Designing Intelligent Machines

Volume 1

Perception, Cognition and Execution

The two volumes of this book were produced as the major components of the third-level undergraduate course *Mechatronics: Designing Intelligent Machines*, written by a Course Team at The Open University, UK. They are:

Volume 1: ***Perception, Cognition and Execution***

Edited by George Rzevski

Volume 2: ***Concepts in Artificial Intelligence***

By Jeffrey Johnson and Philip Picton

Designing Intelligent Machines

Volume 1

Perception, Cognition and Execution

Prepared by an Open University Course Team
Edited by George Rzevski

MECHATRONICS

Butterworth-Heinemann in association with The Open University

OXFORD LONDON BOSTON
MUNICH NEW DELHI SINGAPORE SYDNEY
TOKYO TORONTO WELLINGTON

MILTON KEYNES

BUTTERWORTH-HEINEMANN LTD, Linacre House, Jordan Hill, Oxford OX2 8DP, England, UK

A member of the Reed Elsevier plc group

OXFORD LONDON BOSTON
MUNICH NEW DELHI SINGAPORE SYDNEY
TOKYO TORONTO WELLINGTON

in association with

THE OPEN UNIVERSITY, Walton Hall, Milton Keynes MK7 6AA, England, UK

First published in the United Kingdom by the Open University in serial form for Open University students and staff 1994.

This edition first published in the United Kingdom 1995.

Edited, designed and typeset by The Open University.

Printed and bound in the United Kingdom by Redwood Books, Trowbridge, Wiltshire.

This text forms part of an Open University course. If you would like a copy of S*tudying with the Open University*, please write to the Central Enquiry Service, P.O. Box 200, The Open University, Walton Hall, Milton Keynes MK7 6YZ, United Kingdom.

British Library Cataloguing in Publication Data

A record is available from the British Library

ISBN 0-7506-2404-3

Cover: Computer art created by Dr Paul Margerison using a Sun Sparkstation as part of his PhD research in the Design Discipline at the Open University. The image is a detail from a dynamic fractal, subsequently manipulated using *XV* software.

PREFACE

George Rzevski

This textbook is aimed at undergraduate and postgraduate students and those working in industry who wish to learn the fundamentals of a branch of engineering called ***mechatronics***.

The name was coined in the 1970s to acknowledge an urgent need to integrate two engineering disciplines – ***mecha***nics and elec***tronics*** – with a view to developing and manufacturing mechanical machines controlled by means of electronic circuits. Since then the control and communication technologies have advanced beyond recognition and are now dominated by the software and hardware of digital computers and by embedded artificial intelligence. The name, therefore, may now be considered to be somewhat restrictive. It is, nevertheless, widely used.

The second part of the title – ***designing intelligent machines*** – emphasizes that this book covers new aspects of mechatronics, that is, how to specify and design machines capable of smart sensing, planning, pattern recognition, navigation, learning and reasoning.

The book consists of two independent volumes. Volume 1 covers the fundamentals of mechatronics and discusses the design of machine perception, cognition and execution. Volume 2 is concerned with the concepts of artificial intelligence needed for the design of machines with advanced intelligent behaviour.

Each volume has an 'Overview' which provides the reader with the orientation needed when approaching the study of an unfamiliar and multidisciplinary subject, and provides the rationale for the inclusion and ordering of the topics.

These two volumes were written as the major components of a package of distance learning material for the Open University undergraduate course *Mechatronics: Designing Intelligent Machines*. The contributors to these two volumes were part of an interdisciplinary Course Team, brought together to integrate the disciplines and techniques underlying mechatronics. This Course Team has also generated complementary components of the course, which include video tapes, software, a home experiment kit, study guides and course assessments. More detailed information on the course is given overleaf.

The Open University 'Mechatronics' Course

The two volumes of this book were produced as the major components of the undergraduate third-level course *Mechatronics: Designing Intelligent Machines* by a Course Team at the UK Open University.

Complementary components of the undergraduate course include video tapes, a home experiment kit, software, study guides and course assessments. Video tapes provide students with an opportunity to watch state-of-the-art mechatronic systems in action, to listen to interviews with leading designers of intelligent machines, and to use visual aids to clarify more advanced concepts. The home experiment kit consists of a scanner and a small vehicle connected by an infra-red link to the student's own personal computer. A variety of computer programs and programming environments enable students not only to simulate and experiment but also to design new vehicle behaviours, including autonomous navigation.

If you would like a copy of *Studying with the Open University*, please write to the Central Enquiry Service, P.O. Box 200, The Open University, Walton Hall, Milton Keynes, MK7 6YZ, United Kingdom. Enquiries regarding the availability of supporting material for this and other courses should be addressed to: Open University Educational Enterprises Ltd, 12 Cofferidge Close, Stony Stratford, Milton Keynes, MK11 1BY, United Kingdom.

Course Team Chair and Academic Editor

Professor George Rzevski, The Open University, UK

Authors

Chris Bissell
Chris Earl
Jeffrey Johnson
George Kiss
Anthony Lucas-Smith
Phil Picton
Joe Rooney
George Rzevski
Alfred Vella
Paul Wiese

Supporting staff

Geoff Austin (Academic Computing Service)
George Bellis (Project Officer)
Pam Berry (Text Processing)
Jennifer Conlon (Secretary)
Roger Dobson (Course Manager)
Ian Every (Academic Computing Service)
Ruth Hall (Graphic Designer)
Garry Hammond (Editor)
John Newbury (Staff Tutor)
Christopher Pym (Course Manager)
Janice Robertson (Editor)
Ian Stevenson (Staff Tutor)
John Stratford (BBC Producer)
John Taylor (Graphic Artist)
Helen Thompson (Academic Computing Service)
Bill Young (BBC Producer)

External assessor

Professor Duc-Truong Pham, of University of Wales, College of Cardiff.

Acknowledgements

The Course Team wishes to acknowledge the contributions made in the development of the course by: Mike Booth of Booth Associates; Professor John Meleka; and Dr Memis Acar, Professor J. R. Hewit, Paul King and Dr. K. Bouazza Marouf of Loughborough University of Technology. The Course Team is also indebted to Stuart Burge, Douglas Leith, Don Miles and Peter Steiner, and the many students who participated in the piloting exercises for the course material.

Contents of Volume 1

CHAPTER 5 Cognition
Chris Earl

CHAPTER 6 Planning
Chris Earl

CHAPTER 10 Design
Paul Wiese

OVERVIEW OF VOLUME 1

George Rzevski

Volume 1 provides the fundamentals of perception, cognition and execution. Two later chapters discuss design approaches and architectures of mechatronic systems.

Perception

The role of perception in an intelligent system is to collect information about the system and its environment using sensors and to organize the information collected with a view to minimizing the uncertainty about the world in which the system operates. The term 'perception' is used only when the world in which the system operates is so complex that a substantial post-processing of sensory information is required to determine what exactly goes on there. In considering simpler systems it is customary to talk about sensors, smart sensors (sensors with some post-processing of signals) or instrumentation, rather than perception.

To clarify these issues, consider a practical example. A quality control system for visual inspection of printed circuit boards (PCBs) is located at the output end of an automatic component-insertion machine. The system operates as follows. A video camera observes PCBs as they leave the insertion machine and, based on signals received from the camera, an internal representation of the observed object is constructed by means of rather involved computational feature-extraction and pattern-recognition methods. Thus the computed representation of the actual board is then compared with representations of correctly assembled boards, and if the results of the comparison suggest that the work has not been done correctly, or if the judgement is inconclusive, an instruction is generated to remove the board from the production line. All boards that are not passed are analysed with a view to resolving the ambiguities, if any, or to diagnosing the fault, and a report is produced summarizing the results of this analysis. The reliability of decision-making of the inspection system is less than 100% because the speed at which PCBs pass in front of the camera is very high and therefore, occasionally, the collected information is incomplete and the resulting representation is ambiguous.

The above example will serve to establish rules for demarcation between the perception, cognition and execution functions, and to suggest an overview of the technologies required to construct perception.

Let us start with the demarcation problem. There are no generally agreed ways of determining perception–cognition–execution boundaries. For the purposes of this book we shall always assume that the boundaries are determined by the *function* of each subsystem and that the functions are as follows:

- The function of *perception* is to provide information on the current state of the machine and its environment.
- The function of *cognition* is to plan and initiate machine activities, taking into account information provided by perception.
- The function of *execution* is to start, control and terminate machine actions, based on instructions received from cognition or perception.

I shall now review the description of the PCB inspection system, taking into account the above boundaries.

- The *perception subsystem* observes a PCB at hand by means of a video camera and builds up an internal representation of the observed object which normally contains a sufficient amount of information on critical PCB features to enable a decision to be made as to whether the component-insertion process has been completed correctly.
- The *cognition subsystem* compares the representation received from perception with representations of correctly assembled boards. If the differences suggest that the work has not been done correctly, or if the judgement is inconclusive, it sends an instruction to the execution subsystem to remove the board from the production line. For all boards that are not passed, cognition attempts to resolve the ambiguities or to diagnose the fault, and sends the results of the analysis to execution to produce a printed report.
- The *execution subsystem* removes 'no-go' boards from the production line and compiles, stores, displays, prints and distributes reports.

Although the suggested demarcation rules appear to be straightforward, in practice when one observes a working machine the partitioning may not be that easy. The difficulty is caused by the fact that different functions of a machine are not necessarily implemented as separate physical components or assemblies. An actual machine may have the hardware and software responsible for building a map of the world in which the machine operates connected in a seamless manner with the decision-making circuits and algorithms. A machine may be equipped with grippers that have touch-sensors and actuators which belong to the same closed-loop control system and are placed in close physical proximity to each other and thus give all the appearances of a self-contained unit.

The skill of distinguishing functional concepts from their physical embodiments is particularly important to designers. If a machine is designed first in terms of concepts, and then for each concept a suitable embodiment is sought, the chances are higher that an original and high value-added solution may emerge.

The inspection machine described above represents a typical intelligent machine under development at the time of writing this text. The key technologies required to build its perception subsystem are: (1) the technology of *information processing*, which is concerned with placing information onto carriers in a way that enables it to be stored, filtered, compressed, combined in various ways with information obtained from other sources, packaged, transmitted and distributed;

(2) the technology of *pattern recognition*, which deals with ways of building representations about reality and analysing these representations with a view to reducing uncertainty about the universe of discourse; and (3) the technology of *sensors*, which are devices capable of collecting critical information about a system and its environment without unduly disturbing the world in which the system is located. Elements of these three technologies are given in Volume 1. More advanced topics such as knowledge representation, inference, search, Bayesian approaches to uncertainty, fuzzy logic, neural networks and genetic algorithms are covered in Volume 2 *Concepts in Artificial Intelligence*.

Cognition

Cognition of an intelligent machine is primarily concerned with making decisions on actions that the machine should undertake to accomplish specified goals. These decisions cannot be made without information about the past and current states of the machine and its environment, supplied by perception. In general, at every point in time, cognition is faced with a variety of feasible next states to which the machine could move, and it aims to find the transition which is likely to provide the maximum long-term benefit. The maximum benefit may be expressed in a variety of ways – for example, as the minimum expenditure of the specified resources (usually time or money, and occasionally energy), the minimum risk of failure, the shortest route to a destination and the maximum utilization factor of a given machine. There are, however, many practical tasks that are so complex that it is not practical to seek the optimal sequence of actions. In such cases it is quite sufficient for cognition to find a satisfactory, rather than optimal, solution.

Although every machine equipped with a closed-loop control system, including a boiler controlled by a thermostat, makes autonomous decisions about its future actions, we reserve the use of the term cognition only for those functional subsystems of machines which are charged with decision-making under conditions of considerable variety and some uncertainty. Variety, of course, means that there are many feasible next states to which the machine may move, and uncertainty indicates that it is not clear which one of these potential next states will, in the long run, bring the greatest benefit, or even any benefit, to the mission.

Consider cognition processes taking place in an intelligent aircraft whose task is to reach a target in an urban area and take photographs of specified buildings. I shall assume that: (1) the aircraft's perception system supplies to its cognition system a frequently updated three-dimensional map of the area of concern showing current and past locations of the aircraft; (2) that the mission instructions given to the aircraft cognition by human controllers are, suitably paraphrased, as follows: 'Reach the specified coordinates and look for a wide avenue on the North–South axis. Reduce height to about *X* ft and fly along the avenue northwards until you reach a star-like intersection of six wide streets. Turn into the second street on your left …'; and (3) that cognition has a rapid access to a knowledge base with emergency instructions, listing ways of handling typical unplanned events and emergencies (for example, 'if you miss the intersection, increase the height to *Y* ft, turn around, reduce the height to the previous value and try again'). Cognition processes would then consist of a continuous scanning of the three information sources (the map, the mission instructions and the emergency instructions) searching for hints as to which action to take next. During a mission, cognition will often have to make decisions under conditions of uncertainty caused by the inability of perception to supply complete and reliable information, by ambiguous mission instructions and/or by the occurrence of events that were not foreseen at the time of the development of the emergency knowledge base. A well designed cognition system would be able to learn from experience and thus modify its decision-making behaviour.

Volume 1 provides elements of knowledge which support the design of simpler cognition processes capable of, for example, planning sequences of actions which a robot must follow to rearrange a set of boxes. The key elements of knowledge required to design advanced cognition subsystems capable of supporting the decision-making processes described above belong to computer science and include knowledge representation, search, inference, fuzzy logic, reasoning, associative pattern recognition and learning. These topics are covered in Volume 2.

Execution

Execution is concerned with carrying out sequences of mechanical actions which constitute the interaction of the machine with its environment. Actions are normally triggered, controlled and terminated by an execution subsystem based on information received from perception or cognition subsystems or from the machine operator.

Consider the following examples of activities of typical mechatronic systems:

- A robot arm is reaching for and gripping an object, triggered by a message from its vision system.
- A conveyer belt is moving pallets to a specified destination, repeating the cycle at regular time intervals.
- A system of electro-mechanical links is moving the wing flaps of an aircraft, following instructions from the pilot.

Key elements for understanding execution processes in all these examples are: (1) *dynamics* of machines, the subject concerned with principles of translational and rotational movement of machine components; (2) *control* of machine movements, the role of which is to interpret triggering signals, to initiate appropriate sequences of movements, to monitor execution processes, to decide on corrective actions, if necessary, and to terminate the activities when the objective has been achieved; and (3) *actuators* that generate, transform and transmit the required forces and movements. Fundamentals of control and actuators are covered in Volume 1. Intelligent control, including fuzzy control, is covered in Volume 2.

Design

The activity of designing new artefacts is one of the highest forms of intelligent behaviour. The mental mechanism that supports the capacity of humans to *invent* new artefacts and to work out, in their minds, how these artefacts are going to behave in a variety of situations, is very effective in some individuals and rather mediocre in others. Since we do not understand how this mechanism functions it is not surprising that there are no agreed methods for training all prospective designers to attain the same levels of creativity.

Apart from inventiveness, which may or may not be acquired, there are many important competencies which designers can learn. For example, designers can learn how to:

- identify levels of demand for the new artefact and specify its critical features, that is, features on which its commercial success depends;
- specify product quality, performance and cost, taking into account its technical, economic and social feasibility;
- choose appropriate design methods, technologies, materials, components (for example, sensors and effectors) and manufacturing processes;
- create appropriate shapes;
- generate appropriate configurations of components;
- apply appropriate qualitative, quantitative and experimental methods to assess design hypotheses.

Currently there exists quite a variety of design methods. At one end of the range are those that are distilled from the experience of practising designers and at the other are methods constructed by design researchers from elements of personal observations and the experiences of others, as reported in the design literature and, sometimes, based on hypotheses not yet corroborated in practice. Methods discussed in this book are based on the experience of the individual authors.

The first question always asked is, where to begin? The received wisdom is that the design of a new product should begin only when there exists an identified demand for the product (the so-called *market-driven design*). No doubt this is a reasonable starting point but there are perfectly good examples of new mechatronic artefacts that created a demand by their very presence on the market, the Sony Walkman being perhaps the most striking. In fact, by definition, an *innovative design* will have to create a new demand rather than satisfy an existing requirement.

Irrespective of whether the starting point is a well defined market requirement or a decision to create a product capable of breaking new ground, the time will arrive for the designer to make the first design step. My experience suggests that this first step should be lifting oneself, metaphorically speaking, up on a platform from which one can see a grand picture of the design problem at hand without being distracted by details. The design problem must be digested, vital additional information collected and absorbed, and the problem at hand should be compared with analogous problems for which there are known design solutions. An important aspect of this process is to see directions in which it would be possible to develop the product in the future. A strategic plan for the evolution of the product that is about to be designed will prevent the product being trapped in a technological *cul-de-sac*. Many legal, social, environmental and human-factor issues must be clarified at this stage, including questions such as who is responsible if an autonomous intelligent machine endangers life or property.

Once the designers feel reasonably comfortable with the design situation they can begin making design decisions. Although the precise sequence of decisions does not appear to be important, many practising designers I have interviewed advise starting with the selection of effectors, sensors and the information processing architecture. It stands to reason that one should start with effectors. After all, the main output of intelligent machines is a mechanical activity, such as moving, gripping, pulling or cutting, and therefore the appropriate effectors are of paramount importance for the design success. Simultaneously, one must address the problem of the management of the future machine. Which decisions on a machine's actions will be made autonomously by the machine and which ones by its users? What information will need to be collected through sensors to support these decisions? Which types of sensors are likely to be the most effective, given the world in which the machine will operate? What level of intelligent behaviour will be required to cope with prevailing uncertainties? These are the questions that will need to be answered to clarify the roles of perception, cognition and execution functions.

A very important design decision, which has to be made as early as possible, is the selection of an information-processing architecture. The architecture defines ways in which system components will be related to each other (for example, hierarchically or in layers). In addition, some well developed architectures provide a standard for component interfaces (for example, the internationally agreed Open Systems Architecture) which ensures that all information processing components will fit together.

To proceed further there are basically two alternative design strategies open. In the case of a compact design problem, such as the design of a camera, a CD player or a welding robot, the preferred approach is to tackle the whole problem and to design *top-down*, from the general conceptual solution towards the concept embodiment, following the so-called step-wise refinement approach. Whenever possible though, and in particular when the design problem is very large, it is advisable first to partition the design problem into several smaller problems that can be solved reasonably independently of each other, integrating each new solution with the previous ones. The overall solution thus emerges in an *incremental* manner. Layered architectures are particularly suitable for this approach.

Preliminary design steps, such as problem analysis, legal, social, economic and technological feasibility study, problem partitioning, the selection of architecture and strategic product evolution planning require a very broad socio-technical knowledge. Conventional engineering specialists in mechanical or electronic engineering are not well prepared for these tasks. A broad selection of topics assembled in this book gives an indication of the kind of knowledge that the new breed of mechatronics designers must acquire to prepare themselves for such responsibilities.

For commercial reasons it is of paramount importance to reduce concept-to-market lead time to a minimum, and therefore a wide variety of design activities must take place concurrently. In acknowledgement of this practice, design now often takes place in the context of *concurrent engineering*. A large number of co-designers with specialized skills must contribute to the joint effort in the early stages of design. Design task forces include experts in areas such as: marketing, product design, industrial design, human–machine interface design, legal, social and environmental factors, process planning, manufacturing systems design, costing, testing, reliability assessment, safety, quality, packaging and distribution. They take part in group decision-making and cooperative problem-solving under the guidance of generalists with a mechatronics orientation capable of intelligently assessing numerous trade-offs between different technologies and methodologies.

CHAPTER 1
FUNDAMENTALS

George Rzevski

1.1 Introduction

Some robots, machine tools, vehicles, appliances and cameras have been designed to incorporate advanced information-processing capabilities and are considered to be 'intelligent'. The intelligence is built into the artefacts to improve their performance and to make them more reliable, environment friendly and easy to use. This book is about the design of such 'intelligent machines'.

As an illustration, let's look at current developments in the automobile industry. As a result of a combination of economic, social and environmental pressures, designers of cars are now trying to incorporate the following features: reduced emission of harmful exhaust gases, lower fuel consumption, friendlier human–machine interfaces, better reliability, increased safety, higher levels of creature comfort, easier parking, and improved handling and road holding. The desire to improve performance has always been on the agenda of car designers, but means for achieving the improvements were, until recently, part of the portfolio of mechanical technology. Now, however, many of the improvements have been implemented with the help of *information technology* – the production, storage and communication of information using computers and microelectronics.

Consider an example. In the late 1980s European car manufacturers launched a joint research programme, Prometheus. Its aim was to improve, by the year 2010, car safety by 30% and road traffic flow efficiency by 20%. Within the first two years the following prototypes had been developed through the use of information technology:

an enhanced vision system for the driver, based on the use of ultraviolet and infra-red light and capable of improving visibility in the dark and in fog;

a sensor-based cruise control capable of maintaining a safe distance between cars;

a collision-avoidance system capable of taking over the control of a car in an emergency;

Figure 1.1
Prototype of a self-parking car.

a cooperative driving system which enables drivers of cars and heavy trucks to communicate with each other in difficult traffic conditions;

an automatic route guidance system;

a comprehensive fleet management system for fleets of commercial vehicles which includes the ability to determine vehicle positions;

a 'head-up' display of information (projection of instrument readings on the windscreen), based on holograph technology;

a steer-by-wire car with an aircraft-type joystick instead of the conventional steering wheel;

a self-parking car (Figure 1.1).

The practice of improving product performance, and thus adding value, by building into manufactured objects some form of intelligence is not new. It has emerged slowly over the years, as shown in Table 1.1. Early artefacts had only passive control of performance. For example, the control of temperature in early buildings was achieved by thick walls and small windows. The second generation of designs supplied 'open-loop' controls, where users were expected to provide the necessary feedback and close the control loop (e.g. window shutters). A considerable breakthrough in performance was made when it became possible to incorporate 'closed-loop' control systems into products, which operate automatically without the need for users to be involved (e.g. thermostats). When digital technology in the form of microprocessors arrived on the scene, system controllers could be implemented in computer software. This made them reprogrammable and consequently much more versatile and powerful. At about the same time, some new products began to be considered as machines with a primitive form of intelligence, and the term 'intelligent machines' appeared in the literature.

The current trend is to build distributed information systems into products. This means that by using data communication networks all intelligent activities can be integrated within the product. For example, car management systems currently under development integrate all previously independent monitoring and control functions, such as fuel injection, anti-lock brake control, power-steering, cruise control, monitoring of engine temperature, detection of obstacles and control of electronic power transmission. Many functions previously executed by purely mechanical means are now being implemented by electronics and software – perhaps the most striking examples being four-wheel steering and active suspension systems for cars. The software content in cars changes rapidly. In 1980 there was no software; in 1990 an advanced car had about 60 kilobytes of software; and it is estimated that in the year 2000 the quantity of software will be measured in megabytes.

Elements of intelligence are now designed into a wide range of artefacts, including domestic appliances and leisure equipment, with a view to improving their performance, quality and reliability, simplifying their use and eliminating the need for complex operating instructions. For example, in 1990 Matsushita Electric launched a washing machine which selected an appropriate programme based on reasoning about quantity and type of dirt, size of load and type of detergent used, for a price not much above that asked for a similar conventional appliance. At the same time, the same company developed an intelligent vacuum cleaner capable of adjusting the suction power to suit the prevailing room conditions. A year later, Akai built intelligence into video recorders to achieve a

TABLE 1.1 INCREASING VALUE ADDED BY INCREASING INTELLIGENCE

	Building	**Vehicle**
Passive control	thick walls and small windows	suspension
Open-loop control	shutters	carburettor
Closed-loop control	thermostat	battery charger
Embedded information systems	distributed building management system	distributed vehicle management system
Embedded artificial intelligence	anticipative environment control	route advice, monitoring distance, self-parking

uniformly high quality of images over a wide range of different tapes, and Sharp launched a camcorder capable of learning from its user how to adjust the aperture to avoid underexposure of objects illuminated from behind (see Section 1.6).

Until recently, automation of production relied almost exclusively on rigidly structuring the manufacturing environment, as exemplified by production lines, and using machines (robots) programmed to perform repeatedly exactly the same operations. Changes in patterns of demand created pressures for shorter concept-to-market lead times and for more frequent improvements of products, and, coupled with the availability of advanced information processing resources, these changes began to tip the balance towards increasing use of intelligent machines. For many applications the cost of rigidly structuring manufacturing environments is now prohibitively high in comparison with the cost of developing flexible manufacturing systems, populated by intelligent machines and capable of coping with unpredictable events.

The increased content of information technology in mechanical products has an additional advantage: it enables products to be interconnected into systems capable of providing new services to their customers. Computer manufacturers were the first to realize that they could dramatically increase value added (the difference between total revenue and total purchases) by delivering complete information systems instead of individual computers. All major computer vendors have by now made a further step: they have developed expertise for solving their clients' information problems, and thus became providers of services. In a similar way, some home security vendors developed a similar strategy: in addition to selling security products and security systems, they now offer a complete security service, which includes checking whether an activated alarm is genuine and informing police if this proves to be the case.

Some car manufacturers have collaborated with electronics producers to adopt an identical strategy. They are developing road-automobile communication systems (RACS), which are capable of offering to drivers on-board facilities such as traffic information, weather forecasting, route display and selection, guidance to selected destinations, parking reservation and payment, and road toll payments.

The strategy of at least one large organization is to expand into travel services by installing interactive terminals in the homes of their RACS customers (in addition to terminals installed in their cars). The new facilities offered by such a service would include selecting the most appropriate mode of transport before starting a journey, given the prevailing traffic and weather conditions, and making all necessary reservations and payments. It is only a small step further for the provider of these services to offer holiday packages to their clients.

It is important to remember that the strategy of switching from supplying products (e.g. cars or communication equipment) to installing systems (e.g. road traffic systems) and further to providing a service to customers (e.g. travel service) is feasible only if products have a high information technology content and are thus capable of integration (Table 1.2).

TABLE 1.2 FROM PRODUCTS TO SYSTEMS AND SERVICES

Product	**Car, bus, coach, van, truck**	**Aircraft**
System	*Road traffic control system* On-board computer: traffic information weather forecasts route selection guidance parking reservations payment	*Air traffic control system*
Service	*Travel service* Home computer: mode of transportation selection air/road transport reservations and payment holidays	

The subject of this book, ***mechatronics***, is a new one which crosses the conventional boundaries between the disciplines of mechanical, electrical, electronic and computer engineering. Since each well-established discipline has its own vocabulary, terminology, techniques and world view, it would be surprising if students and practitioners of these disciplines found it easy to communicate with each other when working on mechatronic projects. I shall try therefore to build in this chapter an intellectual platform which, although by necessity modest, should enable readers with different backgrounds to establish a common language for communication. Terms which are regarded as key concepts in the book are picked out in bold type where they are first introduced.

The first sections of this chapter give some background to what we understand as machines, and then I go on to discuss what is needed to make a machine intelligent and how these machines fit into a system. Finally, I draw together these various strands and define what we mean by mechatronics.

1.2 Machines

The term *machine* is used in this book to describe a manufactured object that interacts mechanically with its environment. For the purposes of this interaction, it imports from the environment the required energy, usually in chemical or electrical form, converts it into mechanical energy, and disposes of the surplus energy as heat, sound, mechanical vibrations or chemical substances. Machines may also handle information and materials.

Working parts of a machine are supported by a mechanical structure which will be called in this text the *housing*. Mechanical interaction is accomplished by means of *effectors*, such as grippers, arms, wheels, shutters, wings and jets.

The energy conversion is usually done by means of an internal combustion engine (cars, aircraft) or batteries (cameras), unless the required energy is supplied continuously from a suitable electricity supply system (machine tools, appliances).

Important concepts associated with machines are function, behaviour and performance.

A ***function*** of a machine is a capacity to behave in a specified way. Every machine is designed with a certain *functionality* (a set of functions) in mind. For example, a robot would be designed to carry out welding of car body panels and a laser printer would be used to print documents of a certain format. Machines must be designed to carry out a number of supporting tasks in order to achieve the desired functionality and, at the same time, maintain themselves in good working order. These tasks include provision of power supply, disposal of waste created during operation, housing of physical components, communication with the environment and control of movements.

A ***behaviour*** of a machine is a particular interaction of the machine with its environment over a period of time, defined by a particular set of inputs from and outputs into the environment over that period. Only very simple machines like clocks always behave in the same manner. Other machines will vary their behaviour in response to inputs. For example, an automatic camera will adjust the shutter speed and aperture to suit the prevailing light conditions.

We need to compare the usefulness of various machines, and to do this we should have some way of measuring their behaviours. A set of parameters which provides such measures is called the ***performance***. For example, the performance of a car may be specified in terms of its maximum speed and fuel consumption at a given speed.

Examples of the machines we have been discussing include:

industrial machines, e.g. machines for packaging food, metal-cutting machines, automated assembly machines, welders and industrial robots;

transportation machines, e.g. cars, ships, trains, aircraft and conveyors (Figure 1.2);

appliances and leisure equipment, e.g. dishwashers, washing machines, compact disc players and cameras;

medical equipment, e.g. brain scanners and surgical robots.

Figure 1.2
Sea–land system for handling bulky loads.

1.3 Intelligent machines

The operation and maintenance of machines has to be planned and their behaviour controlled. In other words, machines have to be *managed* (the term 'management' is used here as a synonym for planning and control).

The simplest machines are managed by their operators, i.e. their users. More advanced machines are *programmed* by computer to behave in a particular manner. The pattern of behaviour of such machines will change only if their control program is replaced. An example is a conventional computer-controlled machine tool: whenever it is to be used for processing a different part, the operator has to change the 'part program' to change the behaviour.

There are, however, machines with built-in self-management systems that are capable of modifying their own behaviour, at least to a certain degree, whenever there is a need to accommodate an unexpected change. These will be referred to throughout this book as ***intelligent machines***.

For the purposes of this book, ***intelligence*** is defined as:

> A capability of a system to achieve a goal or sustain desired behaviour under conditions of uncertainty.

The usual sources of uncertainty are:

- the occurrence of unexpected events, either internal, such as a component failure, or external, such as an unpredictable change in the world in which the system operates;
- incomplete, inconsistent or unreliable information available to the system for the purpose of deciding what to do next.

The word 'intelligence' has emotional connotations and some people resent the term being applied to machines. In this book the term is used in a pragmatic manner as a means of distinguishing machines that are capable of making decisions under conditions of uncertainty from machines that are built, or programmed, to carry out repetitive tasks and are capable of changing their behaviour only if instructed by their operators to do so. Such usage of the term 'intelligent' is now generally accepted, although the details of definitions may vary.

The intelligence as defined above may be considered basic. Higher levels of machine intelligence would include an ability to learn from interaction with the world in which the machine operates, an ability to shape its environment with a view to achieving specified goals and, possibly, a capacity for formulating its own goals. Capabilities such as forming new concepts, creating new systems or self-expression are not discussed in this book.

Intelligent machines are not always required. Under deterministic (i.e. predictable) conditions it is much more effective to have a deterministically programmed, rather than an intelligent, machine. Major strengths of programmed behaviour are precision and repeatability. The major weakness is its inability to cope with unexpected events.

Intelligent machines must have a considerable capacity for collecting, storing, processing and distributing information. An intelligent machine receives information from its environment through sensors (such as thermometers, pressure pads, laser bar-code readers, photosensitive cells, video cameras, radars, sonars) and sends information to its environment by means of signalling, broadcasting, printing or display equipment (such as flashing lights, sirens, printers, display panels, screens). The acquired information is, as a rule, stored in digital memories, processed by digital processors and transmitted through digital communication networks.

An intelligent machine interacts with its environment through a set of inputs and outputs.

In general, *inputs* into an intelligent machine are information, energy and, possibly, materials entering the machine, and also mechanical actions that the environment exerts on the machine. For example, inputs into a moving automated train include: electromagnetic signals from coils monitoring track occupancy (information), electric current (energy), and resistance to train movement due to the air pressure and friction (action).

Outputs from an intelligent machine are information, energy and, possibly, materials exiting from the machine, as well as actions exerted by the machine on its environment. For example, outputs from an automated vehicle powered by an internal combustion engine include: light waves that signal its presence and are generated by its warning lights (information), heat, noise and vibrations (energy), exhaust fumes (materials) and its movement along a path (action).

There are many varied application areas open to intelligent machines. Let me list some of them:

- manufacturing
- tunnelling and underground work
- civil engineering and construction work
- transportation by road, rail, sea and air
- fire-fighting and emergency rescue
- nuclear engineering and operations (Figure 1.3)
- underwater operations
- space missions
- processing of materials that are hazardous, toxic, pure, expensive or unstable

***Figure 1.3* Autonomous vehicle for working environments inaccessible to human operators.**

preparation, handling and packaging of food
agricultural work
collecting and monitoring of environmental data
off-shore oil and gas exploitation
microtechnology
medicine and health care
leisure.

Intelligent machines may operate in a self-contained mode or be linked into systems. Examples of systems of interconnected intelligent machines include:

intelligent manufacturing systems
road, rail and air traffic control systems.

The important feature of a system is that its internal links improve performance of its elements so that the overall system performance is better than the sum of the performances of individual elements – a phenomenon called the *emergent property* of a system. For example, a flexible manufacturing system is able to manufacture a range of mechanical components whereas no one machine in this system has this capability.

1.4 Self-regulation

The most elementary behaviour that could arguably be classed as intelligent is *self-regulation*. It denotes the capability of a machine to achieve and sustain the desired behaviour when working in an environment which changes in time in a limited way. The characteristics that change, the range of measurable changes, and the way in which the machine should respond to any particular change are known in advance. Only the timing and magnitudes of changes are not known. Essential components for self-regulation are:

- sensors that collect information about the actual state of the machine and its environment;
- a decision mechanism which selects an action required to achieve or maintain the desired behaviour;
- effectors that move the machine to the desired state.

To understand how self-regulation is achieved let us consider, as an example, an industrial furnace. Let's assume that the task of the furnace is to maintain a given temperature profile over a certain period of time. The temperature is, however, subject to fluctuations due to factors such as imperfect insulation and loading of cold ore which cannot be predicted precisely. Self-regulation of temperature can be achieved by:

- monitoring temperature at various points in the furnace using pyrometers;
- comparing the actual with the specified temperature;
- (based on the result of this comparison) deciding whether to change the level of heating;
- implementing the decision by operating furnace controls;
- obtaining feedback on the results of the intervention.

In general, for the purposes of self-regulation, a machine may monitor one or several measurable physical characteristics, called *variables*, such as position, distance from a given object, direction of movement, speed, acceleration, pressure, liquid level, thickness, and composition.

Whatever the variable or the set of variables, the mechanism of self-regulation is always the same: it follows a *feedback loop*. The steps to be carried out are as follows:

1 Measure values of variables selected to represent a machine behaviour by means of sensors.
2 Compare the measured values with the desired values.
3 Decide which action to take in order to eliminate the observed difference, if any.
4 Implement the selected action.
5 Go back to step 1 and obtain feedback on results of the action.

Such a feedback loop is shown in Figure 1.4.

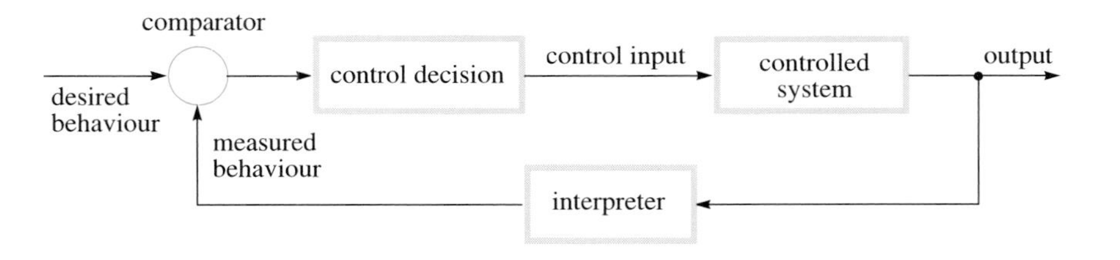

Figure 1.4
A feedback loop.

Additional examples of self-regulation are autofocusing of a camera and autopiloting of an aircraft. Self-regulation is further discussed in Chapter 8.

1.5 Reasoning and knowledge

A higher form of intelligent behaviour than self-regulation is required to cope with an *unstructured* environment. This is, for example, where variable characteristics are not measurable (e.g. the difference between a person and a piece of furniture in a room in which a mobile robot must operate), where several characteristics change simultaneously and in unexpected ways, or where it is not possible to decide in advance how the machine should respond to every combination of events. The machine in this case needs to be capable of *reasoning*. Reasoning can only be done if the machine has access to ***knowledge***, including:

- knowledge about its goals and tasks
- knowledge about its own capabilities
- knowledge about the environment in which it operates.

The machine then has to search through its knowledge space for that element of knowledge (e.g. a rule) which helps it decide what action it should take to compensate for, or take advantage of, the change. Reasoning is also required for planning future behaviours when knowledge about the future is incomplete or unreliable.

Machine designers use various sources to collect some elements of the knowledge required for generating intelligent behaviour. These include human operators who have performed similar tasks successfully, and whose expertise has been written into a *knowledge base* using an appropriate computer-readable language. The rest must be acquired by the machine using its own sensors and information-processing resources.

To illustrate reasoning let us consider an example of an autonomous vehicle transporting goods along the corridors of a store. The vehicle is likely to have a model of its own sensors and effectors and a map of the world in which it operates. This may include knowledge about all available corridors, their lengths, widths and permissible speeds, and even about kinds of objects that may represent obstacles for the vehicle. As it travels, the vehicle observes its world, perhaps by means of video cameras, sonars and laser rangers, and compares what it 'sees' with what is on its internal map. It *navigates* through passages and among obstacles. If it detects an unexpected object within its path (e.g. a person or another vehicle), it may attempt to identify the object by comparison of its observed characteristics with characteristics of objects that it knows; or it may attempt to determine whether the object is moving and, if so, to find out in which direction, or with what speed. If collected information is incomplete, the machine will make an informed guess about the detected object and will engage in planning an appropriate sequence of action, such as sounding a warning, slowing down, or changing direction. During planning, when it has to select a suitable action, the machine may search through its knowledge base for all available alternatives and, if the situation allows, will give preference to actions that minimize the disruption of the schedule. Finally, the selected action will be implemented and a feedback message on the result of the selected action will be obtained, thus closing the control loop.

The similarity with self-regulation is obvious: there is a closed loop which includes monitoring, comparison with a standard, decision making and implementation, followed by a feedback. There are, however, several major differences:

- Raw sensory information received from a variety of sensors must be analysed and interpreted.
- The internal model of the world must be continuously updated.
- Decisions on the next action must be made under conditions of uncertainty caused by the occurrence of unpredictable events or by incomplete, inconsistent or unreliable information.
- Future actions must be planned with a view to minimizing the disruption of the schedule.
- Often, communication with other machines operating in the vicinity is required.

Architectures for machines capable of reasoning are discussed in Section 1.8. Methods for making decisions under conditions of uncertainty, such as fuzzy logic, are covered in Volume 2.

1.6 Learning and pattern recognition

Learning behaviour may be created by making use of artificial neural networks. An *artificial neuron* is an information-processing unit which models the behaviour of brain cells called neurons. A human brain has approximately 10^{10} neurons, each connected to a large number of other neurons. Biological neural networks process, in parallel, electrical signals generated by chemical processes. They store information by modifying strengths of neuron connections, which means that information is stored in a distributed fashion. Neural networks are capable of learning patterns of information and recognizing them even if incompletely specified. Thus, artificial neural networks may be used to provide machines with capabilities to learn as they operate and to generate intelligent behaviours through the process of *pattern recognition*. This eliminates, or at least reduces, the task of collecting knowledge relevant to the operation of the machine and entering it into the knowledge base.

An example of a machine capable of learning is a camcorder developed by the Japanese company Sharp. When a strong light shines behind a person, the subject's face is in the dark, and conventional camcorders (those equipped with self-regulation) cannot obtain a good quality image automatically. The user must intervene and manually adjust the appropriate control. In contrast, a neural network camcorder learns, over a period of time, how to adjust controls automatically by associating patterns of lighting and patterns of manual adjustments carried out by the user. This camcorder learns how to cope with various filming conditions by monitoring the behaviour of its own user.

Neural networks are covered in some detail in Volume 2.

1.7 Autonomous intelligent agents

Intelligent behaviour of a machine may be created through the interaction of *autonomous intelligent agents* resident in the machine, each having its own independent goal. The overall behaviour of the machine is then a result of a synergy of the behaviours of constituent agents.

For example, a machine tool may include two independent intelligent agents, one charged with a goal of achieving the fastest practical speed of cutting a work piece and the other with a goal of maintaining the machine in the best possible working order. Under conditions of normal operation, the first agent will monitor speed of cutting and maintain it at the highest acceptable level while the second agent will be monitoring the tool wear. When the tool wear reaches a certain limit the second agent may decide that unless the cutting speed is reduced, the tool will break in the middle of the current operation and will send a request for slow-down. An independent arbitration system will resolve conflicting demands generated by these two agents.

We will come back to this subject in Volume 2.

1.8 Architectures of intelligent machines

The emergent properties of systems are created by linking system components in particular ways (through *interfaces*). A whole subject of study of these system interfaces has recently sprung into existence, termed the study of system architectures.

The ***architecture*** of a system is a set of rules that primarily defines interfaces between the system and its environment and between system elements. For example, the architecture of a building defines the way in which windows and the roof interface with elevations and the way in which the whole building fits into its urban environment. Similarly, the architecture of a computer defines interfaces between major hardware components as well as the way in which hardware is interfaced with the operating system.

A machine's overall intelligence depends on the way in which its elements are interconnected, and its behaviour differs accordingly. Three main types of architectures are hierarchies, networks, and layered architectures.

A ***hierarchy*** is an architecture which consists of elements linked as 'parents' and 'children'. Each parent can have one or more children. Each child may be a parent of other children. In this way multilevel hierarchies are constructed. The hierarchy always has one element which is a parent but not a child, and this element is at the top of the hierarchy. Unless a hierarchy is infinite, it always contains elements which are children but not parents, and these elements form the lowest level of the hierarchy. A typical hierarchy is shown in Figure 1.5. The important feature of hierarchical configurations is that the contribution of an element to the overall system behaviour critically depends on the hierarchical level to which it belongs. Hierarchies are used whenever it is necessary to reduce the perceived complexity of a system caused by its scale (size). Many human activity systems are arranged into hierarchies, e.g. organizations such the army and the church, or manufacturing systems. In a typical manufacturing system a plant is divided into interlinked manufacturing 'cells', each consisting of mutually interfaced workstations which in turn are made of closely integrated manufacturing modules (units).

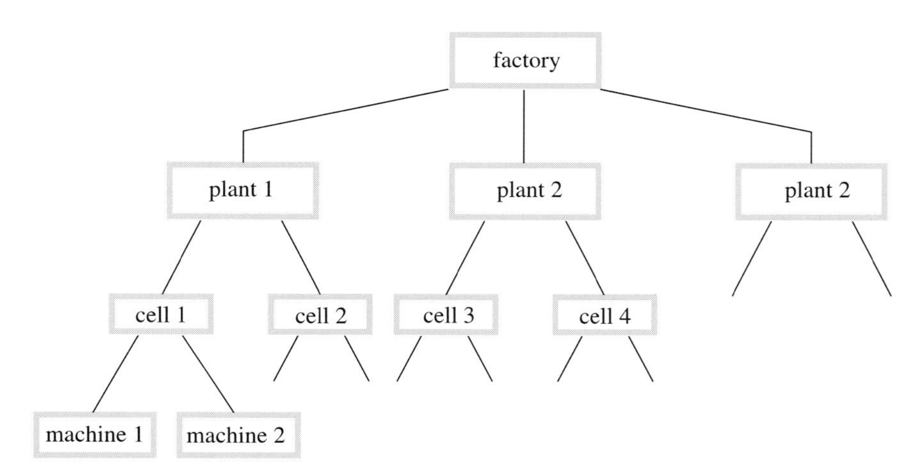

Figure 1.5
A hierarchy.

In *networks,* in contrast to hierarchies, there are no levels of importance and all elements may be connected to each other, as in an 'old boy network' or, alternatively, into rings or stars, as in communication networks (Figure 1.6 over page).

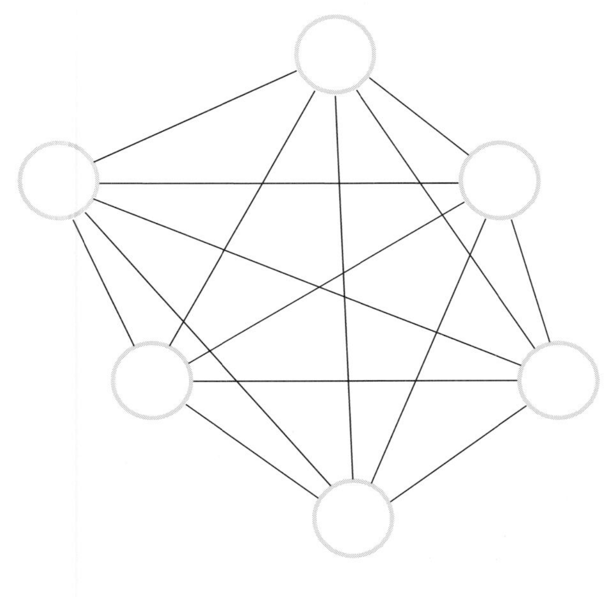

(a) fully interconnected network

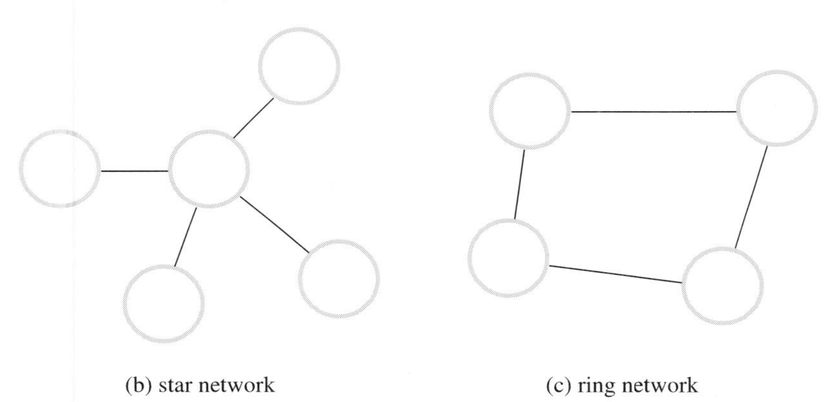

(b) star network

(c) ring network

Figure 1.6 ***Networks.***

Network architectures are used when there is a need for cooperation between units that are equal in importance but different in terms of skills or capabilities. They may be used in distributed computer systems, or, if there is a requirement for parallel processing of information, in neural networks.

Layered architectures consist of self-contained elements, called *layers*, each connected to a set of inputs and outputs and thus each capable of creating a system behaviour (Figure 1.7). Since behaviours generated by different layers may be in conflict with each other (as in the example given in Section 1.7), the layered architecture may include an arbitration mechanism responsible for ensuring a proper overall system behaviour. An important characteristic of layered architectures is that they are ideally suited for incremental development of systems. Because each layer has an independent input and output, a new layer may be added to a system without too much disturbance to previous layers.

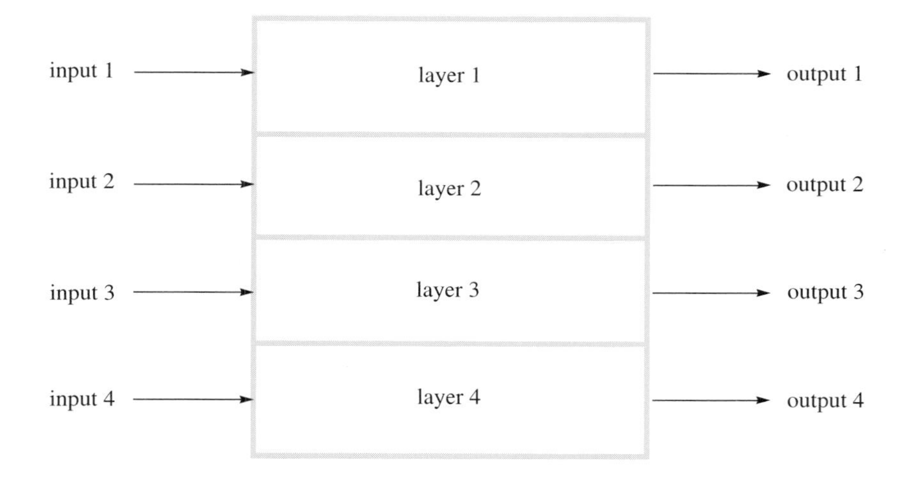

Figure 1.7
Layered architecture.

Machines have different maintenance requirements depending on the architecture used. For instance, modular and integral architectures have contrasting designs and may be chosen for different reasons of maintainability.

In *modular architectures*, constituent elements are interfaced with each other in such a way that each element is as self-contained as possible. Intelligent machines are often designed to be modular so that faulty modules may be rapidly replaced and the system reconfigured. Although modularity is useful from the point of view of maintenance and ease of modification, it carries with it some redundancy of parts and therefore a penalty in terms of weight, bulk and initial cost.

In contrast, *integral architectures* aim to bring together constituent elements of a system so closely that their interfaces tend to disappear. This type of configuration is used, for example, when there is a need to prevent elements from being disconnected or when the prime aim is to miniaturize the housing.

An intelligent machine has to carry out a complex set of tasks and therefore requires an effective architecture. Several possible architectures have been tried by mechatronics designers, including hierarchies, networks and layers. I have selected a network architecture as the most appropriate to discuss various functional aspects of intelligent machines. This architecture is shown in Figure 1.8 (over page) and comprises the subsystems of perception, cognition, execution, self-maintenance, and energy conversion.

The role of the ***perception*** subsystem is to collect, store, process and distribute information about the current state of the machine and its environment. To do that effectively the perception subsystem may be required to construct, store and update models of the machine and its environment.

The ***cognition*** subsystem is responsible for evaluating information collected and processed by perception and for planning actions that will take place. The implementation of the cognitive function is time consuming and requires considerable computational resource; therefore many types of intelligent machine are designed to operate without the capability for planning their own behaviour.

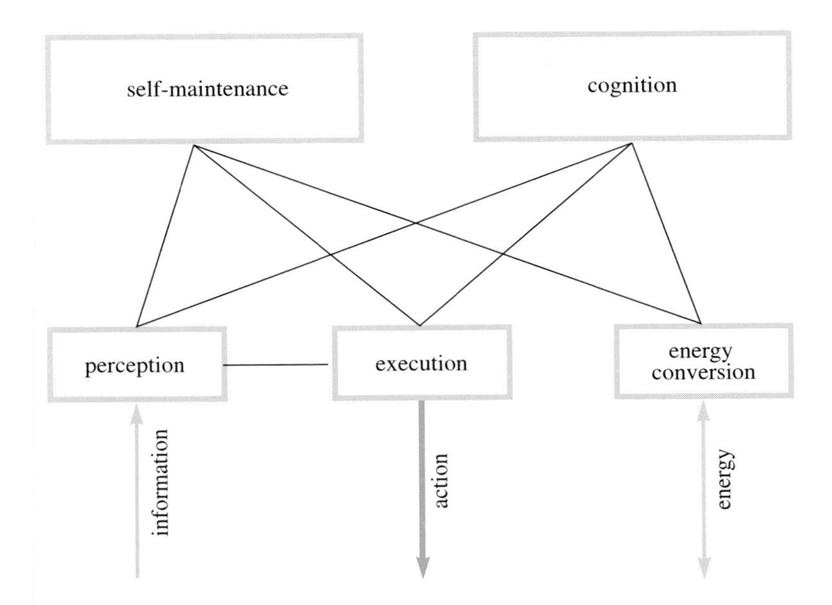

Figure 1.8
An intelligent machine architecture.

The ***execution*** subsystem is responsible for controlling all activities of a machine, based on instructions normally received from cognition (purposeful behaviour) or, in situations when there is a need for a rapid avoidance action, directly from perception (reactive behaviour). The flows of information underlying the purposeful and reactive behaviours are depicted in Figure 1.9. In current intelligent

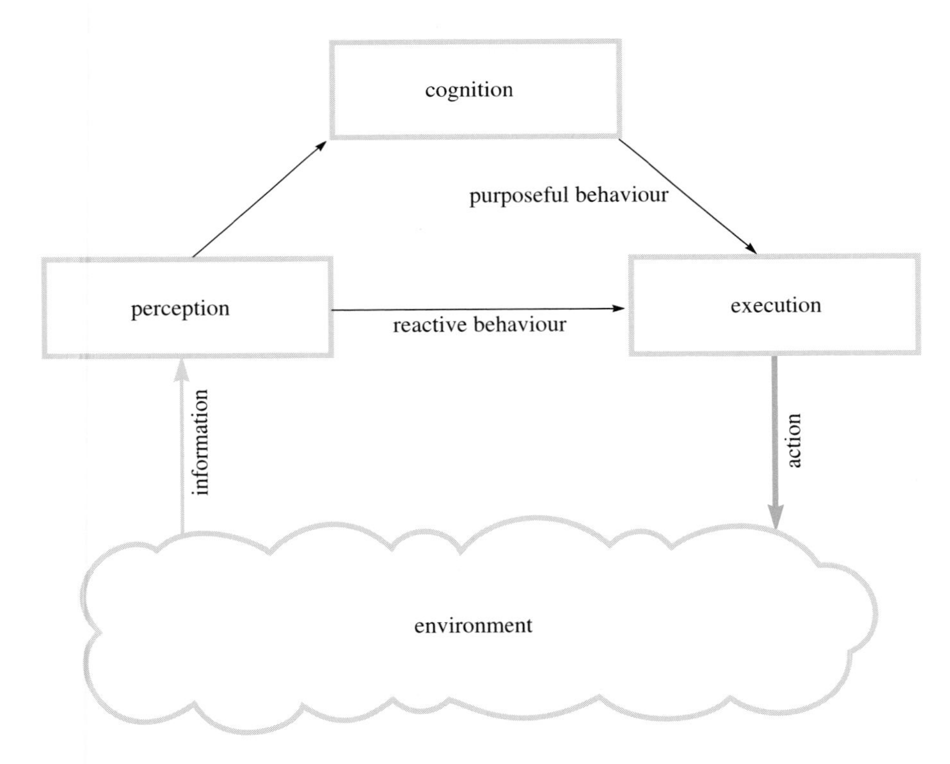

Figure 1.9
Purposeful and reactive behaviours.

machines, as a rule, reactive behaviour is predominant and cognition plays a small role in comparison with perception and execution.

The *self-maintenance* subsystem is responsible for maintaining the machine in good order during its normal operation. This includes intermittent monitoring of the behaviour of key parts with a view to discovering a fault before it occurs (preventive self-maintenance), or immediately after it occurs (self-diagnosis). It may also carry out self-repairs by means of reconfiguration aimed at removing the faulty part, and in very advanced machines it may have a responsibility for continuously improving machine performance. The following are examples of behaviours which are considered to be self-maintenance:

> when a car management system monitors the wear of brake pads and, based on this information, makes decisions on how to compensate for wear, and signals to the driver when to replace the pads (preventive self-maintenance);
>
> when a printer signals the presence of a fault and identifies it (self-diagnosis);
>
> when a spacecraft reconfigures one of its control systems to eliminate a faulty component (self-repair);
>
> when an autonomous transportation vehicle learns from experience which routes through a factory are least congested at a particular time of the day and selects its trajectory accordingly (self-improvement).

To carry out its main function a machine must have an appropriate energy supply and, to maintain itself in good order, it must eliminate undesirable waste, such as heat. These tasks are carried out by the *energy conversion* subsystem.

The choice of energy supply and requirements for disposal of waste products is determined by the nature of the mechatronic machine. For example, the energy supply for a rail locomotive may be from an external source, such as in most electric railway systems, where the energy is converted from chemical via thermal into electrical energy in power stations and transmitted to the locomotive via an electricity supply network and overhead equipment. Alternatively, a self-propelled locomotive may be used, with an energy conversion plant on board, such as a steam, diesel or diesel-electric locomotive.

The choice is also governed by factors such as economic availability of natural resources and environmental and social considerations. Cars are at present almost exclusively powered by internal combustion engines, but pressure from environmentalists may force the use of electric vehicles on a much wider scale in the near future. In some cases there is less choice. An aircraft cannot be powered from an external source, and it would be inconvenient to use an external source for a hand-held camcorder. Nor would we try to supply a process plant with energy from batteries.

The elimination of waste energy often poses equally difficult environmental choices. Disposal of waste heat in very large plants can cause thermal pollution of rivers and streams, and noise may be extremely difficult to control without expensive noise reduction devices.

It is important to understand that all subsystems described in this section are conceptual rather than physical. In other words, although it may be useful, in abstract, to distinguish the task of processing sensory information from that of controlling physical movement of a machine, this does not imply that these two tasks cannot be done by the same computer. Conversely, although it is quite effective to mount tactile sensors on robot grippers, this in no way implies that the functions of these two devices are conceptually the same.

1.9 Mechatronics

Mechatronics is usually defined as an engineering discipline concerned with the *integration* of mechanical, electronic, computer software and other technologies with the aim of designing and manufacturing effective machines. To understand why the integration is necessary, it is instructive to analyse actual mechatronic products, such as mobile robots or camcorders, with a view to identifying the functions that various technologies support. Let's start with the physical components.

The physical machines are constructed from a variety of physical components which may be classified as follows:

- *sensors*, such as sonars and photosensitive cells, which enable machines to collect information about their own state and the state of their environments;
- *input/output devices*, such as keyboards and displays, which enable machines to communicate with their users;
- *information-processing hardware and software*, which provide resources for processing, storing, distributing and utilizing information;
- *effectors*, such as arms, grippers and autofocusing mechanisms, which enable machines to carry out planned actions;
- the *power plant and associated power transmission system*, such as a battery with associated equipment, or a power supply unit which converts electricity from the mains to appropriate power requirements;
- the *housing*, which apart from protecting and keeping machine components together, may be expected to dissipate undesirable heat, protect the environment from noise, control internal temperature and humidity, ensure dust-free atmosphere, eliminate electromagnetic interference, enable ergonomic use of the system and satisfy aesthetic criteria.

In traditional machines (such as steam locomotives, mass-produced cars and high-speed industrial machines) the dominant technology is mechanical. Many timing and control functions in such machines are performed by mechanical means: for example, by steam governors, camshafts and mechanical linkages. Power is transmitted from the power plant to the point of delivery by means of (often long) mechanical transmission systems consisting of clutches, gearboxes, shafts and differential gears. These subsystems of interlinked mechanical components have properties such as inertia and friction which play an important role in the overall dynamics of the machine.

In state-of-the-art mechatronic products, the complexity of information-related tasks (such as sensing, sensor data fusion, world-model updating, planning, learning and control) is such that the role of information technology is dominant. Therefore, distributed digital computers (microprocessors), data communication networks and smart sensors with associated software are likely to represent the most important elements of the machine.

An early example of traditional machines subjected to a radical redesign was the diesel-electric locomotive. The mechanical power transmission subsystem of the conventional diesel locomotive was removed and a generator driven by the diesel engine installed to convert mechanical energy into electrical energy, which could be conveniently delivered to traction motors placed close to driving axles. Control functions were implemented using electronic circuits. More modern examples of a radical redesign of machines to replace mechanical linkages by digital hardware and software are fly-by-wire aircraft and steer-by-wire cars (mentioned in Section 1.1).

How do we distinguish whether a machine is mechatronic or not? Where is the dividing line? Can a diesel-electric locomotive be considered a truly mechatronic vehicle? Do the self-regulatory characteristics of a simple thermostat make a central heating system a mechatronic product? Are conventional programmed robots mechatronic machines?

There is no common agreement at present on mechatronics boundaries. For the purposes of this book we shall consider that the area of mechatronics encompasses machines capable of achieving a given goal or sustaining desired behaviour, under conditions of uncertainty which are caused by the occurrence of unexpected events or by incomplete, inconsistent or unreliable information available to the machine about the relevant events. This definition implies that machines without sensors are excluded from our field of interest. Machines relying on the sensing of well-defined and measurable quantities such as temperature are only of marginal interest.

1.10 Conclusion

In this introductory chapter I have described the economic, social and environmental context within which intelligent machines have emerged as a viable alternative to both conventional mechanical and programmed machines.

I have defined intelligence of a machine in terms of operation: a machine is intelligent if it can achieve or sustain desired behaviour in an environment which is characterized by unpredictable changes. This implies that an intelligent machine can operate to a certain degree autonomously (i.e. without being totally controlled by a human being) in unstructured environments which may be dangerous, uncomfortable or in any other way unsuitable for human operators. Alternatively, an intelligent machine can collaborate with a human operator in environments such as a factory, an office or a home.

I have stressed that in a highly structured world where unforeseeable events do not occur (except in rare system failures), machine intelligence is not required. In such a world the best performance is achieved by machines that are programmed to repeat a particular behaviour precisely and rapidly, such as in an old-fashioned production line. However, current business conditions of never-ending changes of demand and supply, rapidly advancing technology and evolving social and environmental concerns, mean that it is difficult and expensive to structure a working environment in a rigid way. Even in the artificial world of a modern factory there are many potentially disruptive events that could be triggered by external activities. It is very likely, therefore, that, instead of concentrating our efforts into structuring our working worlds, we shall, in the future, take full advantage of our newly developed expertise on how to design intelligent machines that can cope with a certain level of unpredictability.

There are of course worlds that simply cannot be structured, such as civil engineering sites, air space, underwater sites, agriculture and the leisure environment. It is obvious that machines capable of operating in such environments and interacting with humans in a 'friendly' way will be of considerable value.

In this chapter I have considered briefly a number of different levels of intelligent behaviour and several ways of building intelligence into a machine.

The most basic intelligent behaviour, known as self-regulation, may be achieved by arranging appropriate sensors, decision elements and effectors into a feedback loop, as in industrial furnaces, autofocus cameras or aircraft autopilot systems. These systems are capable of coping with a very small degree of uncertainty.

The ability to cope with a substantial uncertainty in a purposeful manner can be achieved by selecting a perception–cognition–execution architecture enabling the machine to reason and to store knowledge about itself and about the world in which it works. The perception subsystem includes, in addition to sensors, considerable information-processing resources which are used to analyse information collected by sensors and make inferences about the actual state of the machine and its environment. The cognition subsystem plans future behaviours of the machine and decides how to react to unforeseen events with a view to minimizing disruption of schedules. The execution subsystem implements decisions made by cognition, turning them into mechanical actions. Large intelligent vehicles and mobile robots are designed in this manner.

A somewhat simplified perception–execution architecture, which generates reactive behaviours, is used widely in intelligent industrial machines and domestic appliances.

The layered architecture may be considered as a possible alternative to the perception–cognition–execution architecture, particularly if computing power is at a premium and the speed of reaction is of paramount importance, but also if the design strategy is to increase functionality of the machine in an incremental fashion by adding capabilities for new behaviours as the need arises.

The emerging concept of autonomous agents is very promising, although there are not as yet many practical applications. Small mobile robots are perhaps the best examples.

An aspect of intelligent machines not much exploited at present, but one that will no doubt play an important role in the future, is the capability for self-maintenance. As the number of intelligent machines in operation increases, it will be necessary to ensure that machines use their intelligence for self-diagnosis, preventive self-maintenance and self-repair.

Finally, the ultimate goal is to design machines capable of self-improvement. At present, considerable effort goes into the development of machines that have the ability to learn from their own behaviour as well as from their users. This ability can be achieved with the help of artificial neural networks. We have already mentioned the camcorder which learns how to handle various lighting conditions by copying the behaviour of its users. Another example of learning behaviour is the airport baggage-checking machine which can be trained to recognize various types of explosive and weaponry.

The rest of this volume continues the theme of providing a general knowledge about intelligent machines and their major functional subsystems: perception, cognition and execution. It also includes an overview of the concept of architectures and describes approaches to designing mechatronic products. The next chapter looks at the important role of sensors in intelligent machines.

CHAPTER 2
SENSORS

Anthony Lucas-Smith

2.1 Introduction

This chapter focuses on the role of detecting devices known as ***sensors***. We will consider how sensors respond to external stimuli, recognizing their presence and measuring selected parameters for input to a system. We need to know the categories and attributes of sensors in order to assess the kind of applications for which they might be exploited. A number of sensor technologies suitable for intelligent machines are identified. The combining of sensors is considered for use in intelligent processing of data where it is impossible for a system to interpret an external situation with data from one sensor only.

We will look at the application of sensors by selecting some case studies. Relatively cheap, simple sensors are used in a prosthetic hand, a sophisticated and complex application, and also in still cameras. The sensors used in guided vehicle navigation are reviewed, and the exploitation of sensors is considered in a case study based on the car industry and the development of techniques to make motoring safer and more comfortable, with engines running more efficiently and in an environmentally acceptable manner.

While reading this chapter you should consider the role of sensors in the context of mechatronics, noting in particular:

- what stimulus or factor is being sensed and for what purpose;
- why a particular sensing technology has been selected and why others have been rejected;
- the need for sensors to be packaged and set in place so that they are unobtrusive, reliable, robust and accurate.

2.2 Sensor functions

Chapter 1 (Section 1.4) refers to the feedback loop in which intelligent machines control their own behaviour as a result of their ability to sense the external environment in which they operate. This implies a linkage between the environment and the recognition function of the system (Figure 2.1a). Such a linkage is provided by the range of devices known as sensors, most of which are examples of ***transducers***, i.e. devices which respond to one physical stimulus and which output a corresponding signal of a different form of physical stimulus. Sensor feedback may be automated or may be handled by human cognition and action. In many applications sensor signals are observed by humans for monitoring purposes and are not part of an automated, closed-loop system.

Actuators are also forms of transducer which respond in most cases to an electrical signal, producing a corresponding output, such as a change in temperature or pressure or a precise movement by a motor (Figure 2.1b).

In many sensor applications, electromagnetic energy such as light is received and an electrical signal transmitted to the control system, though other forms of input signal are commonly encountered. Examples include:

laser bar code reader, used in a supermarket linked with a system to list customers' purchases and provide data for stock control;

photosensitive cell, used in a greenhouse for ventilation control;

photosensitive cells within a video camera, used by a motor manufacturer in a system to enable a robot to insert the windscreen into a car;

pressure sensor, used in steam generator control;

thermocouple, used in a furnace temperature control (Figure 2.1c).

The role of the sensor may be summarized as carrying out the following functions.

1. *Detection* – detects the presence of an external phenomenon, usually represented by a physical characteristic, though chemical and biological properties may also be monitored.
2. *Selection* – selects or filters out and (in most cases) measures a single property of the external stimulus.
3. *Signal processing* – transforms the input signal into a corresponding output signal, expressed in analogue, digital or modulated form. Some ***conditioning*** (i.e. signal interpretation or modification) may be necessary. In many situations the signal generated is very small and first needs amplification. Other types of conditioning include: linearization; limiting the range of values sampled; analogue-to-digital signal conversion; filtering out specified frequencies of an oscillating signal.
4. *Communication* – passes the signal directly to the control system, to a recording system or to a human being.

The sensor can therefore be considered as the agent for *recognition* and a prerequisite for *cognition*. In engineering language the technology of measurement is referred to as ***instrumentation***. *Sensor technology* is a wider concept which includes detection of the presence of a phenomenon.

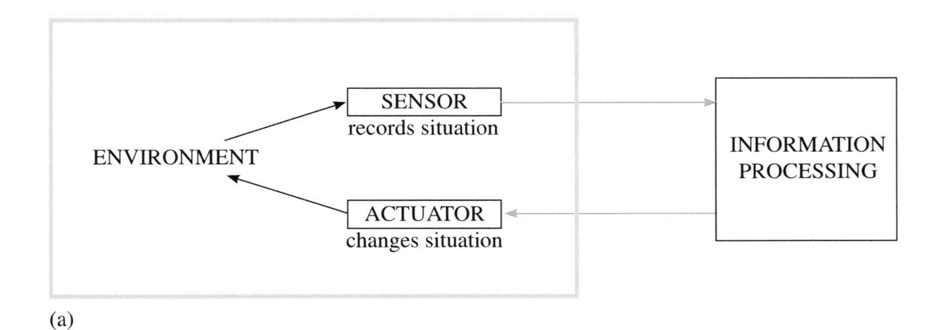

(a)

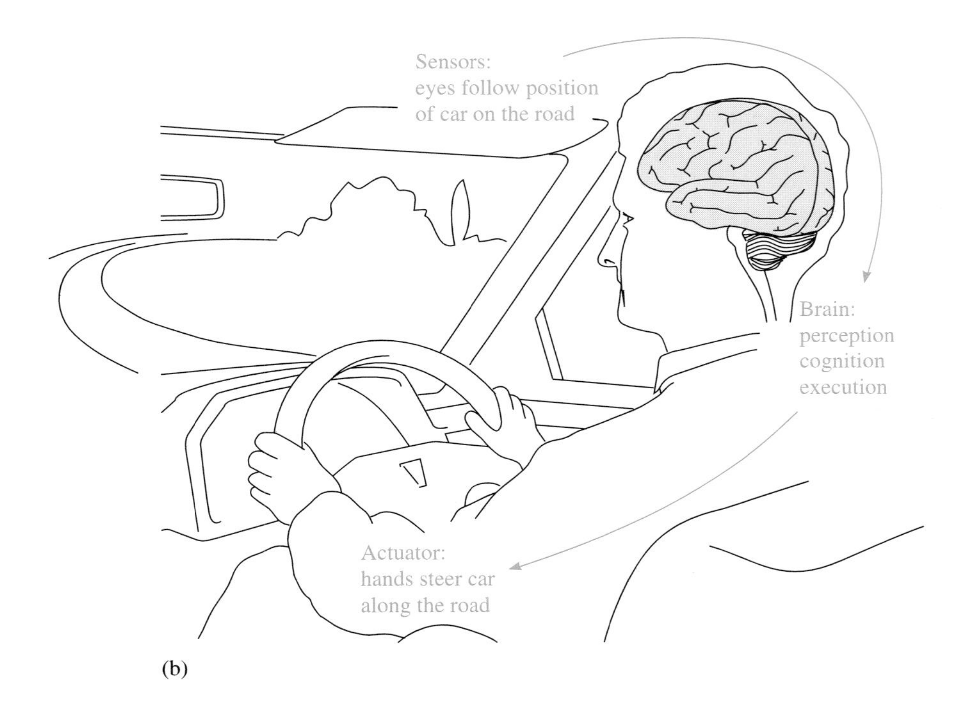

(b)

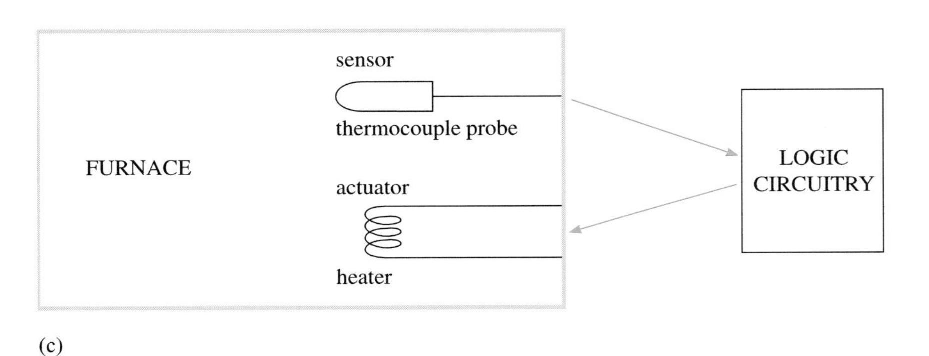

(c)

Figure 2.1
Linkage between sensor and actuator: (a) linkage principle; (b) example 1; (c) example 2.

There is an analogy with human sensors such as nerve cells in the skin which detect the presence of a nearby hot object, sending a signal to the control system which moves the appropriate part of the body away from the dangerous object; or nerve endings in the ear which detect the presence of a sound source, sending a signal to the control system which enables us to identify the location of the sound.

However, the analogy is only partial. Although these human sensors detect external stimuli, no automatic measurement subsequently takes place. Human physiology can distinguish between degrees of 'hotness' and 'noisiness', though in a subjective manner which varies from person to person. Human sensors and brain cells in combination may be capable of estimating values of parameters in the light of experience and by reference to known standards, but have not evolved primarily as measuring devices.

Compared with nature's sensors, manufactured devices can directly detect a much greater range of properties as well as measure them over a wider range of values, within a certain range of accuracy. Nevertheless they cannot currently match the ability of the human brain-and-sensor combination to interpret the significance of combinations of signals received. Mechatronic devices represent an attempt to reproduce such a capability well beyond the power of more traditional, deterministic control systems.

2.3 Components of sensors

It can be a matter of debate on what constitutes a sensor. Its conceptual structure is shown in Figure 2.2. In some cases the components as shown in the figure are not packaged together in the physical structure. It then becomes difficult to identify the sensor as a discrete object. Furthermore, functions such as signal conditioning may be seen by the control system designer as a part of the microprocessor control logic, rather than a physical part of the sensor.

It is also possible to have the sensor and actuator inextricably combined, even though their actions are separately identifiable. Two control devices – the bimetal

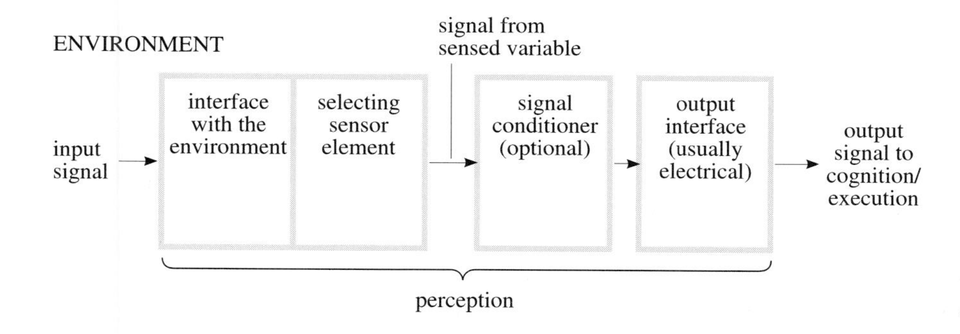

Figure 2.2
Logical structure of sensor.

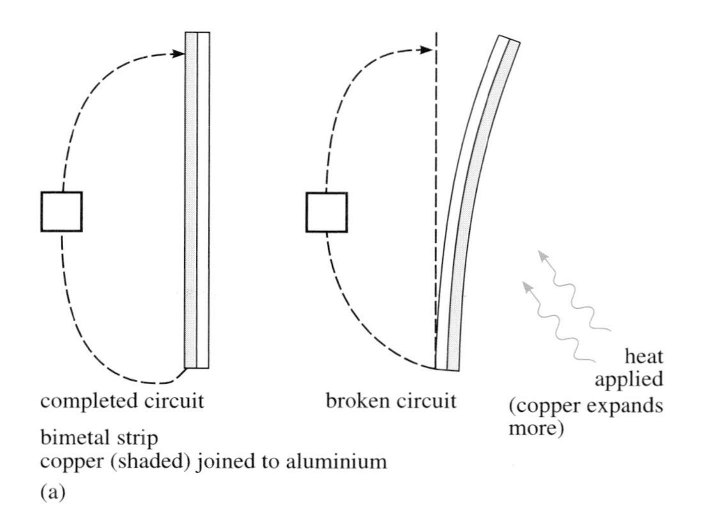

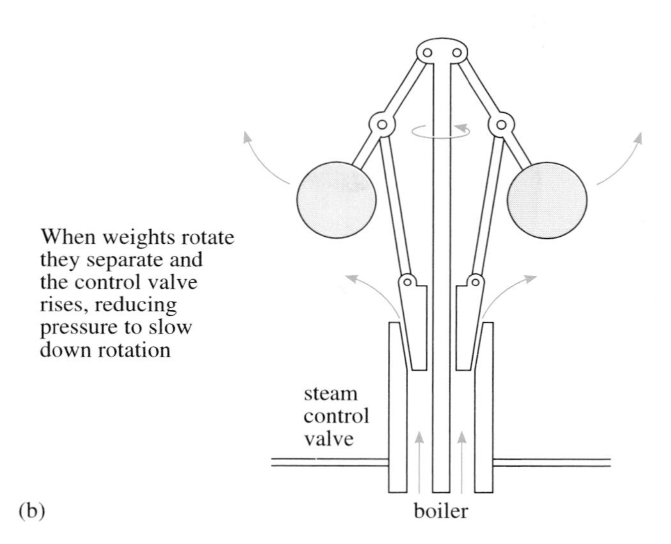

***Figure 2.3* Mechanical transmission of sensor signal combinations: (a) bimetal strip for temperature control; (b) rotary governor for control of pressure.**

strip and the rotary governor (Figure 2.3) – are long-established devices in which the sensor and actuator form a single unit.

2.4 Attributes of sensors

A number of technologies contribute to making possible a bewildering range of sensors to choose from. For example, around thirty distinct, commonly used technologies for sensing simple dynamic properties such as flow rate can be readily identified. This highlights the familiar problem of how to select the most appropriate sensor for an application. It may be useful here to categorize the different aspects or attributes of how they perform, matching the special needs of mechatronics. There is no single, universally agreed classification of sensors, but the following selection of attributes is considered relevant to our purposes.

2.4.1 Phenomena sensed

A wide range of phenomena can be detected and, if appropriate, measured. In this context they are known as ***measurands***.

In many cases a sensor is required only to recognize presence or absence without accurate measurement. For example, many shops have a device for scanning bar codes on products in which a sensor can distinguish between dark and light bands, and collectively recognize them as a bar code. Counting of the bands takes place but is not an inherent measuring function of the sensor.

Measurands can be divided into the following groups:

- positional and dimensional, such as distance from a datum point, linear displacement, angular displacement, liquid level, area, volume;
- proximity, such as with a metal detector which responds to metals within a specified distance;
- static mechanical, such as mass, pressure, torque, density, shear force;
- dynamic mechanical, such as linear and angular velocity, flow rate, vibration, acceleration, pressure oscillation;
- physicomechanical, such as viscosity, humidity;
- thermal, such as temperature and temperature variation;
- electromagnetic, such as wavelength, frequency, intensity, polarization, particularly optical and infra-red;
- electrical electrostatic, such as charge, capacitance, inductance, resistance;
- magnetic, such as flux;
- sonic or acoustic, such as sound frequency, intensity and direction – usually included are ultrasonic waves which are beyond the audible range for humans;
- nucleonic, for the purpose of identifying subatomic particles and measuring their associated parameters;
- chemical, such as infra-red gas analysis, capable of distinguishing between a large number of gases;
- biological/biochemical, possessing the ability to identify living organisms and material.

A distinction should be made between the measurand phenomenon and the sensor technology. For example, a dynamic mechanical phenomenon might be sensed using magnetic or electrical technology, or perhaps a combination of both. The application of a selection of sensor types is described in Sections 2.7 to 2.10 below.

2.4.2 Operating distance of the sensor from the stimulus

This is most frequently encountered in distinguishing between remote and contact (or near contact) sensors. Remote sensing can range from a few metres to millions of kilometres and is clearly the only option when the measurand is always distant, as in astronomical and navigational systems.

When near contact sensing is possible it is often an improvement on remote contact sensing which may be difficult or inconvenient to achieve. In mechanical devices contact sensors might disturb operations by their presence or be easily damaged by friction.

When the phenomenon is distant from the sensor it may be difficult to identify the phenomenon unambiguously and measure its characteristics without being confused by background 'noise' (interference).

As an example, first-generation optical bar code readers required a 'light pen' sensor to be in contact with the bar code and to be moved across its pattern, at a speed within a specified range. The more advanced type of laser reader used at supermarket checkouts operates within a few centimetres of the bar code and does not require fixed orientation (Figure 2.4). Speed and ease of use is improved and could not be matched by a contact sensor for this application. The technology involves a rotating disc with a number of transparent, variable geometry prisms set along the outer rim and a low-power laser light source. A fine beam of light is aimed at each prism and deflected at varying angles so that it scans rapidly through several (non-parallel) planes. Within around half a second the beam is presented with a bar code approximately perpendicular to it and is able to scan through all the lines and spaces. An optical sensor receives the reflected light signals which are conditioned and finally interpreted electronically. Developments are in progress for supermarket laser readers to operate at greater distances and with bar codes distorted, at any angle and even when visually obscured.

2.4.3 Invasiveness of measurement

Invasiveness expresses the degree to which a sensor reacts with the measurand environment in which it is placed. This is a familiar problem to be taken account of in selecting a sensor. No such problem arises in remote sensing. Invasiveness occurs in a number of forms.

1 With close and contact sensing, a form of Heisenberg's 'uncertainty principle' can occur, in which the process of measurement interacts with the

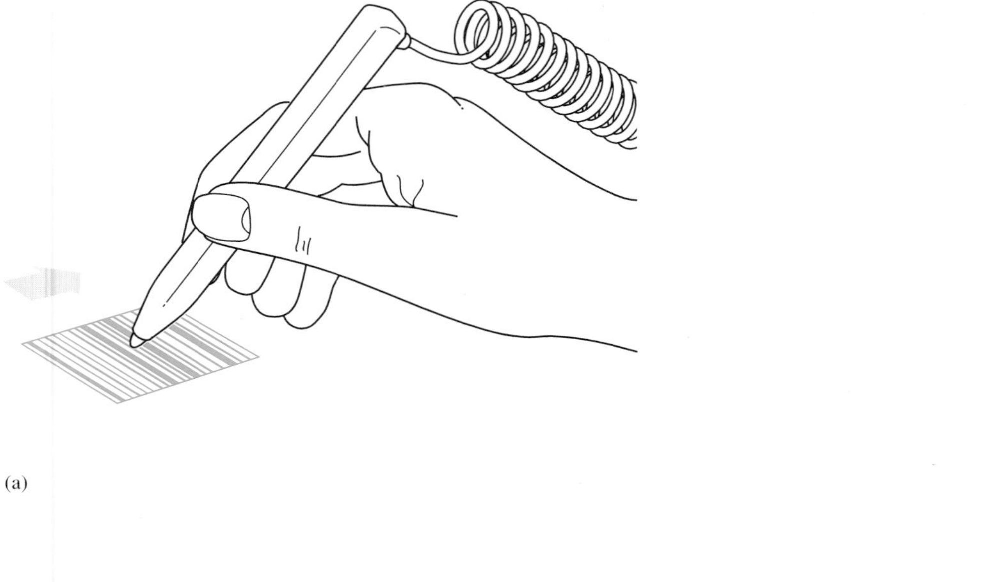
(a)

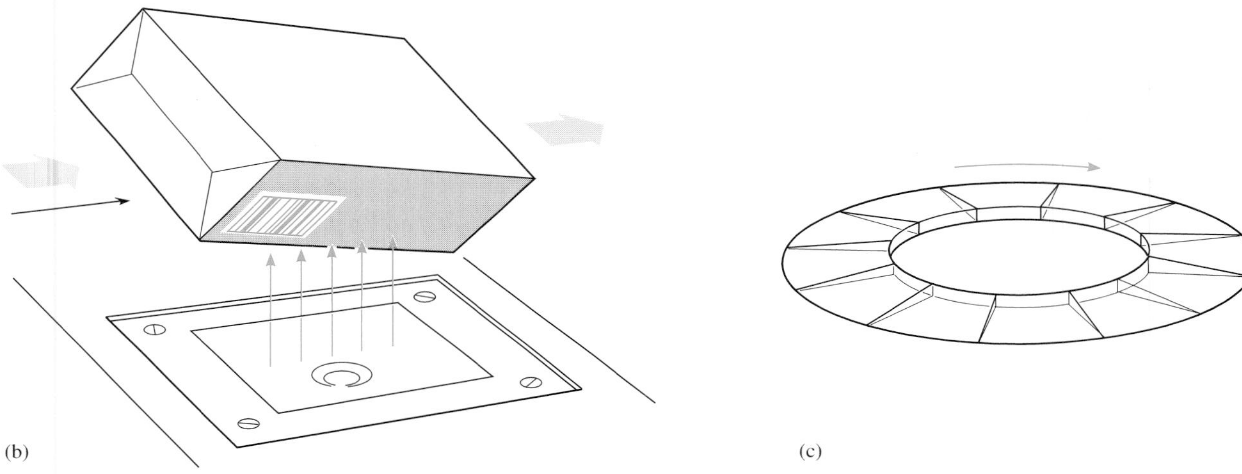
(b) (c)

Figure 2.4
Bar code readers: (a) contact sensor; (b) supermarket laser reader; (c) rotating variable-geometry prism.

measurand and changes its value, causing 'uncertainty' in what was its original value. For example, sensors for very low pressure measurement which sample the contents of a near vacuum will in the process change its pressure.

2 Invasiveness occurs when the sensor's physical presence might affect the operation of the measurand environment. In sampling the pressure of a ball bearing on a flat surface, the contact area would be too small for a bulky pressure sensor. Its presence would so alter the function of the bearing as to invalidate the process of measurement.

3 Invasiveness also applies to sensing within living organisms such as the human body, where sensors might endanger life or be badly shaped for comfort. For example, a blood sugar sensor for diabetic control must be small enough for comfort and capable of maintaining the sterile condition of the blood system.

2.4.4 Invasiveness of the measurand

It also happens that the measurand itself can be a hostile environment and invade the sensor. Typical causes include radiation, temperature, acidity and other chemical conditions such that a sensor within its territory could be damaged or destroyed and therefore be unsuitable for the application.

An important design issue is the avoidance of electromagnetic fields which could affect the process of measurement. Electromagnetic immunity is sought by careful positioning of components and the use of screening techniques.

2.4.5 Form of output signal

In the past, output signals have frequently been mechanical, as with the simple Bourdon pressure gauge (Figure 2.5), and the speed indicator of a rotary governor (see Figure 2.3). Pneumatic output also comes within this category and is occasionally used.

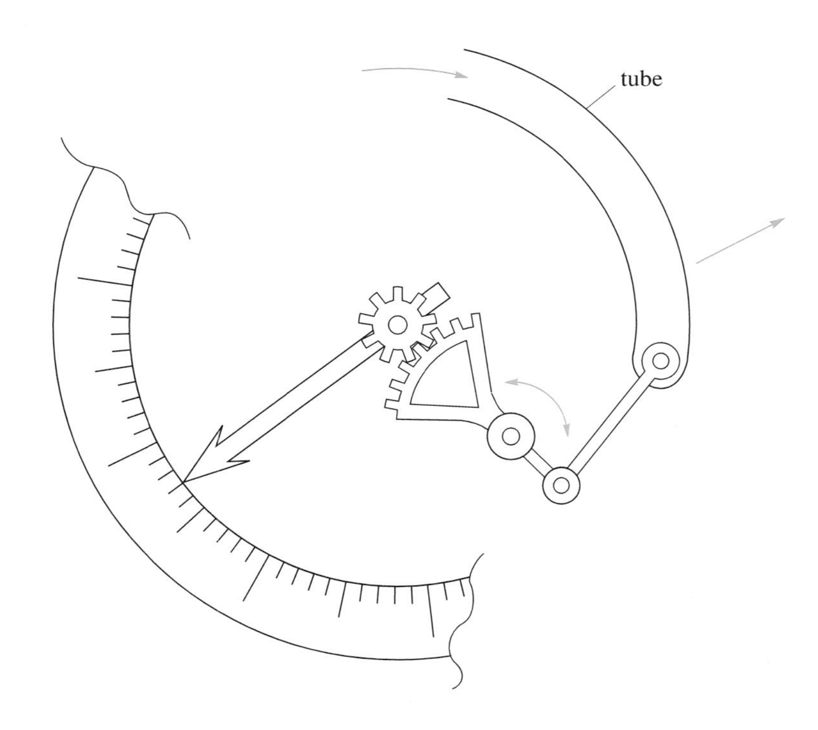

Figure 2.5
Bourdon pressure gauge.

Most output signals are now electrical, and can be analogue or digital. Typical analogue signals are voltage or current dependent or modulated (typically by frequency or amplitude). Digital output is based on the use of encoders, as explained below.

Encoded output

Digital output is acquired through the use of ***encoders*** which contain 'digital' transducers. Although transducer output is analogue in nature, it can be designed to produce output which is easily converted by electronic circuitry to digital form. Consider, for example, a linear displacement measuring device containing a straight line of regularly spaced holes drilled through a flat plate (Figure 2.6). The plate is part of a moving trolley to which is fixed the measurand along the axis of measurement. A light source is placed near one hole with an optical receptor on the remote side. As the line of holes moves with the trolley under the light source, the receptor outputs oscillations corresponding to the number of holes passed and as a function of the distance traversed. The oscillations are analogue but can be converted to a square wave using a 'Schmitt trigger' circuit. The digital signal is then interpreted to calculate the required displacement.

This form of measurement is used by machine tools to achieve accurate machining along an axis. Typically, the movement of a flat bed holding a workpiece being machined can be monitored by turning a gear wheel and recording the displacement on a digital display.

Widely used coordinate measuring devices employ such digital measurement along three orthogonal axes (Figure 2.7). Displacements along the three axes are recorded by moving a small contact sensor (or probe) from one fixed point to

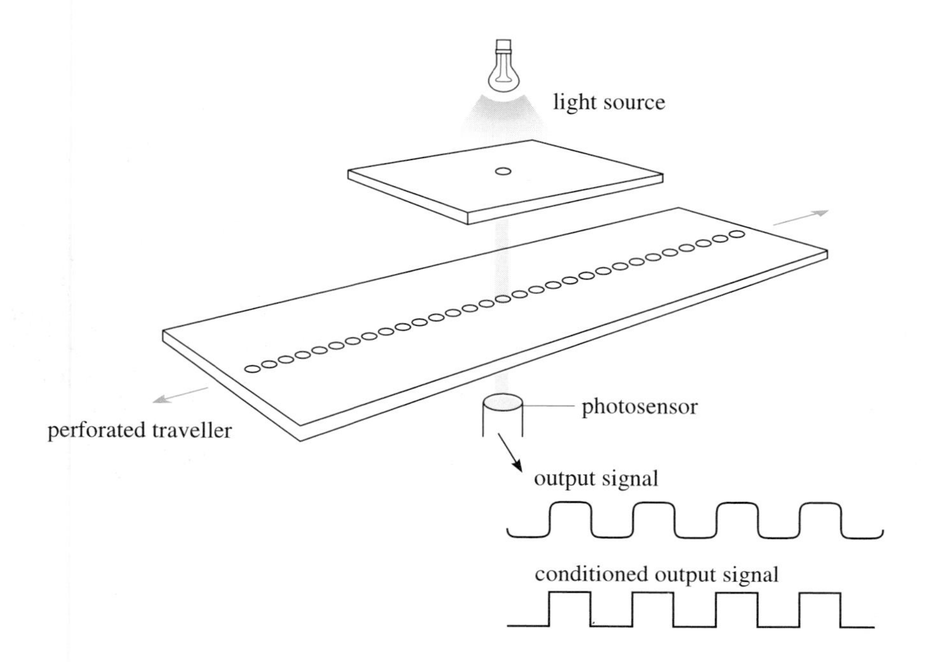

Figure 2.6 Linear displacement measurement.

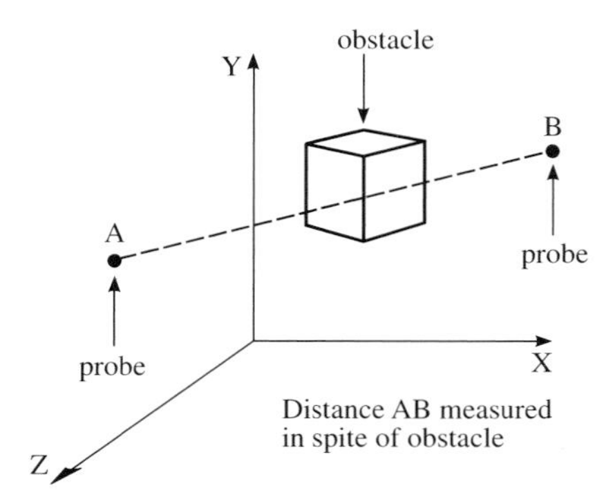

Figure 2.7
Digital measurement using coordinate measuring machine.

another. The linear displacement between the two points is then automatically calculated using coordinate geometry and displayed in digital form. This is particularly useful when the path between the points contains obstacles which make traditional measurement methods impossible.

Angular encoders

Angular digital encoders are a common form of encoder in which a rotating disc has an angular velocity which is a function of the measurand parameter (Figure 2.8). The disc is divided into concentric tracks (typically eight), each of which is monitored by a photocell on one side. Opaque and clear divisions along each track are 'recognized' when light is shone from the other side of the disc. Various alternatives are available for the opaque/clear patterns so that the device can be used to record absolute positions (*absolute encoder*) or relative displacements (*incremental encoder*). Two types of encoding pattern could in theory be used for absolute encoding, shown in Figure 2.8. In practice the *Gray code* version predominates to the almost complete exclusion of *binary code*. Observing the black and white Gray code sectors we can see that when the disc moves from any

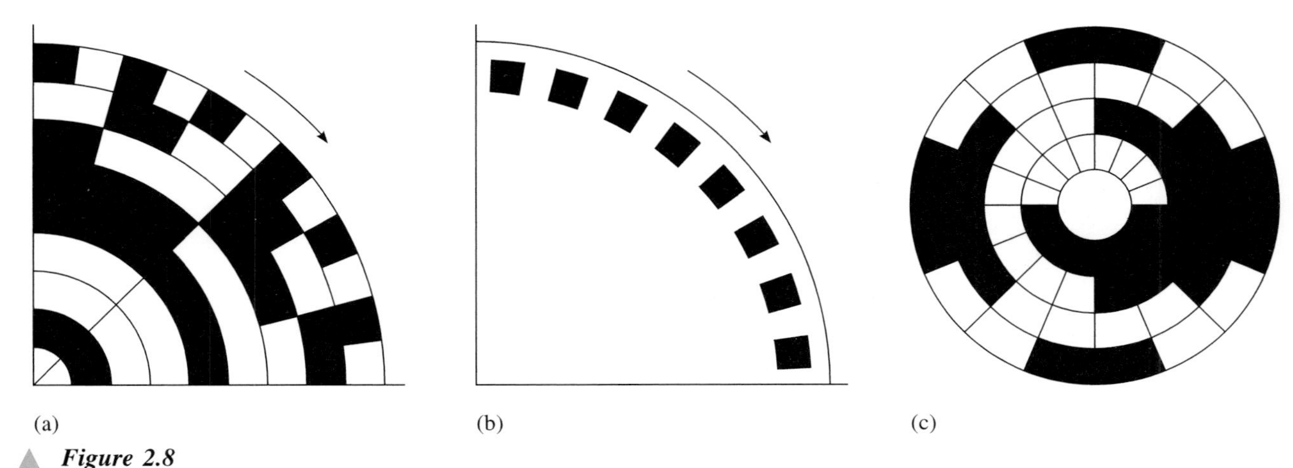

Figure 2.8
Encoder discs: (a) absolute encoder, which indicates position as all sectors are uniquely identified; (b) incremental encoder, which measures angular velocity; (c) Gray code encoder.

sector to its neighbour only one pattern change takes place. This is easy to recognize electronically. With binary code we can identify examples of two or more pattern changes as the disc moves across sector boundaries. Careful alignment of light sensors along the sector and timing of electronic circuits would be necessary to ensure that combinations of pattern changes could be recorded consistently. As with the previous example, an oscillating output can be converted to digital form for convenience of use. A typical use of encoders includes the control of revolving parts in machine tools.

2.4.6 Performance characteristics

Sensors exploit a number of different physical phenomena so that no uniformity of performance can be expected. For example, output signals may be simply related to input signals or may be a complex function of them. The relationship may be well understood over a certain range of parameters but unpredictable outside the range. It is therefore essential to select a sensor in the full knowledge of where its behaviour is understood and can be relied upon. A number of factors must be taken into consideration, as discussed below.

Linearity/non-linearity

The relationship between the input and output signals must be known for the range of applicable parameters. Linear relationships are clearly the easiest to handle, since the direct proportionality allows simple calculation of output. Non-linearity of output may be understood in terms of a mathematical function allowing output signals to be interpreted by calculation. In many cases only an empirically described relationship between input and output exists. If this is expressed graphically, curve-fitting techniques can be used to define the relationship mathematically and store it in a microprocessor as a polynomial or as selected values in a look-up table covering the desired range of values.

Range of operational accuracy

Closely related to the definition of the input/output signal relationship is the more general question of operational accuracy. If the measurand parameter is chosen to vary over a small range (known as the ***operational bandwidth***), inaccuracies due to non-linearity are minimized.

A rule of thumb which design engineers try to follow is to make the operational accuracy around an order of magnitude (i.e. ten times) better than the calculated necessary accuracy. For example, a control system that operates successfully with parameters sensed to an accuracy of ±5% needs a sensor that is accurate to within ±0.5%. This allows for unusual circumstances and possible degradation of the system without the sensor itself affecting performance through its own inaccuracy.

Repeatability

Repeatability of operation is important to considerations of accuracy. Physical changes to the sensor can cause drift in measurement. Regular calibration may be a solution. At worst, the physical changes are continuous (as, for example, with elastic 'creep' and metal fatigue), making the device impossible to calibrate permanently.

Another aspect of repeatability is the common problem of *hysteresis* (Figure 2.9). This is the situation in which inaccuracy is caused by a change in direction of the input signal. For example, in mechanical systems it can be the result of looseness of coupling, often referred to as *backlash*. In systems with large bending forces (or torques), deformation of materials may similarly give misleading values. Hysteresis also occurs in magnetic devices where the magnetizing and demagnetizing processes suffer a delay on change of direction. Such errors can be compensated for by always approaching measurement from the same direction.

Responsiveness to input signal

The frequency of sampling is also an important consideration. Sensors need to respond fast enough for the control system to be effective within its operational timescales. If the response time is not short enough, instability will occur, often seen as *hunting*, in which the control system perpetually oscillates around a desired output. The result is unsteady performance.

Reliability

Reliability includes the concept of repeatability referred to above, but is also concerned with more familiar issues of physical reliability. For instance, some sensor technologies are capable of great accuracy but may be physically delicate and easily damaged, so robustness of a sensor is a significant factor in its selection. It is vital for a manufacturer to understand the environment in which a sensor must be reliable. There are significant marketing and legal factors which influence the design and testing of sensors to ensure that acceptable reliability is achieved.

2.4.7 Ergonomic factors

It must not be forgotten that even highly automated control systems are ultimately controlled by humans. Ergonomic factors cover those aspects of sensor technology where human beings act upon the data output and are responsible for sensor operation. Typical aspects include:

1 *Human–machine interfaces* Sensor outputs must be designed ergonomically, to allow for easy monitoring where necessary. The use of confusing output signals can result in systems being operated inefficiently. At worst

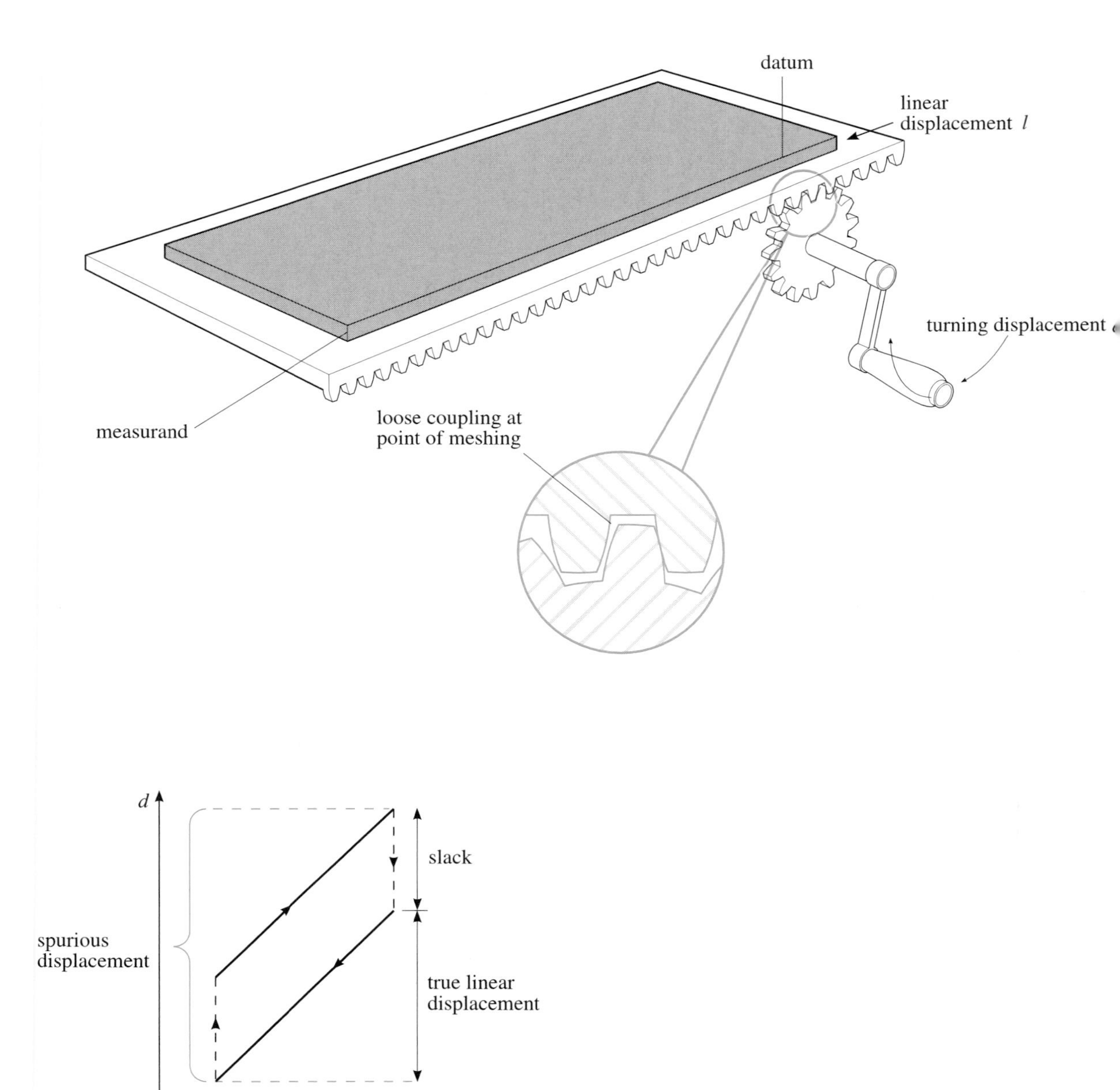

Figure 2.9 Hysteresis in linear measurement.

they can result in damage and danger. Typical issues are whether to produce analogue or digital output; how to group outputs on an instrument panel; how to warn of danger situations.

2 *Ease of fitting and repairing* Sensors must be accessible to allow for ease of connection and disconnection to avoid the expense of lengthy maintenance.

2.4.8 Economic factors

Finally we should not forget the many economic factors that affect the choice of sensor.

1 *Cost* Costs of different technologies vary enormously. Individual sensors may cost from less than a pound to several thousand pounds. Trade-offs are made between the cost of sensors and the number used in a control system. Optimization of costs for complete systems has been a subject of much research. One principle followed has been to process the output signal by electronic circuitry at the earliest possible opportunity. This implies choosing a sensor technology with characteristics that are well understood and appropriate for the application, in order to avoid complex correction. However, an alternative view is that microprocessor power is now cheap enough for conditioning of the signal to be carried out by software, relieving the constraints and costs of sensor design.

2 *Packaging* The ultimate commercial success of a sensor is very much bound up with the quality of its packaging – that is, the way the sensor is encapsulated for practical application. Often, quite delicate sensor technology must be located in hostile conditions, as in automotive and aerospace applications. It must be easy to install and maintain, robust and reliable and yet small enough not to interfere with machine operation.

3 *Market volume* Market volume is usually the overwhelming factor in the cost of manufacture. For anticipated large volume production, considerable savings can be achieved by using automated production. Small volume production using skilled assemblers can be much slower and more expensive.

2.5 Sensors for intelligent machines

Traditional control systems are designed to work in complex, highly variable but nevertheless deterministic situations. Once sensors have been developed to operate effectively within specified limits, following all the criteria described above, it can be assumed that they will produce unambiguous output signals.

Machines possessing artificial intelligence are, however, required to operate under conditions of uncertainty in which individual sensor output will not necessarily provide sufficient data for the control system to take sensible action. In such circumstances ***sensor fusion*** is needed, in which the output from two or more selected sensors are utilized. An analogy would be the use of three human sensors in identifying a situation:

Stage 1. On coming home late at night my *hearing sensors* locate a noise source as being in or near the kitchen. Is it the dog? A burglar? A member of the family or a bona fide visitor?

Stage 2. My *smell sensor* identifies food cooking. Conclusion: not the dog; probably not a burglar, but still a potentially dangerous situation.

Stage 3. My *visual sensor* identifies daughter's suitcase on the floor.

Conclusion: not a burglar; daughter has arrived home late and hungry.

Sensor fusion has provided enough information for intelligent reasoning to take place and a conclusion to be reached. Reliance on information from any one or two sensors would not allow the same conclusion to be drawn reliably.

2.6 The needs for sensors in mechatronics

The following variables need to be detected by intelligent machines, particularly those that move around and have to find their way around a changing environment:

- presence of an object at a distance (technologies available include bar code recognition, optical, ultrasonic and passive infra-red techniques);
- distance (in some cases the detection of an object automatically involves measuring how far the machine is from the object);
- chemical composition;
- temperature;
- pressure;
- air flow;
- acceleration;
- angular velocity;
- speed.

We will examine some of these in a few case studies.

2.7 Case study: use of sensors in a prosthetic hand

This case study illustrates the use of sensors in a mechatronic device in both accepting stimuli from the external environment and in monitoring the internal condition of the device.

2.7.1 Functional needs of the hand

Prosthetic (or artificial) hands are usually cosmetic or, at best, have limited functionality in being hook shaped. There is no widely available prosthesis that looks like a hand and has many of its capabilities. Dr Peter Kyberd and Dr Paul Chappell have developed a mechatronic hand at the University of Southampton in England that fulfils these requirements (Chappell and Kyberd, 1991). It uses three sensor types with an electronic control system.

The basic concepts behind the design are as follows:

- Gripping action is achieved by bringing together the thumb, index and middle finger from a slightly curled starting position. The two remaining fingers do not act independently but follow after a short delay.
- Gripping is either a precision grip, in which the thumb and forefinger come together at the tips of the digits, as in picking a small object, or a power grip where the fingers curl round an object such as a suitcase handle and the thumb follows in a locking position.
- The ability to lift an object requires the sensitivity to hold it tightly enough to maintain a grip on it but not so tight as to crush it. Compare lifting a delicate object and picking up a heavy suitcase. In the prosthetic hand sensitivity is achieved by recognizing slip and tightening the grip until it stops.
- The stimulus to operate the hand is the contraction of muscles located (for preference) in the lower arm. To someone unfortunate enough to have lost a hand (but retained the use of arm muscles) this has some similarity with the original function in both feel and effect.

Achieving these principles was a considerable feat of mechatronic engineering, details of which are not covered here. However, the sensor components, both simple and ingenious, are worth considering here. The requirements are for sensors to recognize:

1. the external stimulus of muscle contraction;
2. the internal state of the device, such as the relative position of the thumb and forefinger, and the pressure being exerted by parts of the fingers;
3. the external condition of a gripped object, whether it is held firmly or is slipping.

Sensors must have: low power consumption to preserve battery life; require little maintenance (at around six-monthly intervals); be easy to service, recalibrate and replace; withstand low-power electromagnetic radiation; and operate within everyday ranges of temperature, pressure and humidity.

2.7.2 Range of sensor characteristics

Myoelectric control of gripping

When muscles in the human arm are flexed to initiate the movement of the thumb and fingers, small voltages known as *electromyograms* are briefly developed. Two signals, from a pair of muscles, are electronically conditioned to instruct the controllers. Sensors in the form of small conducting pads are set within an arm collar so that they are in close contact with the selected muscles. Amplification circuitry enables the electromyograms to be recognized and responded to.

Position sensing by potentiometer

Rotary potentiometers provide a simple method for monitoring the angles between joints of the hand. They are set at selected pivot points on the index and middle fingers (Figure 2.10) corresponding to the equivalent knuckle joint. On the thumb they are used to indicate proximity to the fingers and also the angle of swivel about the thumb axis. This provides the controller with information to

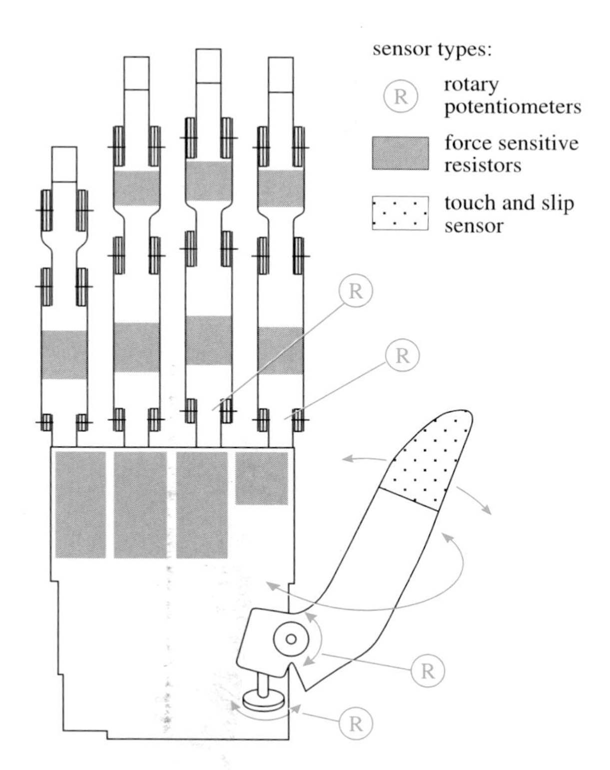

Figure 2.10 Location of sensors in the Southampton prosthetic hand.

distinguish between the various hand states. For example, as the joint angles decrease, the hand approaches the clenched state.

Force-sensitive resistors

Sensitive pads on the fingers and palm of the human hand give feedback on the pressure between objects and the hand. The information is used in determining shape and texture. Analogously, the prosthetic hand uses pressure-sensitive resistors, in the positions marked in Figure 2.10.

From the many technologies available, *conductive inks* were selected. These are impregnated within a fabric base and have the property of changing resistance when compressed. They form cheap, reliable devices of great sensitivity.

Touch and slip sensor

Two sensors are closely located to form a single device used in recognizing that a gripped object is slipping and recording pressure changes so that an appropriately tighter grip can be applied. The structure is illustrated in Figure 2.11. A hearing aid type microphone picks up the slippage vibrations caused by friction between the skin surface and the loosely gripped object. Tightness of grip is sensed as contraction of the rubber tube. A phototransistor emits light which is sensed by the photodiode. Distortions of the tube profile are recorded as changes in light intensity and are calibrated to represent pressure changes.

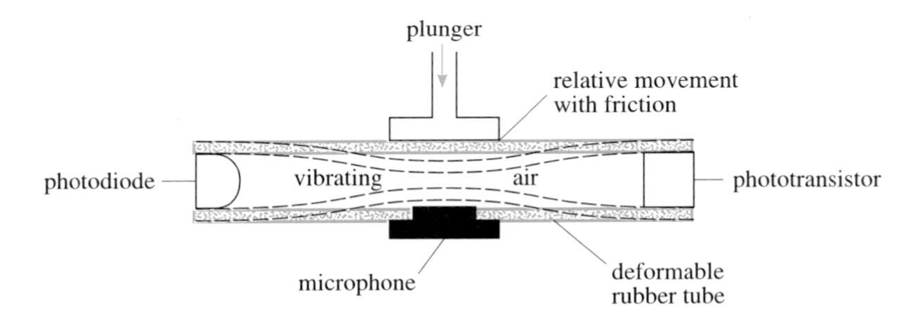

Figure 2.11
Optical/acoustic touch and slip sensor.

The hand in operation

The combination of sensors is used by a microcomputer control system to respond to the user's instructions to operate the hand. In the process the control system must record the continually changing geometry of the hand's components and its reaction to the task undertaken. This is shown in a simplified manner in Figure 2.12. Sensor feedback is used to determine the required action of electric motors to move the digits.

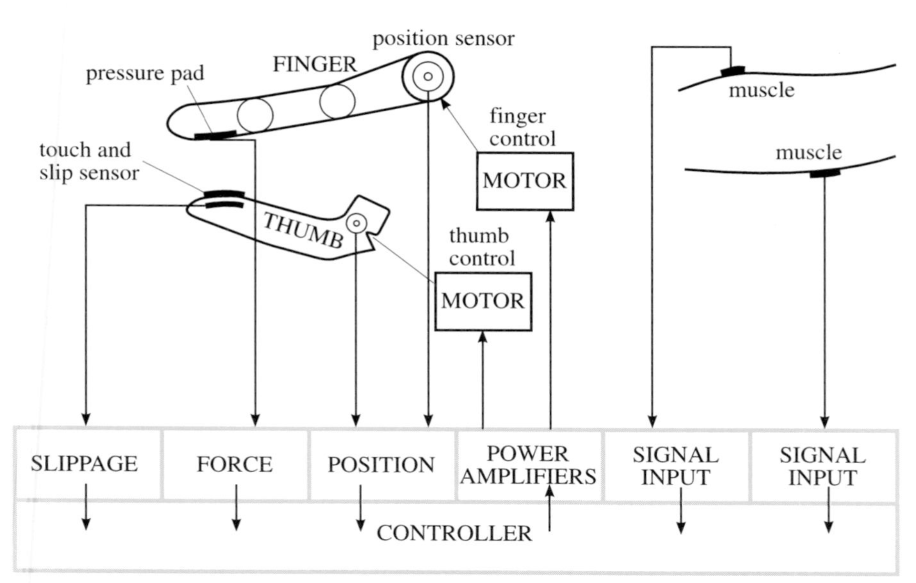

Figure 2.12
Use of sensors in the control of the digits of a prosthetic hand.

2.8 Case study: camera sensors

Modern cameras provide an excellent example of rapid mechatronic developments as they show the progression from precision automatic cameras towards those with intelligent behaviour. Even fuzzy logic (logic capable of handling uncertainty; see Volume 2) can be used to make automatic decisions for the photographer when difficult lighting situations occur. Sensors provide information for the control system to carry out accurate focusing and determine the appropriate exposure for the image photographed. This section refers specifically to features of the Nikon 'professional' F90 camera and acknowledges the assistance of John Clements, Photographic Adviser at Nikon UK Limited. Only selected functions are referred to here and in a simplified manner. The F90 possesses many other advanced features.

2.8.1 Functional requirements

As one would expect, modern cameras must at least be able to focus automatically on objects viewed at the centre of frame, with the option of user-override for focusing on other objects or for manual setting. Additionally the F90 provides a number of refinements, including:

1 Automatic focusing, achieved by measuring the degree of out-of-focus and calculating how far a lens must be moved, rather than by measurement of

distance and setting the lens position. Focus can be wide area or 'spot focus' for a very small area. Also it can be accurately carried out in extreme low-light conditions.

2 Automatic tracking to focus on a moving object and anticipate the subject's movement while taking a sequence of action shots. This results in focus tracking even after the shutter has been released, and a better defined image of a moving object.

3 Automatic metering to control the light exposure on a film, according to the distance of an object, its illumination and its contrast with the surroundings. Fuzzy logic is used to make decisions about which area of a scene is to be taken into account. This is particularly useful when a scene is constantly changing.

4 Intelligent control of flash illumination to take into account scenes with highly reflective backgrounds and shiny objects, and scenes with distant backgrounds but with a prominent foreground. The F90 can also control repeated flashes and vary the precise time of flash during the period of exposure.

2.8.2 Sensor facilities

Light enters through the main lens as a parallel beam and is divided at an angled mirror (Figure 2.13 overleaf). The mirror is semi-silvered and has many thousands of pin holes, which enable incoming light to reach the three main sensors. Around 30% of the incoming light passes through to the autofocus sensor module and flash control unit. The remaining 70% of the incident light is reflected in the opposite direction towards the matrix sensor, which controls exposure.

3-D matrix and advanced matrix metering

The function of exposure metering uses an eight-segment silicon photodiode. The term '3-D' refers to the three dimensions of *brightness*, *contrast* and *distance* taken into account when calculating exposure. Exposure metering seeks the optimum combination of aperture and shutter speed. In matrix metering the camera assesses the picture area in small segments, eight in total. The brightness of each segment is used to determine whereabouts in the picture area differing levels of light are located. The camera has been programmed to adjust the exposure automatically, but subject to the photographer's instructions.

Fuzzy logic enables the camera to arrive at appropriate exposures for a range of different types of composition. Examples include: landscaping with sharp focus across a large depth of field; portraiture with shallow depth of field and a less intense background; sporting action with slight blurring of movement; silhouetting to keep identified features dark against a bright background.

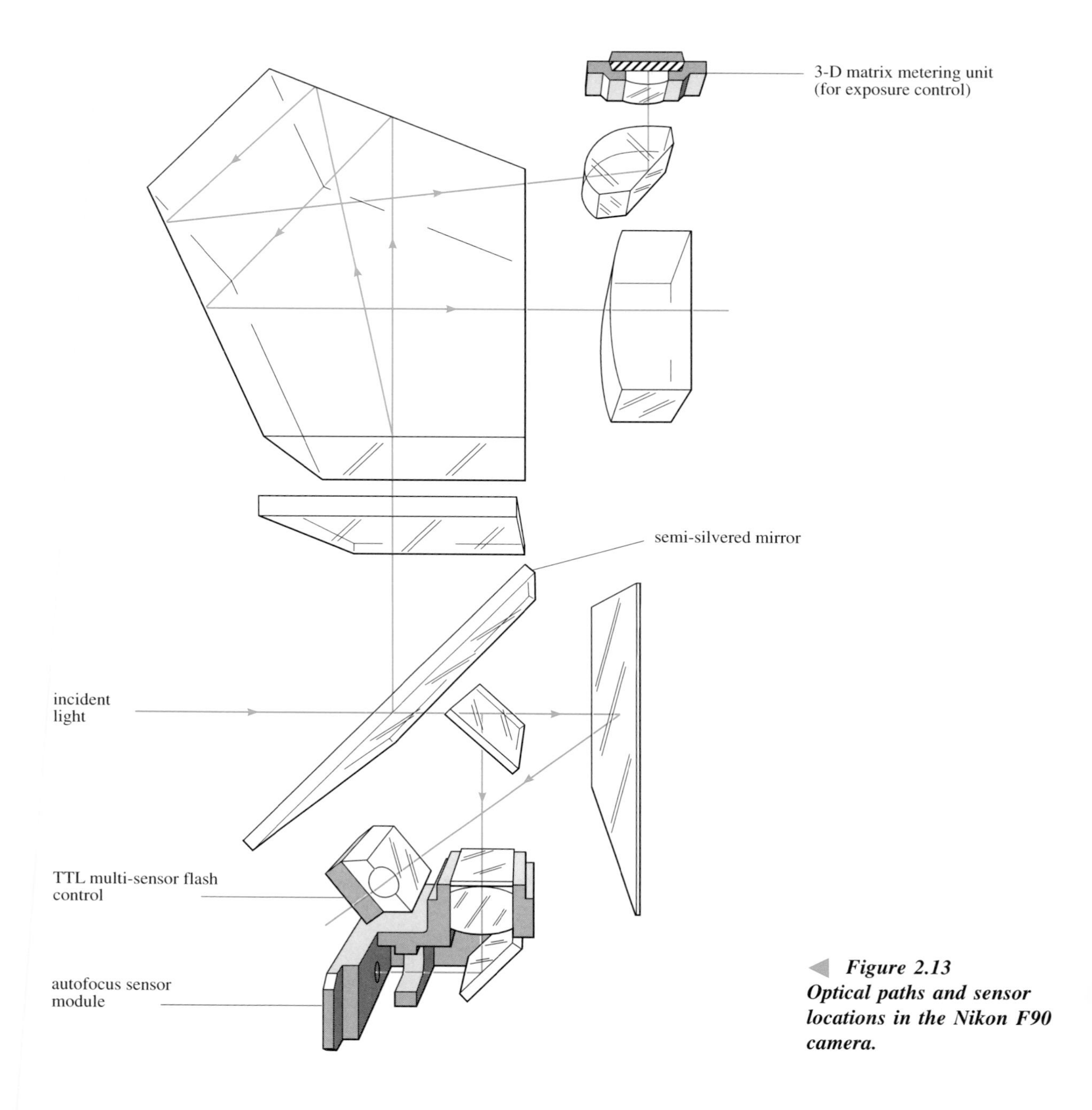

Figure 2.13
Optical paths and sensor locations in the Nikon F90 camera.

Cross-type autofocus

This uses a charge-coupled device (CCD) employing *phase detection.*

Electrical charge is generated in proportion to the light incident on the 246 CCD elements which are located in a cross-shaped mask. Light from a lens is converged through the mask and four separator lenses which project four images of the cross shape onto another CCD-covered surface. The four images are sampled in pairs in order to determine the focus status in both horizontal and vertical directions. If the output is the same then focus has been achieved. Otherwise calculations are made to determine how far the focusing lens must be moved.

In conditions of near or total darkness a light sensor can detect the need for illumination, which is then provided by a light-emitting diode from a flash unit.

3-D multisensor balanced fill-flash

The purpose of this sensor is to record the light intensity for five different segments of the viewed scene. It is therefore referred to as a multisensor and is able to indicate to the control system if there are particularly bright objects in view. This allows the exposure metering to guard against overexposure. The information is also used by the flash control to ensure that closer, darker objects have additional lighting. They would otherwise be underexposed.

The flash unit has the ability to fire off 'pre-flashes' which allow the exposure system to adjust to the artificially raised level of illumination before finally exposing the film.

2.8.3 Integrated sensor system

The three sensor types are combined as an integrated system, as shown in Figure 2.13. Individually, the three main sensor types (plus the illumination level sensor) provide feedback used in the familiar functions of focusing and controlling exposure of the film to light. In such an advanced camera they should not be viewed in isolation. It is in their integration by means of intelligence that sensing fusion takes place. The microcomputer system, with its 1 kilobyte RAM and 32 kilobyte ROM, provides programmed intelligence and a packaged user interface. This allows the user to specify particular qualities required in a picture and leaves the camera to calculate the parameter settings. In case this attracts the criticism that the camera may be too autonomous in imposing its own style of photography, it is worth noting that the photographer can override any functions by setting them manually.

2.9 Case study: sensing for autonomous vehicles

Driverless vehicles encountered in factories include a range of mobile transporters. At their simplest they move along prescribed routes and possess no intelligence. More complex *autonomous vehicles* (AVs) have been developed to find their way among still and moving obstacles and pursue route-finding goals.

2.9.1 Range of sensor technologies

The complex behaviour of AVs requires sensing analogous to human vision to continuously determine their own location and to explore what obstacles are present for avoidance and navigation. Objects encountered may be in contact, nearby (up to a metre or two) or more distant. AVs may need to identify an object, track it and even trace its shape. Sensors may also be used by an AV to record its own relative movements in a 'dead reckoning' form of navigation, i.e. the technique of ***odometry***.

To summarize, the following list covers the wide range of technologies required by AVs, some of which are further discussed in the subsections below.

1 *Internal sensors*:

odometry, which uses optical encoders to measure relative movement in proportion to the number of drive wheel rotations;

inertial guidance systems, which use accelerometers.

2 *External sensors*:

(a) Contact sensors (tactile and non-vision). Sensors may respond to light pressure by contacting and producing a binary output, i.e. either in contact or not. Alternatively, stress sensors are used to measure the force of contact from near zero to very large forces or torques. There are various cheap, simple methods, including:

on/off touch, microswitches;

strain conductive silicon rubber, thin film silicon, conducting inks.

(b) Distance sensors, including:

single photocell

wire guidance and painted strips with optical sensors

infra-red

laser

vision, either binary or Gray code

vidicon television

solid-state television

fibre-optic sensors

ultrasonics.

2.9.2 Proximity sensors

These are needed to monitor the proximity of an end effector to the object approached. Devices include:

optical and infra-red methods;

eddy-current detectors in which an alternating current in a probe tip induces eddy currents in a conducting material located up to 1 mm distant; the resulting change in flux is picked up by the probe tip;

magnetic field sensors, e.g. reed switch;

Hall effect (also see Section 2.10.2) and magnetoresistive devices;

touch-sensitive (capacitive) devices, in which the natural capacitance of a large object changes the resonant frequency of a tuned circuit.

2.9.3 Visual sensors (television)

We mostly associate television with broadcasting cameras, known as vidicon. However, these are slow – 30 pictures per second, produced by a scanning electron beam. Vision as required in the factory situation is more usually provided by solid-state television cameras, in which images can be captured at a much faster rate, in excess of 2000 per second, which is easily comparable to the rate at which images can be analysed by computer. One technique available for speeding up the process of analysis is the use of electronic pre-processing of the image within the solid-state camera.

2.9.4 Ultrasonic techniques

Ultrasonic waves are compression waves of high frequency, typically in the range 500 kHz to 10 MHz, well beyond human audibility. Their use is non-invasive and does not involve the transmission of electric current. They are therefore invaluable in chemical and high temperature environments. The low power emissions are no danger to humans and cannot cause ignition. Normal methods of generation use piezoelectric transducers in which an alternating e.m.f. is applied across the faces of a piezoelectric crystal, causing it to vibrate and produce pressure waves. The detection of ultrasound uses the reverse effect. Continuous or pulsed ultrasound is usable. Pulsed methods (sonar) allow the same device to be used for transmission and detection. Reflection times can be measured so that distance can be automatically calculated.

Ultrasound penetrates solid materials, but to a lesser extent than some electromagnetic waves. Acoustic mismatch can cause attenuation of the signal when it passes through a mix of materials. Ultrasonics is useful in AV navigation in being able to recognize the presence of objects at a distance of a metre or so to an accuracy of 1% or better. For example, an object 1 metre distant can be easily located to within 1 cm, good enough for avoidance, but requiring optical methods if manipulation of an object is required. On comparison, laser range finding is more appropriate for distance measurement from metres to hundreds of metres to an accuracy of around 0.01%. Ultrasonics cannot be used effectively where shape recognition is required, because it cannot distinguish between large and small items in close proximity. Techniques of ultrasonic imaging are well advanced for medical and industrial use (tomography) where images are scanned by humans. Cheap automated methods for use in AV navigation are likely to be available within a few years.

2.9.5 Mobile robots project

The mobile robots experimentation at Oxford University, England, is in response to the need for more intelligent, free-ranging vehicles. It anticipates future mechatronic devices for domestic, factory, hospital, office delivering and cleaning functions.

Robots need sensors to enable them to move around flexibly without damaging themselves or others, i.e. to understand their environment in its ever-changing state. A number of different sensor types are used to achieve more useful knowledge by sensor fusion.

One example of advanced AV is the GEC/Oxford University 'Turtle', which uses infra-red techniques to locate its position. An infra-red laser searches for bar-code targets in fixed positions on surrounding walls. A passive acoustic system is used to pick up and localize a sound source (Brady *et al.*, 1990).

2.10 Case study: automotive application of sensors

In writing this section I gratefully acknowledge the help of Dr Steve Prosser of Lucas, Advanced Engineering Centre, Solihull, West Midlands (see also Prosser, 1991).

Briefly described here is the application of some of the sensor technologies in current use in the automotive industry. No claim is made for comprehensiveness: sensors pervade an enormous range of engineering products and use a wide range of technologies. As an illustration, catalogues of electronic devices typically list over 100 different sensor products readily available over the counter. Let us now consider the range of automotive sensor in current and future use.

2.10.1 The role of sensors in cars

A limited number of passive sensors are familiar to us, particularly speed measuring devices, level indicators for petrol and oil-pressure indicators in engines. Output is normally directed to indicators on the dashboard so that the driver can monitor the situation and, if necessary, take appropriate action such as: slow down; insert petrol; investigate suspected oil leak.

The current trend is for cars to include many more sensors linked to a microprocessor control system, bypassing human control. The intention is for the car to monitor a number of parameters and make adjustments for safe and efficient running. This is the result of legal and marketing requirements, partly inspired by environmental concerns. Broadly, the areas for improvements are in:

fuel economy and gas emission;

steering, braking, accelerating and decelerating safety;

comfort, driveability and performance;

information provision, required in fault diagnosis and vehicle navigation.

The focus of sensor research and development is being directed towards:

- smaller, less intrusive, more reliable and cheaper sensors;
- the use of microprocessors to refine the data received and create smart sensors to contribute to more intelligent control of cars;
- more integrated communication between transducers and control system, achieved through databus techniques where signals share conductors (traditional car-wiring looms require separate connections for each application and therefore use more wiring and junctions);
- the use of sensor data, combined with artificial intelligence, to provide sensor fusion;
- reduced or 'sensorless' applications. Unnecessary monitoring is avoided or carried out by alternative methods. For example, diagnosis of gear-box faults carried out using numerous sensors has been (experimentally) replaced by monitoring the characteristic operational sound patterns in the gear box for sudden changes. A sonic recorder replaces a number of vibration sensors.

2.10.2 Engine performance sensors

Engine parameters have been usefully monitored for some years. These include the following (the sensing method used is in brackets):

manifold absolute pressure as an indicator of engine loading;

camshaft position (Hall effect) – the *Hall effect*, illustrated in Figure 2.14, is encountered when a metal or semiconductor plate is held at right angles to a magnetic field and an electric current passed through the plate; a drift of charged particles to the edges is observed as a voltage across the plate, larger with semiconductors than metals; it can be used to detect either current or magnetic fields and is useful in non-contact devices where a magnet is rotated past a semiconductor;

crankshaft position (variable reluctance sensors);

exhaust gas (ionic conductors);

knock (piezoelectric devices);

air mass flow (hot wire/hot film) used as an alternative to measurement of engine loading, particularly with larger cars;

temperature (thermistors).

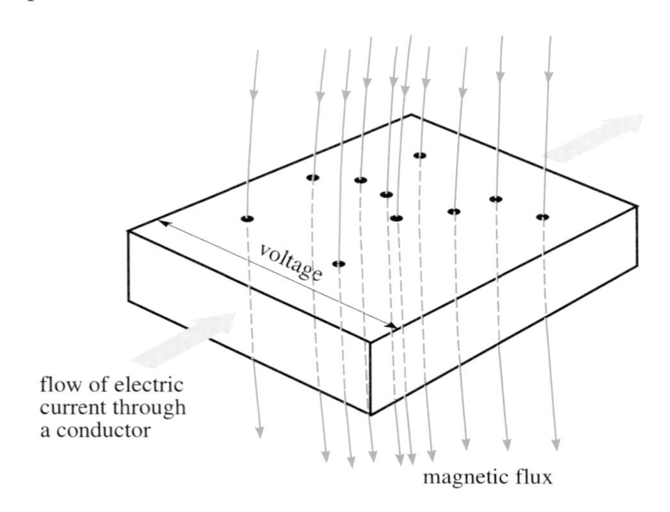

◀ *Figure 2.14*
Hall effect.

Let's look at these in some detail to identify the sensing methods and the rationale behind their choice. The starting point is the requirement for the car engine to respond to the demands of the driver.

Control of engine power

We exert 'driver demand' by pushing or releasing the accelerator pedal. The result can be sensed from either the inlet manifold pressure or from the engine output torque. The latter is preferable but is currently more difficult to measure because torques (twisting forces) are indicated by very small angular displacements along cylindrical components such as a drive shaft. Optical methods aren't easy to implement and contact methods can be too obtrusive, possibly affecting power output. It is therefore more usual to measure the inlet manifold gas pressure.

Gas pressure is measured using silicon diaphram methods in which strain gauges made of piezoresistive material are implanted on the surface of a thin silicon wafer (Figure 2.15). Piezoresistive crystals respond to pressure changes by producing a small electric potential and are sensitive to their alignment. When four are used in a single sensor they are arranged in the form of a Wheatstone bridge, giving an output potential which is directly proportional to the pressure change.

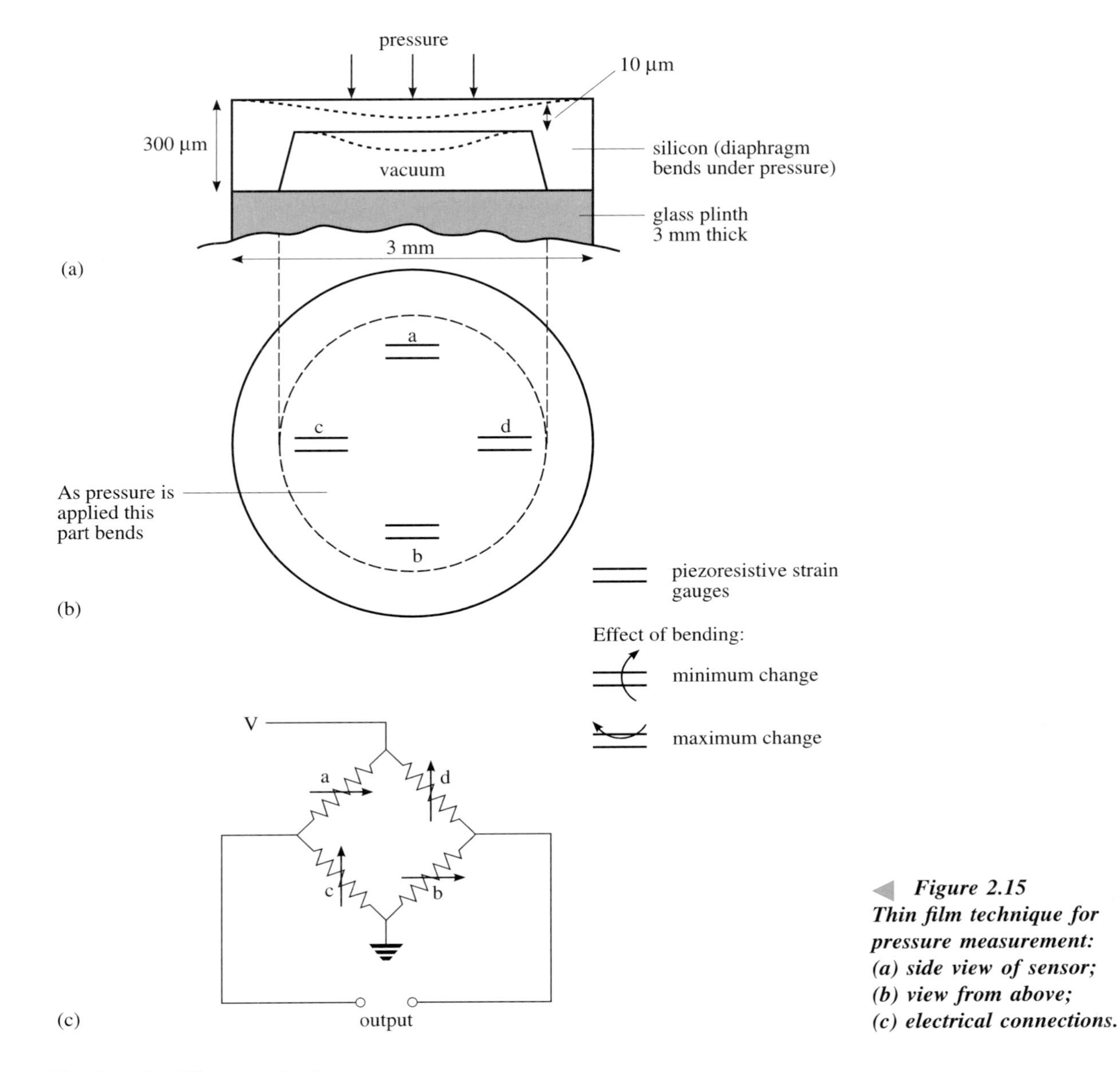

Figure 2.15
Thin film technique for pressure measurement: (a) side view of sensor; (b) view from above; (c) electrical connections.

Engine ignition control

Closed-loop control of fuel ignition has the aim of timing ignition for when the correct mix of oxygen and fuel vapour has been achieved. Lean burn conditions exist when using a higher proportion of oxygen than in conventional carburettor engines. The result is improved fuel economy and the complete combustion of fuel in order to reduce harmful emissions. Timing is therefore critical to efficient operation and is best achieved by taking crankshaft position as a datum point. Hall effect and variable reluctance technologies are commonly used, with magnetoresistive devices being developed for future use. Not all manufacturers

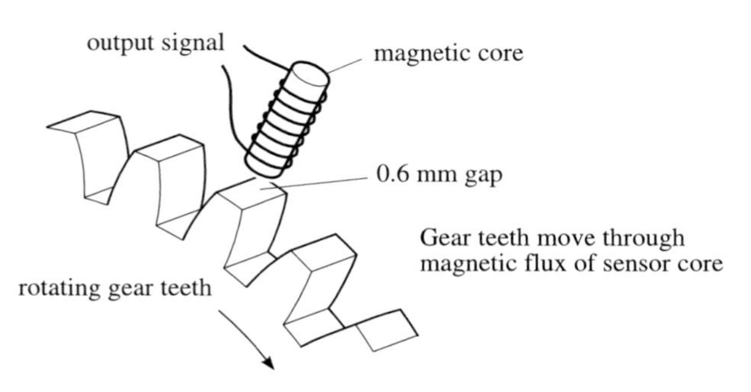

Figure 2.16
Variable reluctance sensor.

are actively pursuing the lean burn approach in isolation. European legislation on car emissions is now encouraging the use of catalytic converters as an arguably more effective, though more expensive, solution to reduce air pollutants in exhaust gases.

Variable reluctance sensors are used for wheel speed measurement in anti-lock braking systems with sensors located around 0.6 mm from toothed gears (Figure 2.16). Motion of the gear teeth changes the magnetic flux in the sensor's ferromagnetic core and causes it to act as an alternating current generator. From the output the wheel speed can be calculated. The current trend is towards the use of Hall effect devices which have a higher accuracy, use a larger air gap (up to 3 mm), and can detect zero speed (variable reluctance effectiveness disappears at around 3–5 mph).

Knock is a condition which must be recognized at an early stage. Drivers are aware of low-frequency *pinking* vibration which indicates the lag of engine timing control after the accelerator pedal has been pushed. The driver has indicated the need for a speed increase but the engine has not yet supplied sufficient power. Sensors can pick up knock at an early stage as a higher frequency vibration of around 8 kHz. Piezoelectric accelerometers can be used, in which physical vibration is applied as pressure oscillations on the crystal surface, resulting in AC electrical output.

Air mass flow

The measurement of air mass flow is significant to the control of lean burn operation. *Vane meter methods* (Figure 2.17) are traditional but decreasingly used. They depend on a lightly sprung vane being rotated by the momentum of air passing into the engine manifold. Movement of the vane is linked to movement of a potentiometer connection. Alternative sensors which are replacing vane meters include hot wire anemometers, in which a platinum wire is heated electrically and then exposed to the cooling effects of the air flow. The additional electric power required to reheat the wire is a function of the air flow. A second upstream heated wire acts as a temperature sensor for calibration purposes.

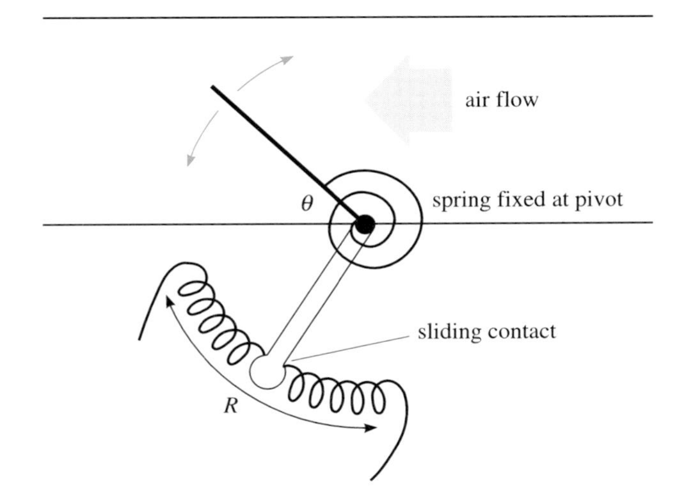

Figure 2.17
Measurement of air flow by vane meter method. Electric potential across resistance R gives angle θ, where θ is inversely proportional to air flow rate.

2.10.3 Movement and position sensors

Accelerometers

Accelerometers to measure horizontal and vertical forces (in units of g, the acceleration due to gravity) resulting from acceleration have wide application, concerned with any kind of car movement. They are of potential use in control of suspension, braking, steering and safety (crash) systems.

A novel form of accelerometer is described by Kellett (1992). It is known as a *servo* or *force balance accelerometer* (Figure 2.18) and has potential for cheap mass production that will allow the development of adaptive suspension for the lower cost end of the automobile market. As acceleration produces deflection of a small cantilever, identified by an optical system, a restoring magnetic force is applied to cancel out the deflection. The electric current providing the restoring magnetic force is proportional to the acceleration. It is an interesting concept that can operate effectively at low frequencies. With a volume of around 4 cm^3 it is larger than thin film silicon devices but quite small enough for automobile applications.

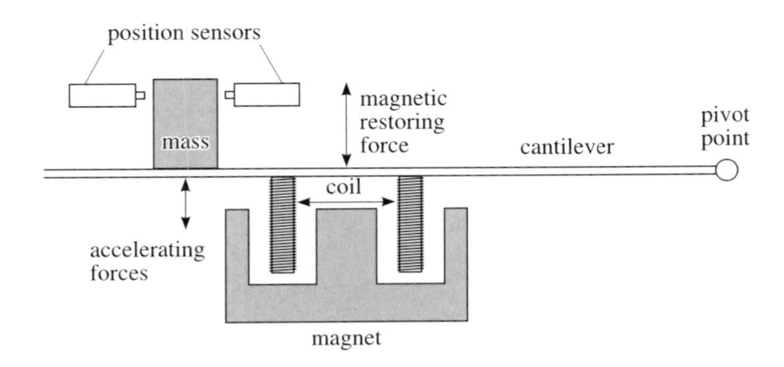

Figure 2.18
Servo or force balance accelerometer.

Wheel speed sensors

A number of applications, including speedometers, tachometers, anti-lock braking systems, transmission and engine timings require wheel speed sensors. In most cases they operate as alternating current generators, so that output frequency is proportional to rotational wheel speed. They are effective at high rotation but less so at low speeds when signal-to-noise ratio is small. There is a growing need for zero-speed sensing capability.

Non-contact position sensors

The need to avoid friction and wear is encouraging the development of non-contact devices, with the ultimate requirement of components which will function perfectly for the entire life of a vehicle. Applications include position sensing of throttle and exhaust gas valves.

The linear variable differential transformer (LVDT) is a commonly used device for measuring linear displacement. It takes the form of a transformer with a central primary coil and secondary coil set as two separate windings in opposition to the primary and on both sides of it (Figure 2.19). A ferromagnetic core is placed symmetrically along the central axis. Electric current through the primary produces no output in the secondary until the core is displaced and disturbs the balance between the opposiing coils. The displacement is related to the output voltage in a linear manner for most of its range. Rotary versions (RVDT) have a similar action and are used to measure angular displacement.

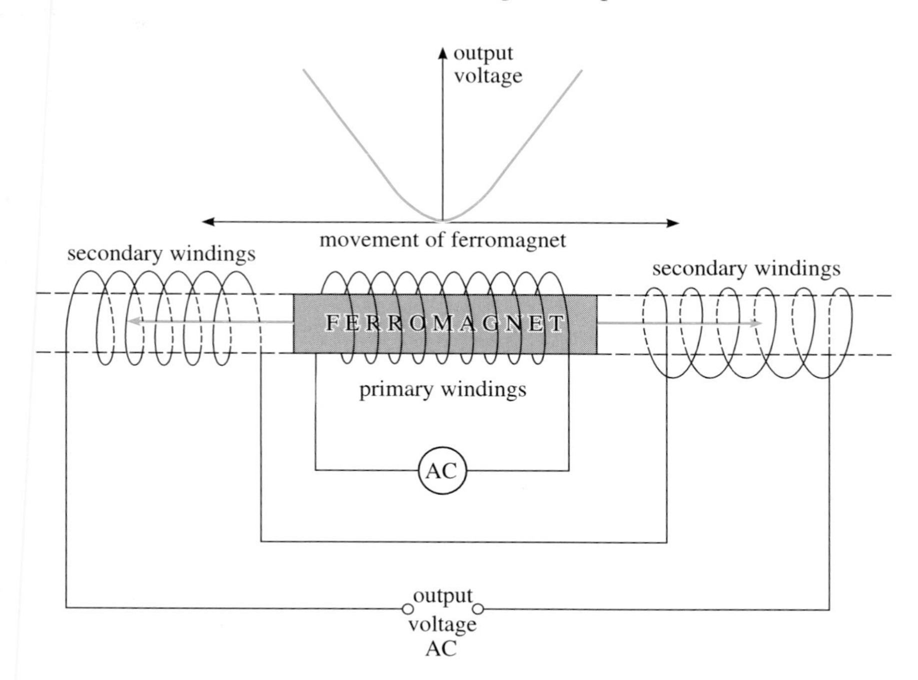

***Figure 2.19* Linear variable differential transformer (LVDT).**

Speed over ground sensors

Traditional systems have assumed that car speed can be calculated directly from wheel speed. However, when wheel slipping occurs, particularly with four wheel drive, true road speed cannot easily be calculated. Methods under test are based on microwave, ultrasonic and optical technologies but are currently expensive solutions.

Angular rate sensors

As well as moving forwards and backwards, round corners and up and down hills (Figure 2.20) cars can have angular motion. This is shown in Figure 2.21 as yawing, pitching and rolling movements. Gyroscopes are used in aerospace for angular motion measurement but are currently too expensive for automotive applications. Other possible technologies such as fibre optic and ring laser gyros, Coriolis effect and piezoelectric devices are under development. All have been used in aerospace applications but are now being adapted for automobile applications requiring lower tolerance and cheaper devices.

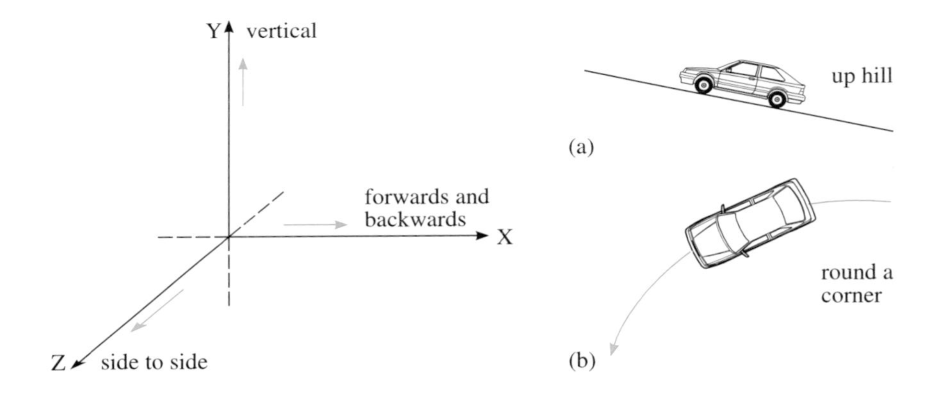

Figure 2.20
Linear motion of a car along three axes: (a) Y and X combined; (b) Z and X combined.

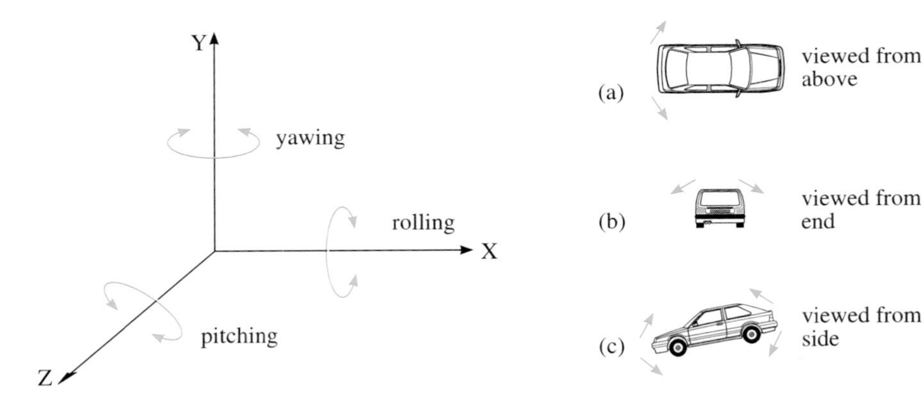

Figure 2.21
Motion of a car about three axes: (a) yawing; (b) rolling; (c) pitching.

2.10.4 Application of sensor fusion

As a result of the provision of a variety of sensors in a car, placed so they can monitor movement along and around three axes, sensor fusion can now be exploited. A few examples under development follow:

1. *Suspension* Conventional suspension is a compromise between soft springing, which isolates the car occupants from road disturbance, and hard springing, which reduces roll and pitch during cornering. An intelligent suspension system which is capable of anticipating road obstacles and of recognizing steering or pitching movements could be used to provide variable suspension according to the needs of the moment.
2. *Yawing, pitching and rolling* Similarly, such conditions of instability and discomfort could be recognized at an early stage and automatically controlled. It is particularly important that sensor fusion be sufficiently intelligent in its approach. Yawing, pitching and rolling are not necessarily dangerous conditions in themselves. Their significance depends on their characteristics and the speed at which they occur. A slow irregular yaw at 40 mph might indicate that the car is steering round bends in the road, whereas an oscillating yaw at 70 mph might indicate a car out of control. A deep pitching movement at very low speed might indicate the brake being applied, whereas rapid pitching at 10 mph might indicate a bumpy road.
3. *Skidding control* Wheel speed and vehicle speed are normally directly proportional to each other. When skidding occurs, the relationship changes. If this can be recognized immediately it starts, automatic avoidance could take place by means of deceleration. In the most complex situation, where skidding and rotating takes place, it is necessary to consider the speeds of all four wheels.
4. *Passive restraint systems* Activation of safety systems such as inflatable air bags can be triggered automatically in anticipation of a crash. Signals from accelerometers are received and interpreted fast enough for the bag to inflate and protect passengers from the impact.
5. *Driver monitoring* A more speculative but quite serious investigation is that of monitoring driver responsiveness. It has been shown that patterns of driver behaviour can be recognized by sensor fusion. The small movements of arm and foot (and eye) that characterize an alert driver are reflected in the speed, acceleration and direction of the car. Sensors that detect changes in the pattern could be used to indicate a drowsy or inattentive driver and trigger a warning signal.

The automotive examples briefly described above are part of the Prometheus programme referred to in Chapter 1. Enormous emphasis at a European level is being put on the development of safer, more efficient, comfortable and environmentally acceptable cars. Mechatronics is seen as the provider of artificial intelligence to control cars as well as many other mechanical artefacts.

2.10.5 Summary of automotive sensor technologies

By way of summary, Table 2.1 outlines the wide range of automotive sensors in current use. They vary considerably from car maker to car maker but are all aimed at monitoring the basic list of variables presented in Section 2.6. More details can be obtained from Westbrook (1988).

TABLE 2.1 SUMMARY OF AUTOMOTIVE SENSOR TECHNOLOGIES

Physical phenomenon	Application	Sensor type
Position	Accelerator pedal position	Potentiometer
	Throttle position	Potentiometer
	Gear selection position	Potentiometer or cam-operated switch
	Valve position	Linear displacement potentiometer
	Throttle closed/open indicator	Microswitch
	Gear-selector hydraulic valve	Optical encoder or position cross-correlation optical device or spatial filtering optical devices
Exhaust gas	Exhaust gas oxygen (stoichiometric operation)	Chemical ceramic composition: zirconium dioxide or titanium discs in aluminium
	Exhaust gas oxygen (lean-burn operation)	Chemical ceramic composition: zirconium dioxide oxygen-pumping devices
Temperature	Ambient air temperature	Thermistor
	Coolant air temperature	Thermistor
	Diesel fuel temperature	Thermistor
	Inlet manifold air temperature	Metal film or semiconductor film
	Diesel exhaust air temperature	Cr/Al thermocouple
Pressure	Inlet manifold pressure	Piezoresistive silicon with strain-gauged diaphragm or capacitive diaphragm
	Exhaust manifold (diesel) pressure	Piezoresistive silicon with strain-gauged diaphragm or capacitive diaphragm

TABLE 2.1 SUMMARY OF AUTOMOTIVE SENSOR TECHNOLOGIES - Continued

Physical phenomenon	Application	Sensor type
	Barometric absolute pressure	Piezoresistive silicon with strain-gauged diaphragm or capacitive diaphragm
	Transmission oil pressure	Differential transformer plus diaphragm or capacitive diaphragm
Air mass flow	Inlet manifold air mass:	
	(a) Unidirectional	Vane meter or hot wire (or hot film) or vortex shedding
	(b) Bidirectional	Ultrasonics or corona discharge or ion flow
Acceleration	Knock	Piezoelectric accelerometer
	Knock and misfire	Ionization measurement or strain gauge devices or force balance devices
Angular velocity	Yaw rate	Fibre optic or ring laser gyros or gyroscopes or Coriolis effect or integrated linear accelerometer or solid-state piezoelectric device
Speed	Crankshaft speed	Variable reluctance or microwave Doppler radar or ultrasonics or laser Doppler velocimetry or reed switch
	Distributor speed	Hall effect or optical digitizer with fibre-optic linkage or eddy current magnetoresistive devices – non-contact linear variable differential transformer (LVDT; Figure 2.19) and non-contact rotary variable differential transformer (RVDT)

2.11 Conclusion

At the boundary between the environment in which the intelligent machine operates and its control system lies the sensor, which must constantly sample required parameters and pass on an appropriate signal. Such devices are hybrid in nature, bridging the physical–mechanical outside world and the digital–analogue domain of the control system. They must be reliable and robust enough to withstand the rigours of the former, yet accurate and responsive enough to input data to the latter.

In understanding the role of sensors in mechatronics we must be able to select them according to a number of technical requirements, while not forgetting the economic and marketing constraints that engineering imposes.

References

Brady, M., *et al.* (1990) 'Sensor-based control of AGVs (automatic guided vehicles)', *Computing and Control Engineering Journal*, March 1990, pp. 64–69.

Chappell, P. H. and Kyberd, P. J. (1991) 'Prehensile control of a hand prosthesis by a microcontroller', *Journal of Biomedical Engineering*, Vol. 13, September 1991, pp. 363–369.

Kellett, P. (1992) 'Accelerometers for suspension control', pp. 75–80 in *Proceedings of International Conference on Mechatronics: The Integration of Engineering Design, Dundee*, Mechanical Engineering Publications Ltd, London.

Prosser, S. J. (1991) 'Advances in automotive sensors', pp. 493–504 in *Sensors: Technology, Systems and Applications*, IOP Publishing Ltd.

Westbrook, M. H. (1988) 'Automotive transducers: an overview', *IEE Proceedings*, Vol. 135, Part D, No. 5, September 1988, pp. 339–347.

CHAPTER 3
INFORMATION

Jeffrey Johnson and Philip Picton

3.1 Introduction

In mechatronics systems, perception concerns the ability of a machine to collect and make sense of ***information*** about its environment and its own behaviour within the environment. It does this through sensors, which have been described in the previous chapter. A good understanding of sensors and the *signals* to which they respond is fundamental in mechatronics.

There is a great demand for machines that are both mobile and can communicate with other intelligent information systems. In 1993 John P. Mello and Peter Wayner wrote:

> Two forces are at work to make wireless data communications one of the most important technologies of the next decade. The first is the trend to untether computers from the desktop. With every improvement in integration, miniaturization, and battery technology, the difference between the performance of desktop computers and portables shrinks, as does the premium you pay for portability.
> The second force driving wireless data communications is the desire for universal connectivity. Computers and their users are more productive when they have access to external data.
>
> *(Mello and Wayner, 1993)*

Extrapolating from Mello and Wayner's view, we can see that developments in communication which permit high volumes of information to be transmitted at low costs will lead to tremendous opportunities for the invention and design of intelligent machines. Certainly machines that can access remote data banks have the potential to perform better. Also this may obviate the need for powerful information processing hardware to travel with the machine. Since mechatronics systems must communicate internally it may be that wireless data communications will be superior to hard-wired carriers, even on quite small machines. For example, the present generation of computers may be the last that have the keyboard and mouse attached by wires. Already the cordless mouse is becoming more common. Who knows what new computer peripherals may exist in the future and how they might be configured when freed from the tyranny of the connecting cable? And who knows how these emancipated subsystems may regroup to form new communities of interacting mechatronics subsystems? An extension of this idea is the situation in which many autonomous mobile machines are communicating with each other. Wireless communication is essential for cooperating machines and for communities of moving autonomous machines.

Apart from representing and communicating information, intelligent machines must *process* information. This can be done by conventional computing, and by the many new approaches to information processing that are emerging within artificial intelligence (AI). This includes new conceptual and software architectures, logic programming languages and object-oriented systems. It also includes novel computational architectures such as parallel processing and neural networks. Neural networks appear to work particularly well in highly unstructured environments in which there is great uncertainty. However, symbolic information processing remains essential for modelling and reasoning about more structured environments.

This chapter presents an introduction to some of the central concepts of information communication and information processing necessary for mechatronics. These include:

1 Information encoding and communication:

analogue and digital encoding
Fourier analysis, filtering, and noise reduction
quantifying information
Shannon's sampling theorem
error-correcting codes;

2 Information processing:

analogue v. digital hardware
microprocessors and digital computers
computer programming
knowledge-based systems
networked and multiple processor computers
non-sequential and neural computers.

3.2 Information, signals and communication

3.2.1 The medium and the message

Suppose you wanted to get some information from A to B, how would you set about it? If the information could be ***encoded*** in words you might shout it: 'left hand down a bit'. If the medium of sound waves through the atmosphere wasn't appropriate, you might encode the information by writing it on paper and posting it into a letter box. But the medium of the postal service is too slow for some purposes, so the speed of communication has to be considered. You could fax your letter: the encoding of information into alphabetic symbols would then be turned into a sequence of 'high' and 'low' voltages corresponding to light and dark dots as the page is scanned, to be carried as an electrical signal along a telephone wire. If A and B are in different countries your electrical signal might be converted to an electromagnetic signal to be bounced off a satellite, before being converted back to an electrical signal on a telephone line.

Clearly then, choosing the most appropriate information carrying medium is crucial to the design of a mechatronics system. The main media available to mechatronics engineers for carrying information include:

- fluctuations in electrical voltages along wires
- electromagnetic waves (see Figure 3.1):

    gamma rays
    X-rays
    ultraviolet
    visible spectrum
    infra-red (near, mid, and far)
    microwave
    radar
    radio (UHF, VHF, HF, MF, LF)

- light along fibre-optic cables
- sound waves through the atmosphere
- forces transmitted by rods and cables
- chemical and biochemical states
- magnetic fields
- atomic structures and radioactive emissions.

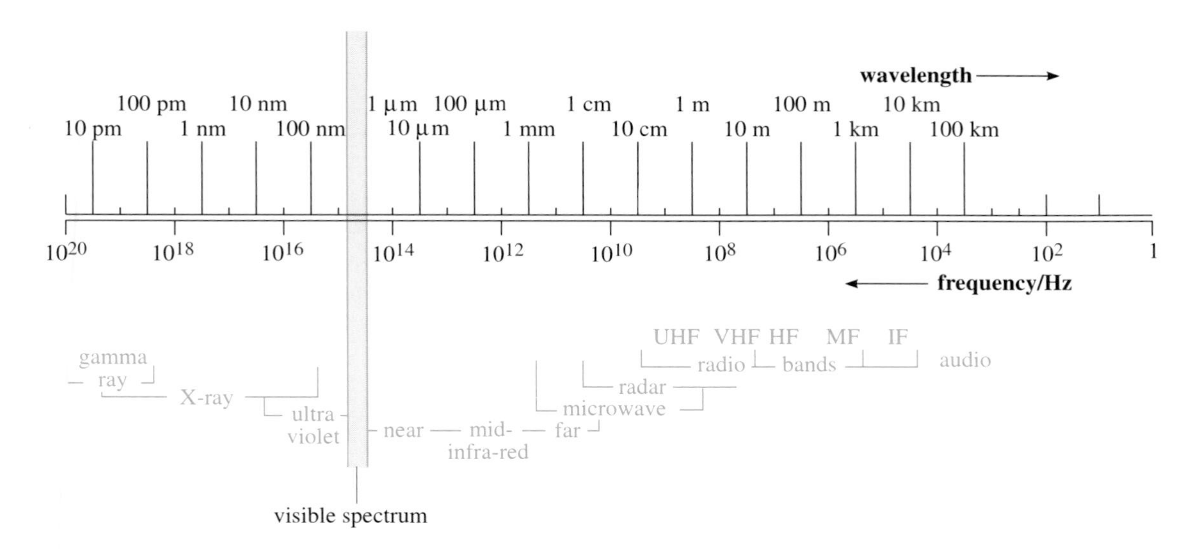

Figure 3.1
The electromagnetic spectrum gives a rich and versatile source of information-carrying signals.

Another aspect that has to be taken into consideration is the fidelity with which the signal can be transmitted and received. Your reader will have to accept that your fax is a degraded version of the original, especially since your original has been converted from its analogue form to discrete binary dot/no-dot form, as shown in Figure 3.2. In addition it will be degraded by *noise* (interference) and distortion in the communication channel. Noise will be added to the signal which will cause small unpredictable changes to the signal.

message (a)

message (b)

Figure 3.2
The effect of transmission on a simple 'message': (a) information encoded in characters before transmission; (b) information encoded in dots after transmission.

Now suppose the dot information of Figure 3.2(b) were put through an *optical character recognition* (OCR) machine. These machines attempt to recognize characters from dot patterns and they are remarkably good at it. Such a machine might recognize all the characters correctly, and so we might obtain a perfect reconstruction of the original message as shown in Figure 3.3(a). However, we must ask what 'perfect' means here. It is possible that the OCR machine knows nothing about italic fonts or character spacing, so it might print out the information as shown in Figure 3.3(b). Does this perfectly convey the information encoded in Figure 3.2(a)? The answer depends on the purpose of the message. If it is to convey the word 'message' then no information is lost. However, if the intention is to show what a particular style of writing looks like (in this example, italics), then all the information has been lost.

message (a) message (b) me88age (c)

Figure 3.3 **Possible output from an optical character reader.**

Of course, the OCR machine may make mistakes in interpreting the distorted information. For example, the tops and tails of each *s* have merged with the middle to produce something that could easily be mistaken for an *8*, as shown in Figure 3.3(c). This could be a problem which might only be solved by having a higher level of processing – in this case a spelling checker might do the job.

This simple example of information being carried between two sources illustrates a number of points:

- the same information can be encoded in different ways;
- the same information can be carried in different ways;
- information can be transformed from one encoding to another without loss of information;
- signals may become degraded by noise, and information may get lost or spurious information may be added;
- some signals carry information in continuous *analogue* form, and some carry it encoded in discrete *digital* form;
- information can be amended by *meta-information* (information about information), such as the tolerances on a sensor, or the prior knowledge that 'me88age' is probably a mis-spelling of 'message'.

We shall return to the subject of information and how to measure it after a brief review of some signal processing techniques.

3.2.2 Periodic signals and the frequency domain

A particularly useful way of analysing signals and waveforms follows from the work of the French physicist and mathematician, Baron Jean Baptiste Joseph Fourier (1768–1830).

It follows from Fourier's work that any periodic waveform, $p(t)$, can be written as a sum of sine waves:

$$p(t) = a_0 + a_1 \sin(1 \times 2\pi f_0 t + \phi_1) + a_2 \sin(2 \times 2\pi f_0 t + \phi_2) + a_3 \sin(3 \times 2\pi f_0 t + \phi_3) \ldots$$

where a_0 is the average value of the signal; $a_1, a_2, a_3 \ldots$ are the *amplitudes* of the ***Fourier*** components; and f_0 is the *fundamental frequency* of the waveform. The other term, ϕ_i, is the *phase*, which is a value in radians between $-\pi$ and $+\pi$, and is a measure of how far to the right or left the component sine wave is shifted relative to the sine wave with the fundamental frequency. Depending on the nature of $p(t)$ there may be a finite or an infinite number of Fourier components.

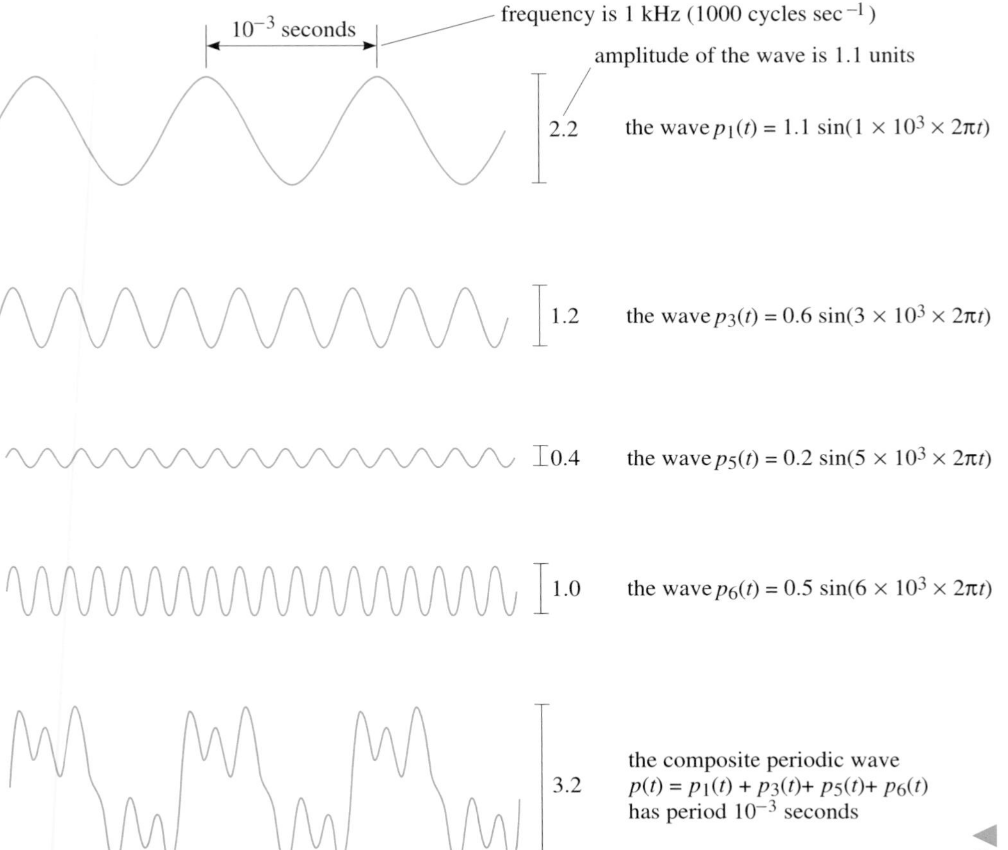

Figure 3.4
Composite periodic wave.

An example of a periodic waveform is illustrated in Figure 3.4 in which the wave $p(t)$ at the bottom is the sum of the waves p_1, p_3, p_5 and p_6. The *period* of the wave $p(t)$ is 10^{-3} seconds, which means that the composite waveform occurs exactly once in that interval. This is called the *fundamental period*. The fundamental frequency, f_0, of $p(t)$ is the reciprocal of the period, namely $1/10^{-3} = 10^3$ Hz or 1 kHz. Each of p_1, p_3, p_5 and p_6 has a frequency which is an integer multiple of the fundamental frequency. Thus our wave could be written as

$$p(t) = 1.1 \sin(1 \times 10^3 \times 2\pi t) + 0.6 \sin(3 \times 10^3 \times 2\pi t) + 0.2 \sin(5 \times 10^3 \times 2\pi t) + \\ 0.5 \sin(6 \times 10^3 \times 2\pi t)$$

We can represent the composite wave by a diagram which shows amplitude and phase against frequency, as illustrated in Figure 3.5. We say that $p(t)$ has been transformed into the *frequency domain* and that this diagram is a *frequency spectrum,* or *line spectrum*. Notice that the phase of each component is zero since the individual sine waves are all 'in phase'. This will be true of all the composite waveforms described in this chapter, so we will leave out the phase component of the spectrum. However, in general, the phase is an important part of the spectrum and should not be overlooked.

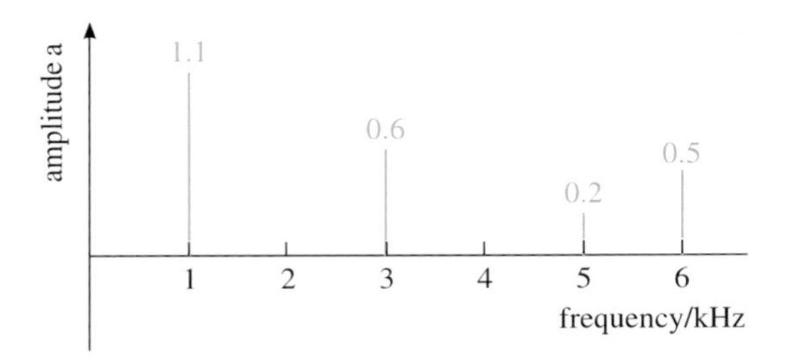

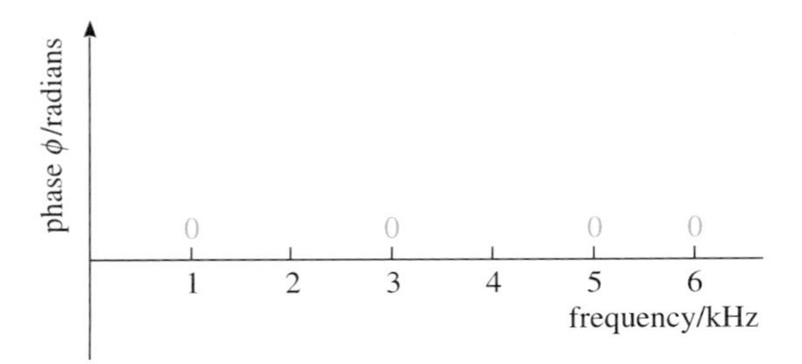

Figure 3.5
Line spectrum of a periodic waveform.

The information content of the frequency, amplitude and the phase information is the same as that of the waveform. But the frequency domain representation is often much easier to draw than the waveform representation, and is easier to analyse and manipulate. A particularly important infinite Fourier series is that for a *square wave*, illustrated in Figure 3.6(a), which is given by the equation:

$$p(t) = \frac{4}{\pi}\left[\sin\,(2\pi f_0)t + \frac{1}{3}\sin\,(3 \times 2\pi f_0 t) + \frac{1}{5}\sin\,(5 \times 2\pi f_0 t)\ \ldots\right]$$

In Figure 3.6(b) the fundamental frequency, f_0, is 500 Hz with amplitude $4/\pi$. The components with frequencies of 1.5 kHz and 2.5 kHz have amplitudes $4/(3\pi)$ and $4/(5\pi)$ respectively.

The first three terms of this series give the approximation to the square wave shown in Figure 3.6(b). Part (c) of the figure shows the line spectrum for the approximation to the square wave obtained by taking the first three terms of the equation.

3.2.3 Non-periodic signals

Non-periodic signals also have a frequency domain representation. Figure 3.7(a) shows a signal which begins at time zero and ends after time T and does not repeat itself. This is the kind of signal that might come from a sensor such as a transducer when measuring temperature in an industrial process.

This signal can be made into a periodic wave as illustrated in Figure 3.7(b), which we know has a Fourier spectrum. If T is, say, 10 minutes, the fundamental frequency is 1/600 cycles per second (Hz). All the other terms in the Fourier

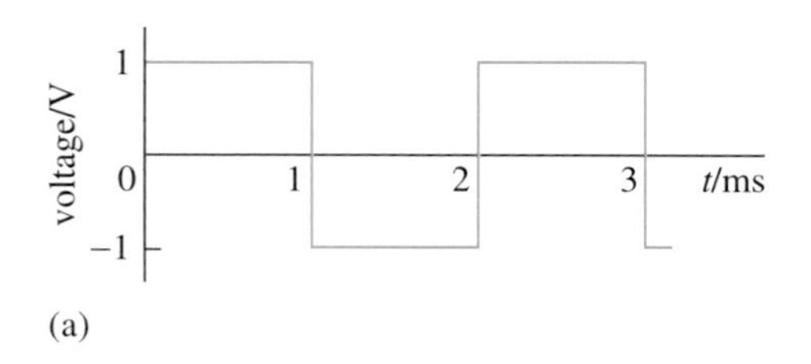

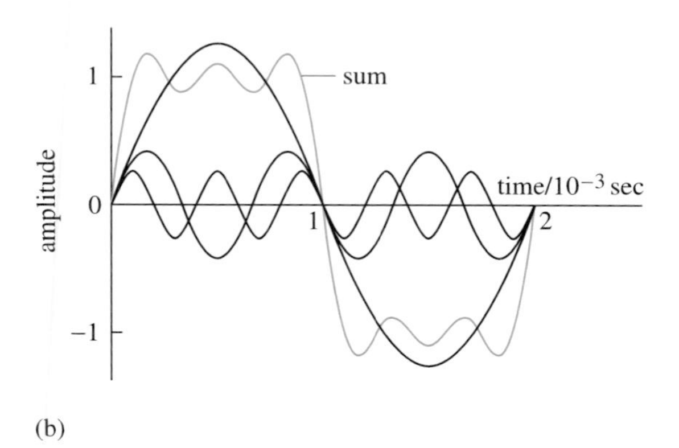

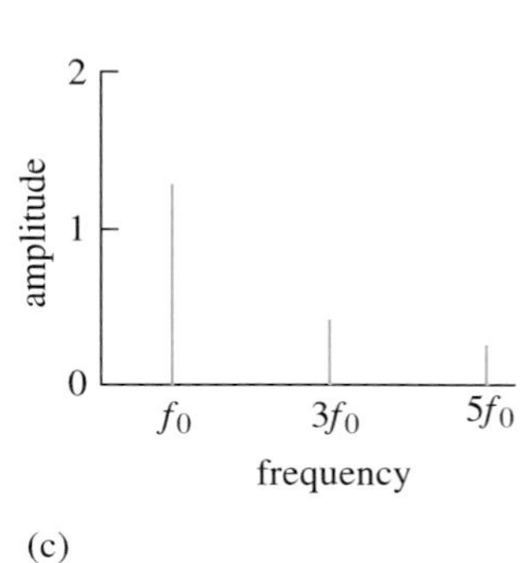

Figure 3.6
(a) A square wave; (b) its approximation by sine waves; (c) the line spectrum of the composite wave form in (b) – adding more terms results in a better approximation.

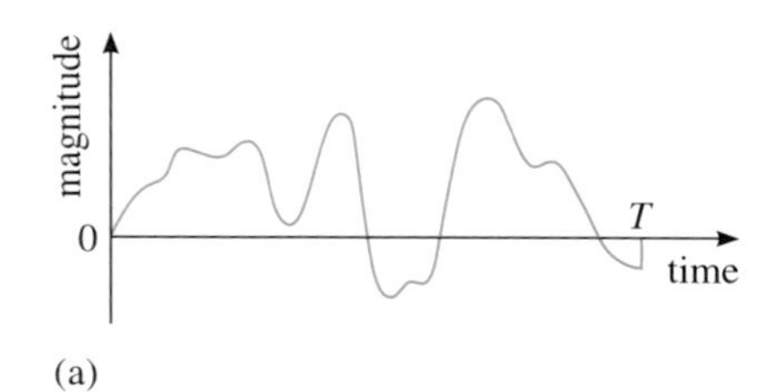

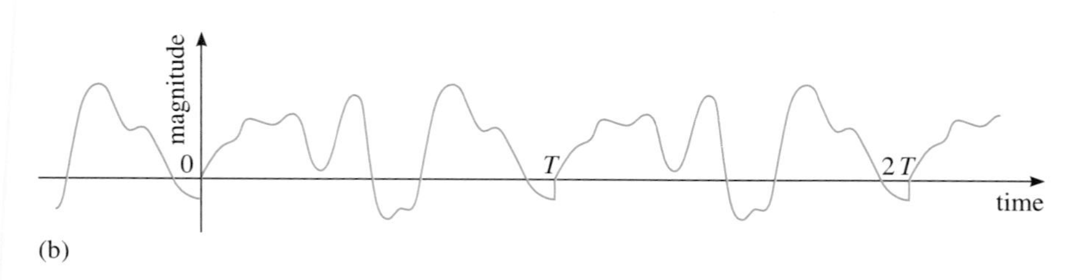

Figure 3.7
Representing non-periodic waves: (a) sample of a non-periodic wave between times $t = 0$ and $t = T$; (b) repeating the wave creates a periodic wave with period T.

spectrum are multiples of this. This means, for example, that between 0 Hz and 10 Hz the spectrum may contain up to 6000 lines, which would of course tend to merge together if drawn. The result is a *spectral outline*, as shown in Figure 3.8. This can be thought of as the envelope of the tops of the component amplitudes.

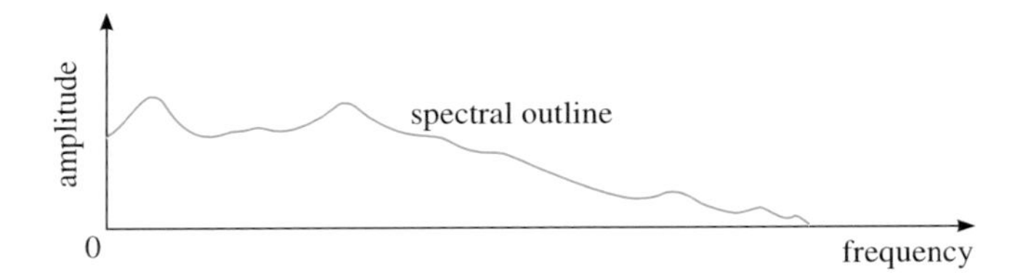

Figure 3.8
A spectral outline 'envelope' of the tops of component amplitudes.

The mathematics of the Fourier analysis of non-periodic signals is beyond the scope of this chapter, but for our purposes we need to realize that all practical signals can be represented in the frequency domain in this way.

An important frequency-domain characteristic of a signal is its *bandwidth*. The bandwidth can be loosely defined as the difference between the maximum frequency and the minimum frequency present in the signal. This is the range of frequencies in which the signal exists in the frequency domain. Figure 3.9 shows two examples of signals in the frequency domain and their bandwidths.

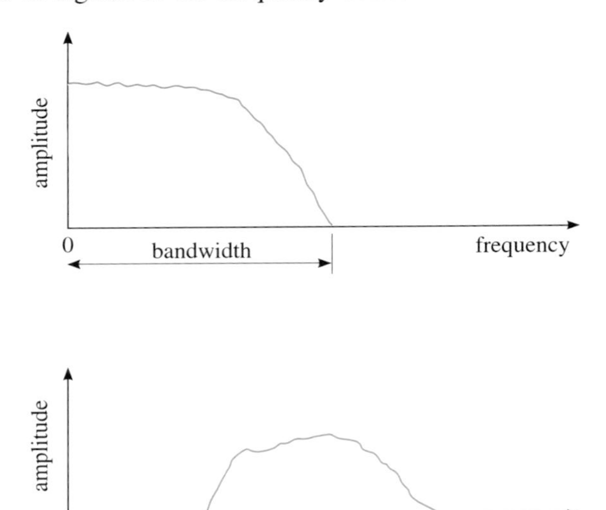

Figure 3.9
Two signals in the frequency domain, showing their bandwidths.

3.2.4 Modulation and signals as information-carrying waves

Much of the theory of signals and information has its origins in the early days of radio when the major preoccupation was the design of transmitters and antennas to overcome atmospheric degradation of analogue signals. The considerable body of theory that has accumulated since those early days is very relevant to information communication in mechatronics.

It would not be practical to transmit sound-wave information by radio waves of the same frequency because many stations have to share the available electromagnetic spectrum. Thus each station is allocated a frequency range for its transmissions. The most common way to transmit information through the

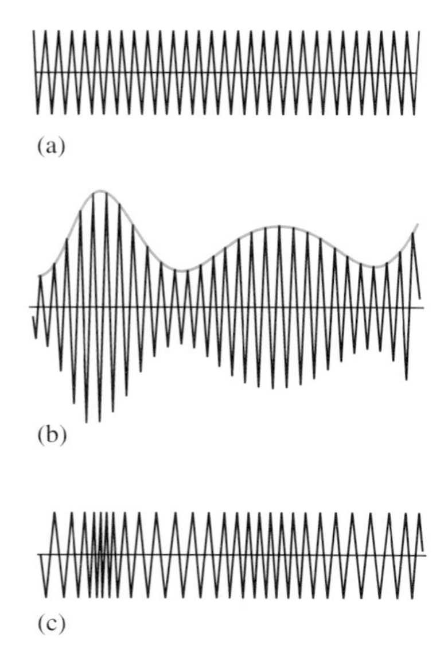

Figure 3.10
Waves can carry information by amplitude modulation or frequency modulation of a carrier wave:
(a) carrier – a structure produced electrically;
(b) amplitude modulation – the amplitude of the carrier has been varied in proportion to the sound signals;
(c) frequency modulation – the frequency of the carrier wave has been varied in proportion to the sound signals.

atmosphere is to modulate the information onto a *carrier wave*. Two types of ***modulation*** are illustrated in Figure 3.10. These modulations each have their own virtues and drawbacks. Amplitude modulation (AM) is more sensitive to atmospheric interference and noise than frequency modulation, so frequency modulation (FM) is usually preferred these days for high fidelity broadcast radio.

One technique of amplitude modulation is to multiply the signal by the carrier wave. The effect is to shift the signal from its original range of frequencies up to a range centred on the carrier frequency. You can see in principle what happens by looking at a single sine wave, $\sin(2\pi f_s t)$, and a carrier waveform, $\sin(2\pi f_c t)$, where f_c is much greater than f_s. The product can be manipulated using a trigonometric relationship:

$$\sin(2\pi f_c t) \times \sin(2\pi f_s t) = 0.5\ [\cos((f_c + f_s)2\pi t) - \cos((f_c - f_s)2\pi t)]$$

The resulting waveform can therefore be considered as consisting of two sinusoidal waveforms (cosine being sinusoidal in shape) with frequencies of $(f_c + f_s)$ and $(f_c - f_s)$ respectively. This is shown in Figure 3.11(a). Also, part (b) of the figure shows the effect of amplitude modulation on a non-periodic signal with a bandwidth of B. Notice that the mirror image of the signal spectrum is also produced to the left of the carrier frequency. For this reason we need to remember that if a signal with a bandwidth of B is to be modulated in this way, then there has to be a transmission channel with a width of $2B$ which is centred on the carrier frequency.

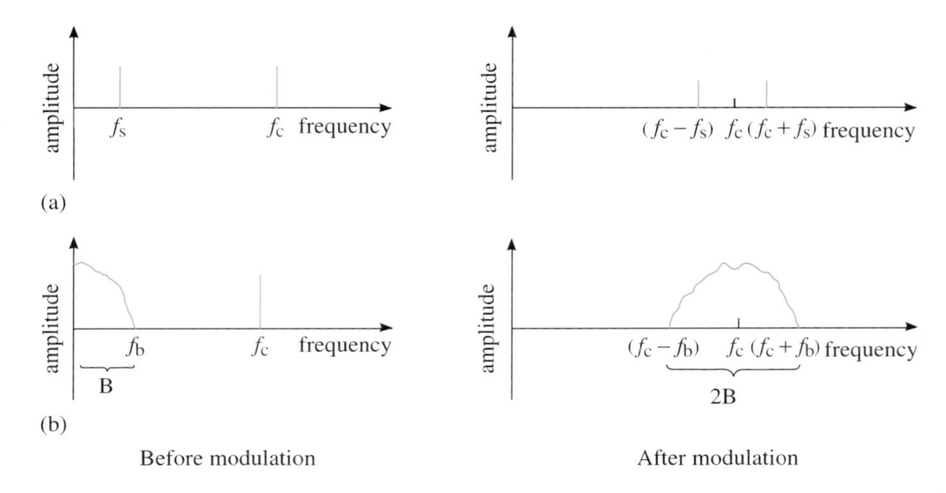

Figure 3.11
The effect of one type of amplitude modulation on the spectrum of a signal.

Once a signal is received, it has to be shifted back to its original frequency range. This process is known as *demodulation*.

3.2.5 Filters

One of the reasons that the frequency domain representation of signals is so important is that we can more readily appreciate how to filter out parts of the signal that we do not want. In general, ***filters*** are electronic circuits which reduce the range of frequencies present in a signal. They reduce or attenuate the signal strength at some frequencies while allowing others to pass through almost unchanged. Figure 3.12 shows the four main types of filter. The idea is simple enough: given a signal, we filter out the parts which are not wanted in order to retain as much as possible of the parts that we do want.

The *frequency response* of a filter is a description of how its attenuation characteristics vary with frequency. Often the term 'bandwidth' is again applied: in the case of a filter it indicates the range of frequencies over which a signal can pass without getting significantly attenuated.

Filtering is extremely useful in communications and signal processing because sometimes atmospheric noise occurs over a range of frequencies, and it can be reduced using filters. Furthermore, by means of modulation the frequency spectrum of a signal can be shifted to avoid areas of noise. A simplified illustration of this idea is given in Figure 3.13.

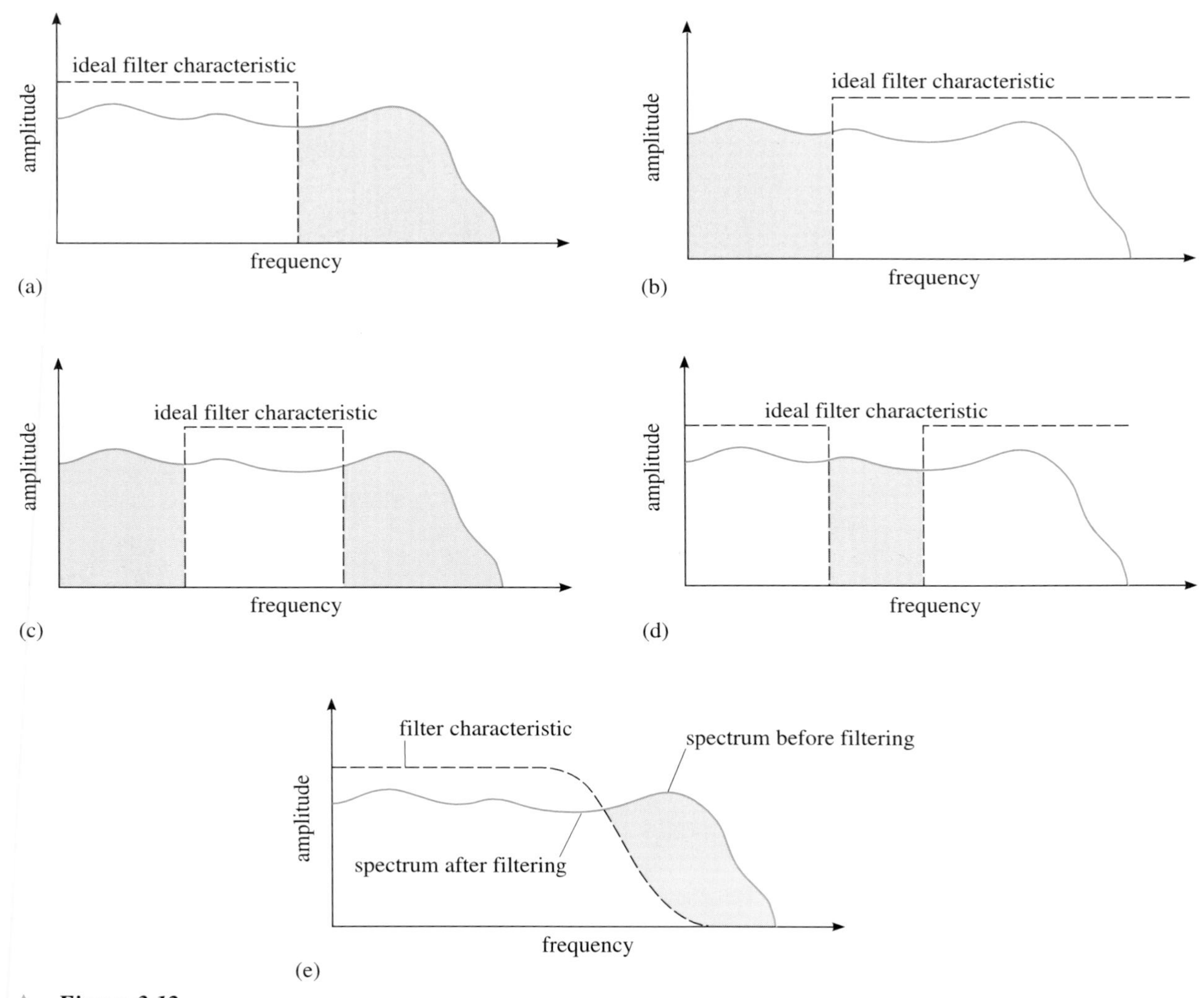

Figure 3.12
The effect of various filters: (a) ideal low-pass filter; (b) ideal high-pass filter; (c) ideal band-pass filter; (d) ideal band-stop filter; (e) in a real filter the signal strength is attenuated over a range of frequencies.

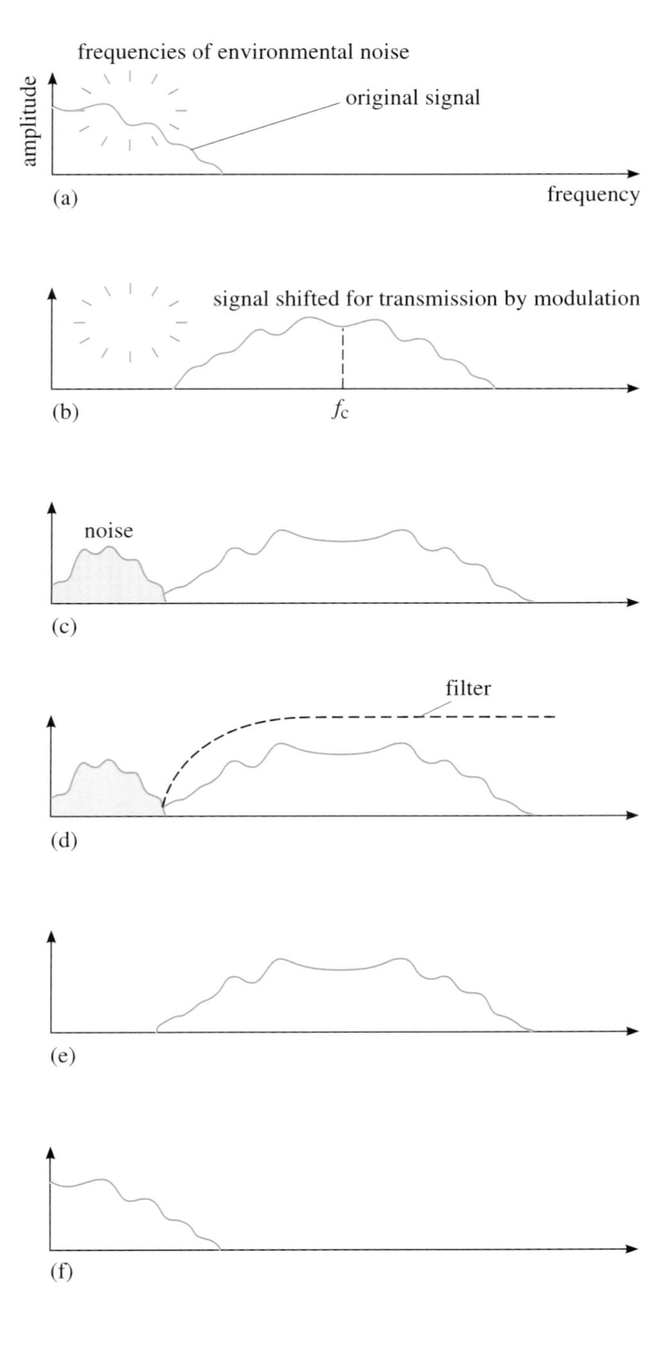

Figure 3.13
A simplified illustration of how modulation, filtering, and demodulation can be used to avoid noise:
(a) the frequencies of the original signal clash with sources of atmospheric noise;
(b) the signal shifted to a different part of the frequency spectrum for transmission (note: this modulation creates a 'mirror image' of parts of the spectrum);
(c) the signal received distorted by noise;
(d) a high-pass filter to remove the noise;
(e) the filtered signal;
(f) the signal shifted back to its original frequencies by demodulation.

3.2.6 Some fundamentals of information theory

In 1948 Claude Elwood Shannon of Bell Laboratories addressed the question of how much information there is in a signal. Suppose you took your driving test yesterday and telephoned a friend to tell her the good news that you had passed: how much information would you have conveyed? Shannon would say that there were two outcomes which we will suppose were equally likely as far as your friend was concerned – either you would pass or you would not pass. That is, there was a 0.5 probability of your passing and a 0.5 probability of your failing. The concept of information can be used to represent this kind of uncertainty. The information I associated with an event with probability p is defined to be

$$I = -\log_2 p \text{ bits}$$

where $\log_2 x$ is the logarithm to the base 2 of x; or

$$\log_2 x = \log_{10} x / \log_{10} 2$$

and

$$\log_{10} 2 = 0.301$$

The message to your friend describes one of two possible events, pass and fail, each with probability $p = 0.5$. Thus the information in your message was

$$I = -\log_2 0.5 = 1 \text{ bit}$$

In other words you conveyed one bit of information to your friend in terms of passing the driving test, 'yes' or 'no'.

The term *bit* is used to measure the information content of a message, but is also used in computing to mean a binary digit. It is more than a matter of convenience that when a message is binary the information content in bits is the same as the number of binary digits that would be needed to represent the information. So, in the above example, the information on passing the driving test is either 'yes' or 'no', which could be coded as a single binary digit with a corresponding value of 1 or 0. The choice of using logarithms with a base of 2 (that is, $\log_2 x$) ensures that this is true.

Suppose you also told your friend how euphoric you felt. How much more information would this be? Well, assuming you are a normal person, your friend would probably be able to guess that passing your driving test would make you feel pretty good. In fact she might think it was a thousand to one that you could feel any other way. If her assessment was correct, the probability of your being elated is $p = 0.999$ and the extra information conveyed by your elated outpourings would be:

$$I = -\log_2 0.999 = 0.0014 \text{ bits}$$

Not very much! The total information content of your signal that you've passed and that you are elated is found by simply adding the individual information contents:

$$\text{Total } I = 1 + 0.0014 = 1.0014$$

In this case the highly predictable message 'I'm very happy to have passed my driving test' conveys very little information. However, human communications are more subtle than telecommunications. Perhaps your friend begins to feel uncertain about your friendship if you don't keep in touch, and the fact that you actually called and are chatting away may have significantly more bits of information in it.

The idea of 'bits' in information theory relates to how much information the signal conveys over and above what could be deduced from the probabilities alone. When information is communicated using a binary waveform, the probabilities that the next part of the wave will be '1' or '0' are equal at 0.5, and by a similar calculation to that above, the information conveyed by one pulse of a binary waveform is one bit in information theory terms. It is possible to use other than binary waveforms, such as quaternary waveforms which have four distinct levels. However, for the rest of this book we will assume that we are talking about binary waveforms, which are the most commonly used.

We can now determine the limit of how much information can be transmitted through a medium or channel using binary waveforms. At one extreme, a signal could contain a string of all 1s, say, which would look like a constant signal, corresponding to a waveform with a frequency of 0 cycles per second. At the other extreme a signal could contain a string of alternating 0s and 1s, which corresponds to a square wave. One period of a square wave contains two bits of information, a 0 and a 1. The frequency of the square wave is the reciprocal of the period and is a measure of the number of cycles per second. Since one period contains two bits, the number of bits per second is twice the frequency of the square wave. Therefore a binary signal can be transmitted over a communication channel between the frequencies of $f_{min} = 0$ and f_{max}, where f_{max} is the frequency of the square wave that corresponds to alternating 0s and 1s.

As we saw earlier, the bandwidth of a communication channel is defined to be the difference between the maximum and minimum frequencies that the channel can effectively carry. The bandwidth of the channel described above is $f_{max} - f_{min} = f_{max}$. The amount of information that can be carried is twice the frequency of the square wave and therefore equals $2f_{max}$, i.e. twice the bandwidth. Many channels do not have bandwidths that go from 0 to some higher frequency, but by using modulation (described earlier) it is possible to shift any signal so that it fits the frequency range of a channel. We can therefore say that if a channel has a bandwidth B, the bit rate is twice the bandwidth, $2B$. For example, coaxial cable has a typical frequency range of 60 kHz to 60 MHz, and therefore has a bandwidth of about 60 MHz. When a binary signal is sent along a coaxial cable the bit rate is twice the bandwidth, namely 120 Mbits sec^{-1}.

To get an idea of the amount of information in sound consider a compact disc. Here the sound signal is *sampled* 44 100 times each second to give an amplitude in the region of 0 to 65 535 or 2^{16} possible sound levels. Thus the amplitude data per sample is stored in 16 binary bits, and the CD stores data to be played at a rate of $44\,100 \times 16$ bits sec^{-1} = 705 600 bits sec^{-1}. Just like the engineers who designed the compact disc system, we must consider the question of sampling a waveform and seeing to what extent it can be reconstructed.

3.2.7 Sampling signals and the sampling theorem

Sampling signals is a major area in communications and signal theory, particularly because of the widespread use of digital computers in signal processing. When a continuous signal is sampled discretely (i.e. by taking a finite number of samples in a given time), it is important to know if the sampled values can adequately represent the original waveform, and if so what is the optimum sampling rate to achieve this. ***Shannon's sampling theorem*** states that

> A continuous signal can be represented completely by, and reconstructed from, a set of instantaneous measurements or samples of its value made at equally spaced times. The interval between the samples must be less than half of the period of the highest frequency component in the signal.

This remarkable result is easy to illustrate intuitively when we consider a single sine wave. If we sample at a rate less than twice per cycle, we will miss it either rising or falling (Figure 3.14). A rate of more than twice the frequency ensures that we will sample the sine wave sufficiently in every cycle. In other words it is clear that more than two samples per cycle are necessary in order not to lose essential information. If we have more than two samples per cycle for the highest frequency component of the sampled wave we certainly have more than two samples per cycle for all the lower frequency waves.

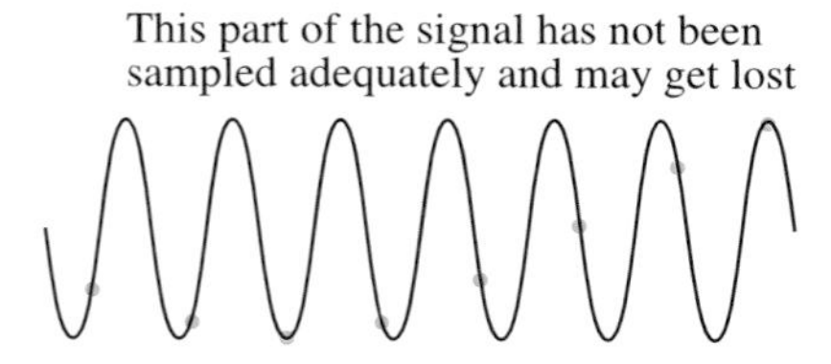

Figure 3.14
Insufficient sampling: at least two samples per cycle (shown by dots) are necessary to ensure that essential information is not lost from a sine wave.

Figure 3.15 shows a number of sampled waves. The third of these (c) shows why it is strictly necessary for the sampling to be at a frequency *greater* than twice the highest frequency component. In the case of equality it is possible to sample the wave everywhere at amplitude zero, and it will not be possible to reconstruct the wave from this.

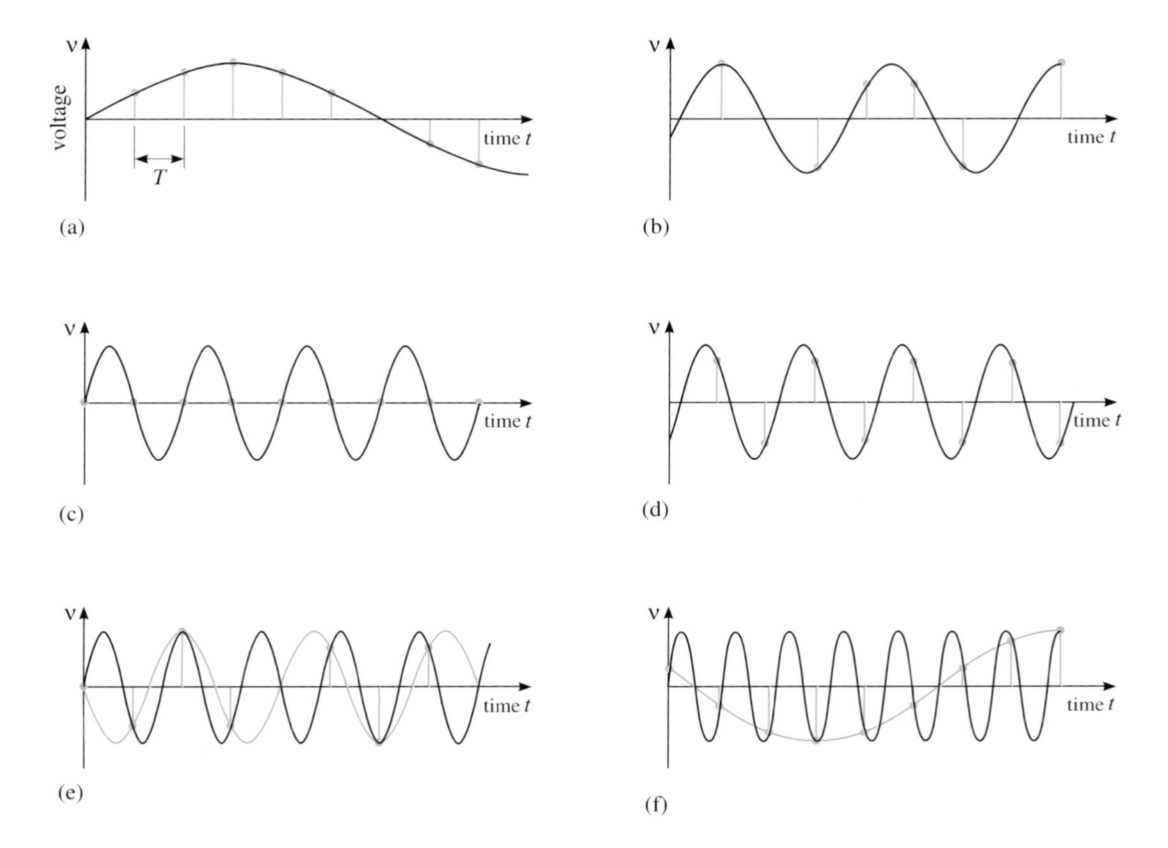

Figure 3.15
Illustrating the sampling theorem: (a) frequency $\ll T/2$; (b) frequency $< T/2$; (c) frequency $= T/2$; (d) frequency $= T/2$; (e) frequency $< T/2$; (f) frequency $< T/2$ showing aliasing.

Parts (e) and (f) of Figure 3.15 show the phenomenon of *aliasing* which occurs with undersampling: two sine waves can be reconstructed from the same data set so that one can be the 'alias' of the other.

Once the sampling rate has been selected, f_S, it is quite common practice to place an *anti-aliasing filter* before the sampler. This is a low-pass filter which allows signals through with frequencies below $0.5f_S$, and attenuates signals above $0.5f_S$. This ensures that no parts of a signal exist above $0.5f_S$ so that no aliasing occurs.

3.2.8 Digital encoding

We have considered examples of signals which represent data in analogue form. However, it is becoming increasingly important to encode signals in digital form. In principle, any medium (which can be in either of two states) can be used to carry digital binary signals. For example, Morse code can use sound to carry a digital encoding of sequences of letters of the alphabet by tapping two distinct

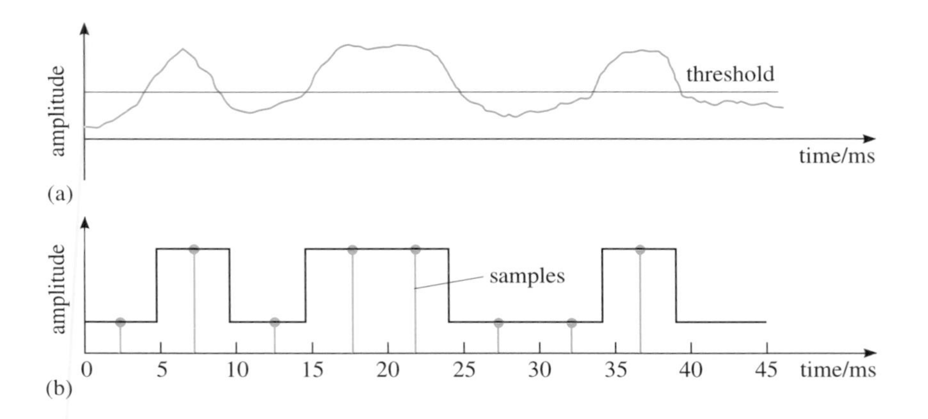

Figure 3.16
A digital waveform: (a) noisy signal; (b) ideal signal.

characters – the dot and the dash. Morse code can also be transmitted down a telegraph wire, or flashed with a torch.

There are many reasons for using ***digital encoding*** of information. The main one is that it provides better immunity to noise in a signal than analogue waveforms. Also digital encoding allows error-correcting signals to be transmitted (see next section) and for digital filters to be used.

Figure 3.16 shows a digital signal being received by a system. Although the signal in part (a) is very noisy, it is possible by using a simple threshold to produce the very 'clean' digital signal (b). Assuming that the signal is sampled every 5 ms and that the first sample represents the left-most digit, the signal could be interpreted as the binary integer:

01011001

which is

$$(0 \times 128) + (1 \times 64) + (0 \times 32) + (1 \times 16) + (1 \times 8) + (0 \times 4) + (0 \times 2) + (1 \times 1) = 89$$

as a decimal number. This number might mean that a sensor has a voltage output of 89 millivolts or a whole variety of other interpretations depending on the application.

Digital encoding is useful when the information to be transmitted is already encoded in symbolic form. For example, the 'message' string of alphabetic symbols that we transmitted over the fax uses characters which are meaningful to human beings. Because of the importance of encoding characters in the computer, the ASCII (American Standard Code for Information Interchange) character codes are now universal. These assign characters to many of the 128 7-bit binary sequences that can be encoded by a binary number such as 01011001 and its decimal equivalent 89 (the extra first bit is included for error checking, as described in the next section). The code for A is 65, that for B is 66, ..., that for X is 88, that for Y is 89, that for Z is 90, ...and so on. Thus we can establish a relationship between the waveform shown above and the letter Y.

3.2.9 Self-correcting codes

When discussing analogue signals in Sections 3.2.4 and 3.2.5, we saw how they could be processed in order to minimize the effects of noise. However, although these approaches minimize the effects of noise they do not allow the receiver to detect or correct errors. With digital encoding it is possible to build *redundancy* into the coding and so enable the receiver to detect errors or even correct them. Redundancy means adding extra bits to the code word which aren't necessary to convey the information content.

A simple example of error detection is the *parity* error. If seven bits of information are sent down a channel, a further (redundant) bit can be sent which is 1 if the number of 1s in the seven bits is odd and 0 if it is even. If the received eight bits do not have matching 'parity' an error has occurred.

This form of parity is referred to as *even parity* because the number of 1s in the whole eight bits including the parity bit is always even. For example, if the seven bits are

0110100, add the parity bit 1 number of 1s = 4 (even)

0001100, add the parity bit 0 number of 1s = 2 (even)

Another way of using redundancy to detect errors would be to send each set of eight bits twice. The receiver could then check that they were equal; if not, it would indicate that an error had occurred. However, this amount of redundancy can often be much better used by employing ***error-correcting codes***. These allow a sufficiently small number of errors to be detected and corrected at the transmitter without the need to request retransmission. When communicating over large distances, say to a spacecraft, this would be very important. This is because communications are constrained by the speed of light, which results in significant delays over large distances. It is also important if the source of the signal cannot be instructed to retransmit (as in the case of one-way communications).

Hamming codes use redundant bits in a way which shows not only that errors have occurred, but also which bits are incorrect. We'll look at a simple example of such a code, which allows one error to be corrected in each code word. Suppose the numbers zero to nine are to be transmitted as 4-bit binary numbers:

Decimal	Binary			
	A	B	C	D
0	0	0	0	0
1	0	0	0	1
2	0	0	1	0
3	0	0	1	1
4	0	1	0	0
5	0	1	0	1
6	0	1	1	0
7	0	1	1	1
8	1	0	0	0
9	1	0	0	1

Consider the groups of three 3-bit numbers, BCD, ACD, and ABD.

Let X, Y, Z be parity bits for each group so that each group has even parity. For example, if we were transmitting the number 2,

$A = B = D = 0, C = 1$

Therefore:

BCD = 010 with parity X = 1

ACD = 010 with parity Y = 1

ABD = 000 with parity Z = 0

The 7-bit codeword ZYAXBCD is now transmitted. This particular ordering of the bits quickly allows us to determine which bit is in error, as we shall now show. With this ordering, Z is bit 1, Y is bit 2 and so on until D is bit 7.

The receiver carries out the parity checks on ABCD again to produce three new parity bits X′, Y′ and Z′. These are compared with the values of X, Y and Z that have been received to see if there is a difference. If there is a difference, it is possible to work out which bit must have been corrupted during transmission.

Suppose that the number that is transmitted is 2 again. The complete 7-bit codeword (ZYAXBCD) is:

0101010

Let's see what happens if one bit is corrupted during transmission: for example, C is now 0 instead of 1. The received codeword is:

0101000

Now carry out the parity checks again.

BCD = 000 with parity X′ = 0 but X = 1

ACD = 000 with parity Y′ = 0 but Y = 1

ABD = 000 with parity Z′ = 0 and Z = 0

Two of the parity checks have changed. We now create a new 3-bit number which corresponds to the number of the bit where the error has occurred, based on the two sets of parity checks. If the new parity check is the same as the old one then a bit is set to 0, and if the parity check has changed we set a bit to 1. So for this example:

		Bit number
BCD = 000 with parity X′ = 0	but X = 1	1
ACD = 000 with parity Y′ = 0	but Y = 1	1
ABD = 000 with parity Z′ = 0	and Z = 0	0

Reading this number from top to bottom we get 110 = 6 in denary. So bit 6 (C) is corrupted and we can now correct it.

If there are no changes to the parity checks then no error has occurred in transmission and the bit number is 0. All this assumes that there is at most one bit error. There are Hamming codes which do not have this restriction, but these are beyond the scope of this book.

3.3 Information processing

3.3.1 Analogue versus digital hardware for perception

In this section we will consider the ways that information can be processed in mechatronics systems.

Sensors can be classified as *active* and *passive*. For example, a bat navigates by 'sonar'; that is, it is constantly sending out high-pitched sounds, and it perceives its environment from the way in which these sounds bounce off objects and return to its ears. Thus, apart from having excellent auditory sensors, the bat is actively generating the signal they will respond to. We will say sensing is *active* when the mechatronics system creates its own signal through a transmitter.

In active perception we will usually have a *transmitter–receiver pair*. For example, in a system which automatically opens a garage door the transmitter will be located in the car and the receiver (sensor) will be located in the garage.

Sometimes a system's environment provides information about itself 'for free', so to speak. For example, a reflex system which switches on the lights when it gets dark is responding to a signal generated by the sun. Here such perception will be called *passive*.

In general, mechatronics designers will not depend on the hit-and-miss possibility of the environment providing information about itself, and active perception will be built into the system. In this respect remember that we expect mechatronics systems to be *processing* perceived information, and the system will usually want the information to be available on demand. So an unmanned vehicle will probably have lights fitted to ensure that its camera has an adequate signal, even when the vehicle sometimes operates in daylight.

Suppose that a sensor has done its work and we have its *output signal* on a wire. What next?

First we must know the way the information is encoded on the wire. The main ways are:

- *analogue* – the information is represented by a signal which varies continuously in time;
- *digital* – the information is represented by a sequence of a finite number of states. Most commonly only two levels are used, in which case the system is binary, e.g. 5 V represents a logic 1, 0 V represents a logic 0.

3.3.2 Processing analogue signals

When information is encoded in analogue form, its processing usually involves signal conditioning which attempts to preserve as much waveform information as possible. Amplification and filtering are the main kinds of processes involved, and are essential to any electronic system which has signals as inputs and outputs.

Some systems are designed so that the analogue signals are used directly to produce a desired response. An example of this kind of system is a feedback control system, where the output is sensed and an analogue signal fed back to the controller. Inside the controller the signal is compared with the desired value of the output and an error signal produced which is then used to initiate some control action. In our terms we would see such control action as *reflexive*.

One alternative is to convert the feedback signals to a digital waveform and to use computing hardware to produce the control action. The output of the controller would have to be converted back to an analogue signal to actually control the system. Although this method uses a computer, the system is still reflexive. However, once a computer has been introduced into the system it can start to take advantage of general kinds of 'knowledge', and can process information in a deductive way. The computer can become a cognitive subsystem. The conclusion is therefore that a signal has to be converted into digital form at some stage in order to be processed 'intelligently'.

3.3.3 Analogue to digital conversion

It is becoming increasingly common these days for information to be processed in digital form, even when the digital numbers represent analogue quantities. Analogue-to-digital conversion devices (ADCs) are major components in information processing.

The simplest form of ADC *thresholds* an analogue signal: above the threshold the output is '1', and below the threshold the output is '0'. This information can then be processed by logical devices as described in the next section.

More sensitive devices convert analogue signal levels to digitally encoded binary numbers. Since the binary number will inevitably have a finite number of bits (usually 8, 12 or 16) the analogue signal will never be totally accurately represented. For example, if eight bits are used to represent a signal which ranges from 0 to 1 V, there will be 256 possible numbers that can represent the analogue signal. The signal is said to have been *quantized* into 256 discrete levels, with each level 1/256 V apart.

Finally, before the signal can be processed by a computer, it would have to sampled as explained in Section 3.2.7.

3.3.4 Logic gates

Digital signals are processed by networks of *logic gates*. Even if computers are used, these are themselves constructed from millions of logic gates. The basic logic gates are AND-gates, OR-gates and INVERTERS. These can be combined to form more complex gates, the most commonly found one being a NAND-gate (abbreviated from NOT AND) which consists of an AND-gate followed by an INVERTER. The symbols for these gates are shown in Figure 3.17.

The principle behind logic devices is simple. Suppose 5 V on a wire means logic 1 and 0 V means logic 0. If both the inputs to, say, an AND-gate are 1 (5 V) the output will be 1 (5 V), otherwise the output will be 0 (0 V). The logic tables for the four devices shown in Figure 3.17 are as shown in Table 3.1. Remarkably, *any* logic circuit can be built from two-input NAND-gates.

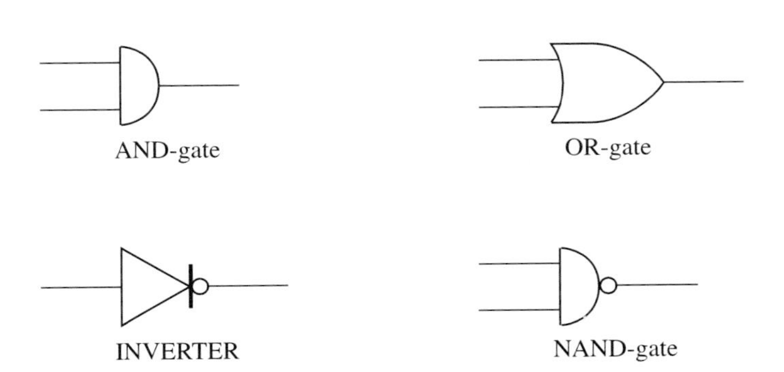

Figure 3.17 Commonly used symbols for logic gates.

TABLE 3.1

Inputs	Output	Inputs	Output	Input	Output	Inputs	Output
00	0	00	0	0	1	00	1
01	0	01	1	1	0	01	1
10	0	10	1			10	1
11	1	11	1			11	0
AND-gate		OR-gate		INVERTER		NAND-gate	

Logic devices can be used to model logical reasoning in hardware, such as:

If A AND B but NOT C then D

where, whatever A, B, C and D actually mean, their *logic values* are represented by voltages on wires. In particular the voltage on the output wire corresponding to D tells us its logic value given the input logic values on the wires A, B and C.

3.3.5 Applications of logical devices

The main operations we perform on digital data are *logical* and *arithmetical*. The simplest logic, which characterizes a wide class of digital hardware, is *Boolean*. The term comes from the nineteenth century logician, George Boole. The Boolean operators are AND, OR, and NOT which physically correspond to the AND-gates, OR-gates and INVERTERS just described.

To illustrate the use of logic devices, consider a vehicle equipped with a sensor which provides a 5 V output when the safety belts are correctly fastened and 0 V otherwise, and a sensor which provides a 5 V output when there is a key in the ignition and 0 V otherwise. The manufacturers may specify the logical requirement that the safety belts must be used properly *and* the key must be in the ignition before the vehicle can start.

The sensor information can be fed through an AND-gate and the output of the AND-gate fed into the ignition circuit, which requires this particular wire to be logic 1 before power will be supplied to start the vehicle, as shown in Figure 3.18(a).

Of course the output of a logic device may be used for more than one purpose by *splitting*, as shown in Figure 3.18(b), so that turning the key in the ignition either starts the vehicle, or causes an alarm to sound.

In this case the logic is such that the signal to the alarm will be 1 (sound alarm is 1) in all cases that the safety belts are undone or when the key is not in the ignition switch. However, since the key switches on the power we will only hear the alarm

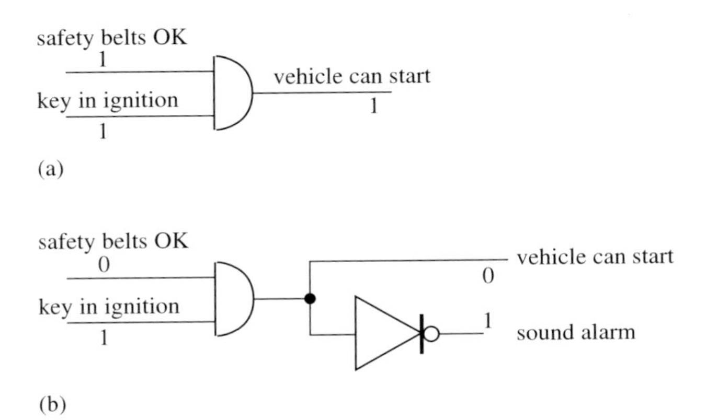

Figure 3.18
Safety feature in a car.

when the key is in the ignition and the seats belts are not done up correctly.

Consider another example with a parking vehicle moving forward receiving information from two sensors, one at the front (F) and one at the right side (S). Suppose these sensors provide binary information 1 when the vehicle is within a metre of an object, otherwise they provide zero. Let F = 1 mean that the front sensor has sensed an object and F = 0 mean that an object has not been sensed. Let S = 1 and S = 0 mean that an object has or has not been sensed at the side of the vehicle. Then we may have *logical rules* such as:

> If F = 0 AND S = 0 then move forward to the right.
>
> If F = 1 AND S = 1 then the vehicle is parked, stop.

In this case we could put the output wires from the two sensors into an AND-gate and take its single output to the control circuit where, for example, logic 0 tells the propelling and steering motors to combine in a forward direction to the right, and logic 1 switches off all the motors. Of course this control strategy is both crude and incomplete, and we will consider more subtle approaches in Volume 2 of this book.

Usually these simple gates are combined to form large networks which perform much more complex logical operations with many inputs and many outputs. If the gates are connected so that there is no feedback in the system, the network is said to be a *combinational logic* network. If feedback is present, there is the possibility of storage of information, and the network is then said to be a *sequential logic* system. In this system the network settles into a stable state, and stays there until there is some change of the inputs. When the inputs do change, the network will eventually settle again into a new stable state. Thus the network can be described as following a sequence of states, giving rise to the term 'sequential'.

All digital hardware consists of combinational and sequential logic circuits. However, rather than design each circuit from basic gates, the designer could use devices which are general-purpose logic circuits. The most popular of these is probably the *microprocessor*. Using a microprocessor replaces one skill, namely designing circuits from individual gates, by another skill, namely *programming*.

3.3.6 Digital computers and microprocessors

The digital computer is a relatively recent invention, with primitive commercial machines being developed in the 1950s and 1960s. A major breakthrough in digital computation occurred in the 1970s when Intel developed the first ***micro-processor*** in a single integrated circuit (chip). This 4-bit processor had all the major components of the modern microprocessor.

A microprocessor is an example of a *von Neumann architecture*. In a von Neumann computer the memory is arranged as a sequence of cells. Each memory cell (usually an 8-bit *byte*) is numbered in the *address space* of the processor. The size of the address space is determined by the number of wires (bits) of the *address bus*. For example, the address bus shown in Figure 3.19 has 16 bits, and the address space is $2^{16} = 64\text{K}$ (K means 2^{10} or 1024). The data bus has 8 bits, allowing data in the range 0 to 255. In most computers a significant part of the address space is used for data storage in random access memory (RAM), i.e. memory which can be written to and read from many times, and read only memory (ROM). However, some of the addresses are 'special' in that the processor can 'write to' and 'read from' external devices such as interface hardware. For example, 'writing' the number 1 to memory location 59013 could switch on a motor if appropriate interface hardware is connected to the address bus, while writing a 0 to the same address could switch off the motor.

If the processor requires larger values than 255 it uses more than one byte. For example, two adjacent bytes can be used to represent unsigned numbers in the range 0 to 65 535, or signed numbers in the range −32 767 to 32 768. Decimal

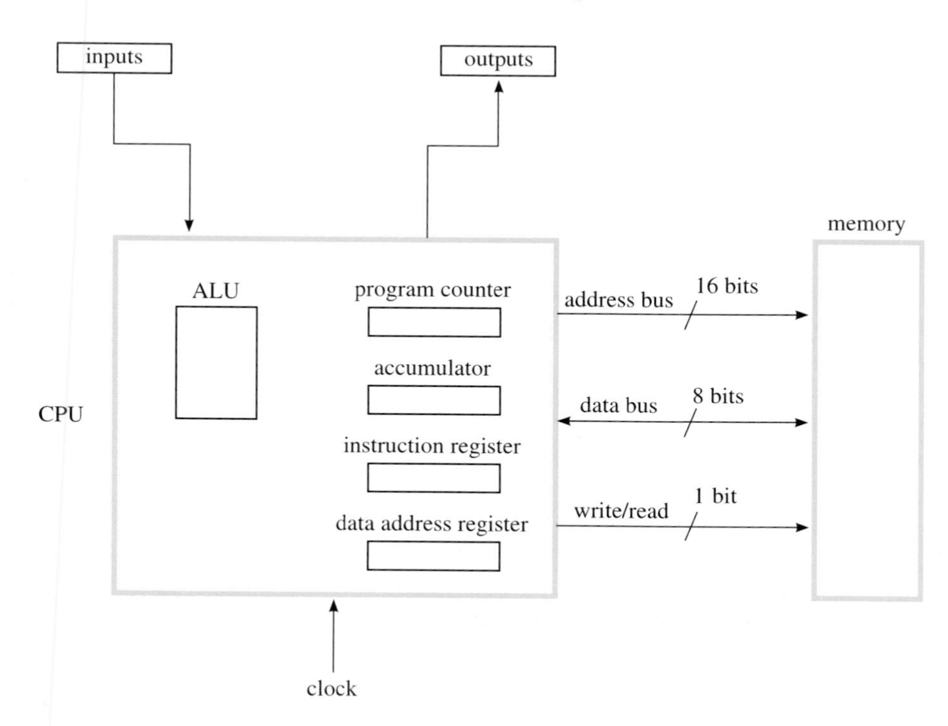

***Figure 3.19* Block diagram of a typical microprocessor.**

numbers are usually represented in *floating point* form. Typically this involves the use of four bytes. The number is represented in three parts: the sign, the value correct to seven significant figures, and a power of ten by which the value must be multiplied. The available 32 bits are shared between these three. With this scheme, very small and very large numbers can be represented, the range being approximately 10^{-38} to 10^{38}.

These options for representing and manipulating numbers and data depend on the architecture of the processor and the supporting software.

Within the *central processor unit* (CPU) the two main items are the *registers* and the *arithmetic logic unit* (ALU). The registers have two main functions. In the first instance they are repositories for data called from memory, and those data can subsequently be operated on within the registers. In some computer architectures different registers have different functions. Those in which arithmetic can be performed are sometimes called *accumulators*. Some registers have special functions associated with calculating addresses for the address bus.

The ALU performs all the arithmetic and logical operations on the data. Logical Boolean operations on single bits are fairly straightforward – just like using individual logic gates. Arithmetic is a little more complicated, but still depends on logical operations.

The remaining components of a microprocessor are involved in the timing, decoding, and general housekeeping of the system.

Typically a microprocessor can perform the following functions:

- load data from a given memory address into a register;
- save data from a register into a given memory address;
- increment and decrement by 1 the number in a register;
- perform arithmetic and Boolean operations on the data in two registers and write the result in a selected register, e.g.

  bitwise AND, OR, INVERT,

  bitwise left and right shift,

  integer addition and subtraction,

  integer multiplication and division.

It is perhaps surprising that such a small repertoire of basic operations can result in the tremendous information processing power of modern microprocessors.

Some microprocessors such as the INMOS transputer have floating point arithmetic as primitive operations in their hardware. This is much more complex than integer arithmetic, and hence many microprocessors are designed to be used in combination with a separate *coprocessor* when intense floating point processing is required. Otherwise floating point arithmetic is usually *emulated* in software, and takes longer.

3.3.7 Hardware, software, and stand-alone microprocessors

When a microprocessor is powered up, various pieces of support hardware ensure it reaches an electrical state at which it is ready to function. The microprocessor expects to find its first instruction at a given location in its memory space. In systems such as personal computers the initial instructions are 'hard-wired' into the system by the designers. Usually there is a large piece of software called the *operating system* which gives programmers easy access to devices such as disk drives, keyboards, printers, screens, and add-on hardware. For example, DOS stands for disk operating system (which is misleading because DOS knows about many other kinds of devices, such as those just mentioned).

However, microprocessors do not require disks, keyboards, screens, or any of the other microcomputer paraphernalia to work. In engineering applications there is usually a minimum configuration made up of the processor, a ROM which contains the program in executable machine code (see below) which the processor can run directly, some RAM, and hardware which interfaces the processor to sensors, effectors, and possibly communications hardware.

Software is usually (and rather ambiguously) classified as *high level* and *low level*. The lowest level of software is *executable machine code*, a sequence of binary code words which are understood by the CPU. In the early days of computing, engineers actually wrote programs in this binary form. A higher level of programming is available using an *assembler* which makes life much easier by employing meaningful mnemonics and names for data addresses, while maintaining a one-to-one correspondence with the machine code it produces. At a higher level still one finds *languages* such as C, Pascal, FORTRAN and BASIC which have constructs such as for–until loops, if–then–else tests, multidimensional data arrays, and so on, but still give access to the registers and lower level functions. The programs which transform *source code* written in these languages into executable machine code are called *compilers* and *linkers*. Some compilers allow software to be developed on one processor to be used on other. These *cross-compilers* are useful because they allow more powerful computers to be used for development work, and because they allow the engineer to work within a familiar software environment.

Assembly language programming requires knowledge of the particular hardware being used, and usually takes longer than programming in higher level languages. The advantage of programming at assembly level comes from the engineer accessing the hardware directly, and therefore more efficiently for a given purpose. Assembly language programs usually run faster than compiled programs.

Higher level languages allow code to be written which is more comprehensible to humans than assembly language code. This code is usually compiled and linked to other software modules to produce an executable module which the processor can run. Most languages can link to assembled modules. Time-critical software is usually written in assembly language, as is software which accesses part of the hardware in unusual and specific ways. In mechatronics applications this may mean all the software.

Software for microprocessors is usually developed in *development systems* produced by the chip manufacturer. These are computer environments which allow software to be developed using high-level languages and cross-compilers, with the resulting software transferred or *ported* to the target microprocessor by disk or by ROM.

3.3.8 Computer programming

Conventional ***computer programs*** consist of lines of code which manipulate numbers or logical values, or control the *flow* of the program (i.e. determine what happens next). Computers interfaced to external devices will also have interface commands.

As an example, the fragment of program below is written in the C programming language. It might be used to respond to a noisy sensor which occasionally gives incorrect responses. When the sensor correctly detects something, the motor is to be switched off. The strategy to cope with noise is to require the sensor to respond at least ten times in a hundred, and therefore to ignore isolated 'blips' (this simplistic approach would not be used in practice). The first line performs the operation of assigning the value zero to the memory location called '`score`' (arithmetic operation). The second line sets up a loop to be executed 100 times (program control). The brackets show that everything within them is to be executed each time round the loop. The fourth line of code reads the value on a port attached to a sensor (interface operation) and stores it in the memory location called `test_value`. The next command executes an if–then test to see if `test_value` is 1 (logical operation) and if it is the next command increments the value of `score` (arithmetic). The next line of code tests the value of score and if it is true (logical) that `score` is greater than ten (arithmetic) then data is sent to an output port which switches off the second motor (interface). The closing bracket completes the lines of command executed each time round the loop: `count` is incremented and if it is less than 100 the program goes round the loop again (program control).

```
score = 0;
 for ( count = 0; count < 100; count = count + 1 )
 {
    test_value = READ_SENSOR( 3 );
    if ( test_value == 1 ) score = score + 1;
    if ( score > 10 ) SWITCH_OFF_MOTOR ( 2 );
 }
```

READ_SENSOR and SWITCH_OFF_MOTOR are examples of subroutines whose lines of code are given elsewhere in the program. When they are encountered, the program jumps to those lines of code (program control), executes them (arithmetic, logical, interface) and returns to the next command (program control).

Programming computers in this way requires great care. First of all we can note that every part of every line has to be absolutely correct, and every line has to be in the right place. Swapping any two lines would make the program function differently. This means the program designer requires a lot of information and must put it together perfectly. This kind of programming is highly skilled and is very prone to errors. Various techniques are used by programmers to circumvent error. They may use names for memory address which are meaningful and make the program easier to 'read'. Similarly they try to break a program into subroutines which cleanly perform some task without upsetting other things. Nonetheless, programming remains error prone and a whole discipline of software engineering has grown up to overcome the common problems of faulty software.

The kind of programming we have described is called *procedural* since knowledge is encoded as procedures or subroutines in a computer programming language. This differs from the *declarative* style of programming in which knowledge is encoded in passive data structures but cannot be executed directly. Conventional computer languages such as BASIC, FORTRAN, COBOL, and C are procedural.

The main languages used by artificial intelligence researchers include LISP, which is functional, and PROLOG, which is declarative. LISP is a language designed to manipulate symbols rather than numbers. Its data elements are *lists* of symbols that represent things. Programs written in LISP consist primarily of collections of independent procedures called *functions*. These are constructed from a small set of primitive functions within the language. PROLOG programs may be viewed as logical clauses from which the interpreter makes deductions. Thus the logic programming approach is appropriate for manipulating knowledge expressed in a way without *a priori* structure.

A further class of programming languages is exemplified by Smalltalk, one of the first *object-oriented* languages. This permits the user to construct hierarchical classes of objects with subclasses 'inheriting' properties. The general idea is that the programmer creates classes of objects which 'encapsulate' their own data, and devises methods for manipulating data. In strict implementations, objects communicate by sending messages to each other. The C programming language has been extended to include object orientation. The result, C++, is becoming increasing popular for programming.

The general thrust behind these AI programming styles is to have programming languages which allow knowledge to be represented and manipulated in a convenient way. They are widely used in robotics research where the computer systems must be able to cope with many objects in a highly dynamic and uncertain environment.

3.3.9 Knowledge-based systems

Knowledge-based systems and expert systems are successes arising from research into artificial intelligence. In the 1980s the limitations of conventional computing became widely recognized after the Japanese announced their multi-million-dollar ICOT programme of research towards the 'fifth generation' of computers. One of the ideas in the ICOT programme was a new information processing paradigm called *expert systems* or, more generally, ***knowledge-based systems***. Conventional computer programs can perform miracles of calculation which can make our human powers seem puny. However, computers have difficulty in processing disparate and incomplete information, something at which we humans excel. Humans learn from others and from experience. As students we learn from knowledge which is encoded in language, symbols and pictures. We tend to become expert in things we do many times. Sometimes this expert knowledge is encoded in rules, but sometimes it is not, and one 'just knows'. Research on expert systems over the last thirty years has attempted to abstract that knowledge from human experts and build it into computer systems. In this we can identify two major types of knowledge: *facts* and *rules*.

Experts know many facts about their domain of expertise. For example, engineers know it is difficult to get sustained power from electrical batteries. Experts also know how to apply rules to obtain new facts from old. Suppose we know that an aeroplane must exceed a given power-to-weight ratio to fly. This can be stated as an if–then rule: *if* the power to weight ratio is lower than *P*, *then* the aeroplane will not fly. Suppose we know that the power-to-weight ratio of batteries is lower than *P* in all cases. A general rule of logic says that if something is true in all cases, it is true in any particular case. Suppose also that the power-to-weight ratio of a battery-powered aeroplane is less than *P*. We can apply the expert rule to this to deduce that a battery-powered aeroplane of a certain weight will not fly.

In general we can *elicit* facts and rules from experts on an *ad hoc* basis, and the order in which the rules are elicited does not matter too much. Also we can abstract meaningful knowledge in small pieces without having an initial overview of the whole domain. In 1984 Marvin Minsky, one of the founding fathers of AI, commented:

> In a sense today's expert systems demonstrate a marvellous fact that we did not know twenty years ago: if you write down if–then rules for a lot of situations and put them together well, the resulting system can solve problems that people think are hard. It is remarkable that much of what we think requires intelligence can be done by compiling surface behaviour rules. Many people in this field are surprised at that.
>
> *(Minsky, 1984)*

John Gero defines expert systems as follows:

> Expert systems are computer programs which attempt to behave in a manner similar to rational human experts. They all share a common fundamental architecture even if the knowledge encoding mechanisms differ. An expert system will have the following components:
>
> *an inference engine:* this carries out reasoning tasks and makes the system act like an expert;
>
> *a knowledge base:* this contains the expert's domain specific knowledge and is quite separate from the inference engine;
>
> *an explanation facility:* this interacts with both the knowledge base and the inference engine to explain why an answer is needed at a particular point or how a question can be answered; further, it is used to explain how a conclusion was reached or to explain why a specific conclusion could not be reached;
>
> *a state description:* this contains the facts which have been inferred to be true and those which have been found to be false during a particular session;
>
> *a natural language interface:* few expert systems have this yet.
>
> *(Gero, 1985)*

As Gero says, expert systems should have an explanation facility so that users can query the advice they are given. This is necessary because expert systems can make mistakes, and sometimes they make the most elementary blunders. Humans take huge amounts of 'obvious' information for granted but unless this information is made explicit within computer expert systems they will not work properly. Thus we expect expert systems to have to justify their reasoning when a human operator sees something suspicious.

Although the terminology is not standard, it is sometimes useful to distinguish between an expert system which models some area of human expertise, and a knowledge-based system which works on the same principles but is not intended to act like a human expert. In mechatronics we are more likely to have a knowledge-based system designed for some particular purpose. The PROLOG language is suitable for writing knowledge-based systems since clauses and predicates are its natural data structure.

3.3.10 Parallel computation

The enormously successful von Neumann machine architecture lies at the heart of the majority of contemporary computers, including the ubiquitous IBM PC compatible types. Individual sequential computers have been ***networked*** into more complex computational architectures for many years. When we speak of *networks* of computers we usually mean autonomous sequential processors where each processor can send information to the others and receive information from the others. This raises the important question of *control* and the possibility of *deadlock*. Without external controls it is possible that two processors will lock each other with both waiting to receive information from the other before proceeding. This is rather like two polite people obeying protocol: 'after you', 'oh no, after you', 'oh no, after you'.... In systems with many independent processors such as an automated factory or an advanced aircraft, the issue of deadlocking is very real and could have disastrous consequences.

In some ways a network system can itself be considered a 'computation system' of greater complexity than its constituent parts. This makes the term 'computer' rather vague. In future in this chapter we will use the term 'processor' for a single device which processes information, with the assumption that processors can be 'computers' within more complex 'computers'.

In this connection it is interesting to recall the robotics scientist Rodney Brooks from the Massachusetts Institute of Technology who asserts that the machines of the future will have 'intelligent' behaviour as a result of the autonomous action of many independent processors. As far as we know, biological systems are like this. For example, your brain has billions of processors called *neurons*, but no 'master neuron' has yet been discovered.

Networked computers are most commonly encountered in institutional settings such as banks, retail distribution systems, and motoring organizations. However, information processing lies at the heart of mechatronics, and we expect the number of processors in machines to increase greatly by the turn of the century. The existence of many autonomous or semi-autonomous processors in machines raises new research problems in information engineering.

In mechatronics we can expect to find the trusty von Neumann machine employed for many information processing tasks into the foreseeable future. These processors are relatively inexpensive, and they have well tried and tested software and software engineering procedures. But other approaches to computation are emerging, as we shall see in the next section.

3.3.11 Non-sequential computers: neural computers

The networked systems discussed in the last section were made up of communicating von Neumann machines. There is a whole class of processors which work on entirely different principles in a non-sequential way, called ***neural computers***. Put somewhat simplistically, *artificial neurons* receive information and process it by *firing* or not firing. When these neurons are configured in networks, with the outputs of some neurons being the inputs of others (Figure 3.20), they possess some remarkable information-processing capabilities.

Your own biological neurons appear to be firing or not firing all the time in response to stimuli from your sensors. For example, in the children's game you shout 'snap' when two cards with the same picture are displayed. We don't quite know how it happens, but your eyes take in the information of the current card and your brain remembers the last card. Your 'snap comparison processor' takes in

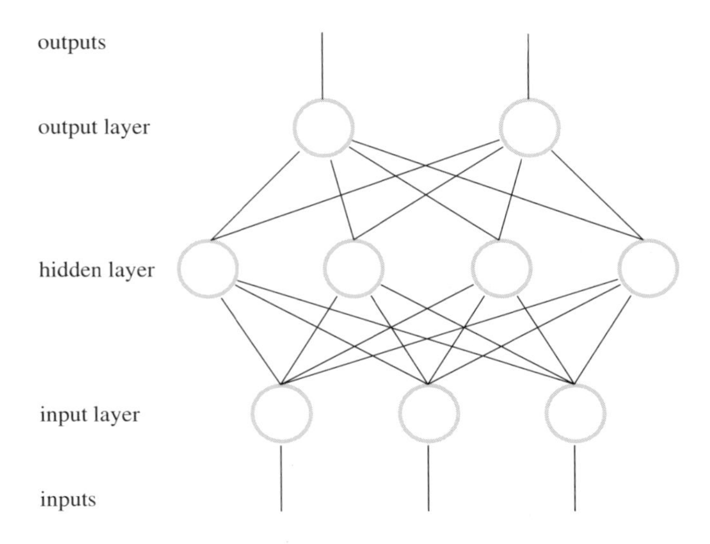

Figure 3.20
A typical neural network configuration.

this information, and when two cards are the same it 'fires' and results in you calling out. Of course, sometimes it fires incorrectly, and you shout 'snap' when the cards are different.

The idea of implementing neuron-like information processing into electronic hardware goes back to the 1940s and the work of the mathematician Walter Pitts and a neurologist Warren McCulloch. Their idea was that the artificial neuron would receive information of various *weights* on its inputs, sum this information, and if the sum exceeded a given threshold the neuron would fire. When the neuron fired its output it would be a value such as 1, but when it did not fire the output would be zero (Figure 3.21).

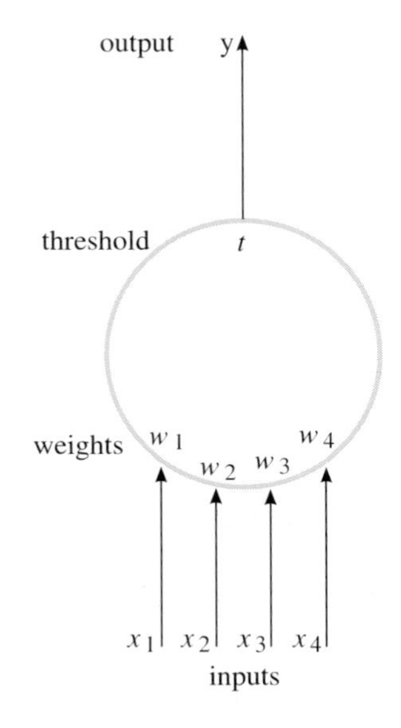

Figure 3.21
A four-input artificial neuron.

Marvin Minsky and Seymour Papert showed in a seminal work (Minsky and Papert, 1969) that such a neuron is severely limited in its information processing capabilities. In particular a single neuron cannot perform the EXCLUSIVE-OR operation (this is like the OR function described earlier except that when the two inputs are both 1 the output is 0). However, they showed that by configuring many neurons into networks the whole system can indeed perform the EXCLUSIVE-OR.

The details of how neural networks process information are given in Volume 2. In general terms a neural network will look something like that shown in Figure 3.20. The main difference between a neural network and a sequential processor is that the former can be 'programmed' by learning from examples. There is a set of inputs for values to be entered to the network. Depending on how the neurons are connected (the *network topology*) each neuron will receive information, process it (usually by adding the values and applying a threshold), and output zero or some other value such as one. In a wide class of networks, each connecting link has a weight which is applied to the value it carries. These weights lie at the heart of *feedforward* networks since they are adjusted when the network is learning, using a process called *back propagation of errors* (details of this are explained in Chapter 4 of Volume 2).

Neural networks can be implemented in hardware or simulated in software on sequential computers. The main differences between neural networks and sequential computers usually cited include those in Table 3.2.

TABLE 3.2 SOME DIFFERENCES BETWEEN SEQUENTIAL COMPUTERS AND NEURAL NETWORKS

Sequential computers	Neural networks
Are programmed by humans.	Learn from examples.
Data errors cause unpredictable failures.	Are tolerant of data errors.
Small malfunctions cause catastrophic failures.	Degrade 'gracefully' on being damaged.
Provide precise logical and arithmetic processing.	Produce classification with occasional error.
Can implement rule-based symbolic information processing.	Information is distributed as numbers over the network links.

It is interesting to note that the pioneer of sequential computing, John von Neumann, was also a pioneer of neural computing. The digital computer began life at least as far back as the nineteenth century with the work of Charles Babbage and his supporter Lady Ada Lovelace. It took about a century for sequential computing to reach its present state, so one should not be surprised that sixty years after its invention, the artificial neural system remains an area of intense research and immense potential.

Like expert systems before them, neural networks have been overhyped in recent years and there are occasional claims that they are 'artificial brains'. Most artificial neurons are not in the least brain-like in their functioning, and it is important to keep things in perspective and realize that the term 'neuron' is more a metaphor than a simile. Nonetheless, artificial neural networks provide a powerful new information processing paradigm for the mechatronics engineer. They are especially useful in collecting together noisy data from different sensors to produce a recognized state which can be symbolically encoded and fed into the cognition and execution subsystems of an intelligent machine.

3.3.12 Which information processing paradigms can be applied when?

Although there are no hard and fast answers to this question, there are some guiding principles. Neural networks are particularly useful for pattern recognition in the context of interpreting sensor data. They perform well in these circumstances because they can train on real data which reflect all the uncertainty and noise of the sensor data. Neural networks can also be used for recognizing good options in scheduling problems such as motion planning.

Symbolic information processing is appropriate to aspects of mechatronics systems which are more structured and behave more reliably. We usually take a model of the subsystem and apply rules and reasoning to make deductions. Sequential computer programs may be appropriate for complex systems for which there is an explicit symbolic or numerical model. This approach is also appropriate for software whose performance is time critical. Knowledge-based systems provide appropriate information for systems whose behaviour can be captured in terms of statements of fact, and rules. Expert systems may be especially appropriate for machines which take over human functions for which knowledge already exists and can be elicited from the human expert.

The distinction between the various ways of symbolic information processing can become blurred when we look at the details. For example, both sequential computer programs and knowledge-based systems make use of if–then rules; here the difference lies in the information-processing control mechanisms and the way the systems are built. However, the explanation facility expected of expert systems suggests that such a feature should exist in systems which interact with people. Ultimately *all* machines are supervised by humans, and machine decision-making must be comprehensible to the human controller. In summary:

1 Neural computing is particularly suited to processing information from less structured environments, including those which produce noisy data, and those for which there is no explicit model. Neural networks *interpolate* well within their training data, but cannot *extrapolate* successfully from it.

2 Symbolic computation is well suited to more structured environments in which there is less uncertainty about the input data.

   (a) Sequential programming, in the context of software engineering, is appropriate for very complex systems with a known model, and for time-critical functions.
   (b) Knowledge-based systems are appropriate when there is knowledge about the system expressed as facts and rules.
   (c) Expert systems are appropriate when there is human knowledge and expertise which can be elicited into facts and rules. The explanation facility of expert systems may be important in safety critical applications and when interfacing to overall human control.

3.4 Conclusion

In this chapter on information in mechatronics we have addressed some fundamental issues:

- What is information?
- How can information be encoded?
- How can encoded information be communicated?
- How can information loss be minimized?
- How can noise be minimized?
- How can signals be sensed and their information abstracted?
- How can information be processed?

  Sequential information processing by von Neumann machines.

  Parallel information processing.

  Information processing architectures such as expert systems.

  Non-programming paradigms such as neural networks.

Inevitably in this broad chapter we have only been able to skim the surface of communications, signal processing, information theory and computer architectures. Mechatronics engineers must be aware that there is a wide body of theory and practice relevant to perception in machines, and that there are many areas of specialism.

References

Gero, J. (1985) 'Editorial: Expert systems in CAD', *Computer Aided Design*, November 1985, Butterworth & Co. (Publishers) Ltd.

Mello, J. P., Jr. and Wayner, P. (1993), 'Wireless mobile communications', *Byte*, February, pp. 147–154.

Minsky, M. (1984) 'The problems and the promise' in Winston, P. H. and Prendergast, K. A. (eds) *The AI Business: the commercial uses of artificial intelligence*, p. 244, MIT Press, Cambridge, Mass.

Minsky, M. and Papert, S. (1969) *Perceptrons: an introduction to computational geometry*, MIT Press, Cambridge, Mass.

CHAPTER 4
PERCEPTION

Jeffrey Johnson

4.1 Introduction

Perception is the process by which information enters a system and is transformed into a useful form. For example, the human system is well equipped to receive information through its 'senses' of sight, hearing, touch, smell and taste. Information can come from the system's environment, or it can come from within, as when humans experience pain or when a machine is monitoring its internal performance. Such internal monitoring may be related to ***goal-directed behaviour***, or it may be motivated by self-diagnosis. For example, a vehicle may count the number of times it thinks its wheels have revolved when using dead-reckoning to estimate its position, or it may have sensors to check that none of the wheels has fallen off.

When speaking of perception I will make a distinction between *sensing* and *processing* the information that is sensed. We sometimes complain that someone does not 'listen' to us, that what we are saying 'goes in one ear and out the other'; our exasperation comes from the fact that they have certainly heard what we said, but their brain appears to choose not to process the information in any active way. ***Perception*** in mechatronics systems is the process by which data on a *system* and its *environment* are *received* by appropriate *sensors*, *decoded*, and *processed* to give useful *information* which can be transmitted and used elsewhere within the system.

The grey area begins when we talk about 'decoding' and 'processing' the perceived signal, since this may require no further processing (as in the case of breaking a light beam and triggering an alarm), or it may require a great deal of processing (as is the case in a robot's vision system).

There is a fundamental distinction between symbolic and non-symbolic ways of representing information. Symbols are cultural artefacts which are used to 'label' observables and derived concepts. The words in text such as this book represent information in symbolic form whereas radio waves represent sound information in non-symbolic analogue form. An important special case of symbolic form comes when the mechatronics system has a *model* (defined in Section 4.12) of its environment or some process. In this case the system can manipulate information by *reasoning*. We will say that perception includes delivering information encoded in symbolic form, but further information processing which is model based will be defined to be part of ***cognition***.

Suppose you hear something as you pass a room. Your sound sensors (ears) detect sound waves in the air. They convert them into analogue mechanical and then analogue electrochemical signals in your brain. This part of the process is perception. You may instantly 'know' you are hearing one of Mozart's symphonies. We could debate whether this instant ***pattern recognition*** remains part of perception or cognition. The words 'Jupiter Symphony' may come into your mind. By this time you have moved from the analogue signal of the sound to information encoded in symbolic form (words), and so you are in the realm of cognition. You may engage in some model-based reasoning and decide you have time to pause a while to hear more, and so the perception–cognition process will have resulted in analysis and effected a consequent action (stopping to listen).

This chapter will concentrate on the perception side and Chapter 5 will pick up the theme of cognition and show how we close the loop in Chapter 1, Figure 1.6.

4.2 Analogue and symbolic representations in perception

Consider a scene which contains a chessboard and pieces such as that in Figure 4.1. In computer graphics the screen is divided up into a grid. Each cell in the grid is called a picture element, or *pixel*. Suppose this scene has been scanned with a monochrome black-and-white camera and digitized to give a 768×576

Figure 4.1 Photograph of scene containing a chessboard.

pixel computer image with one byte per pixel representing grey levels between 0 (black) and 255 (absolutely white).

Could you tell from the photograph in Figure 4.1 if the game has begun? If you knew nothing about chess you could not answer this question. Indeed, for the question to be meaningful it is necessary to have a 'model' of the game of chess in your mind.

The starting position for chess can be represented *symbolically* as shown in Figure 4.2(a). Thus the white pieces in Figure 4.1 are all in their positions at the beginning of the game. But what about the black pieces? As it happens, some of the black pieces cannot be seen because they are hidden by others, or they have merged into other pieces or black squares in the poor light, or both.

Figure 4.1 shows an ***analogue*** image (photograph) of the chessboard, Figure 4.2(a) shows an *iconic* representation of the chessboard, and Figure 4.2(b) shows a ***symbolic*** representation of the chessboard. These represent increasing levels of abstraction from the real chessboard. The photograph *looks like* the real thing, the iconic representation uses symbols which *approximate* the real thing, while the alphabetic representation uses symbols which are *associated* with the real thing.

The association of Figure 4.2(b) is not absolutely straightforward. Although we use the first letters of most of the pieces of rook, knight, bishop, king, queen and pawn, we use the symbol N for knight since we want to reserve the symbol K for

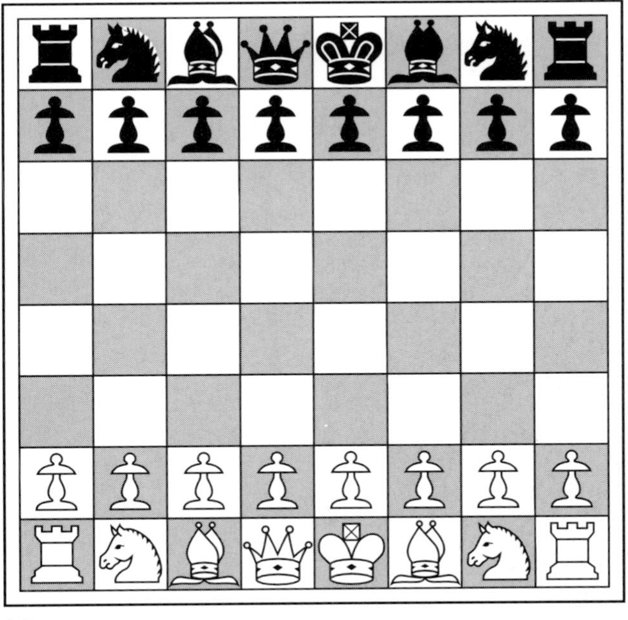

(a)

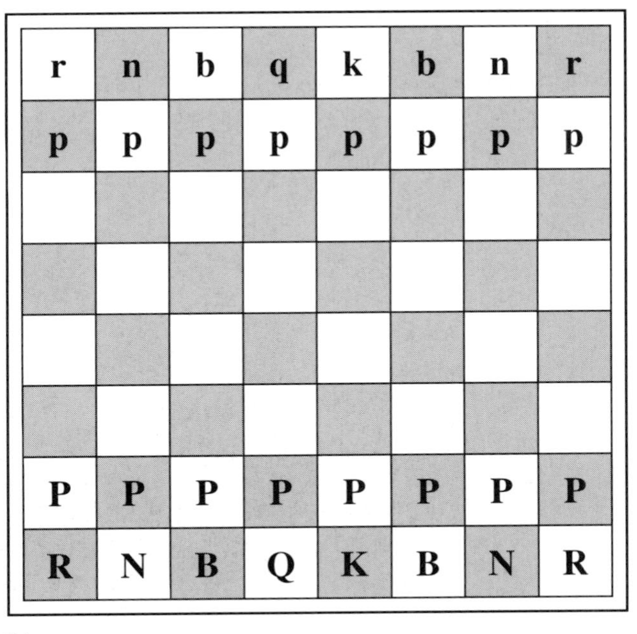

(b)

Figure 4.2
Symbolic representations of the starting position in chess: (a) iconic representation; (b) a symbolic representation.

the king. Children often use the term 'horse' for the knights of a chess set since that is what they look like. However, many chess players hate this incorrect usage and would much rather misspell knight as 'night' than accept the symbol H to represent 'horse'. Which all goes to show that it does not matter what the reason is for using a particular symbol for representing something, as long as the symbol represents one, and only one, thing and that this convention is universally adhered to.

The iconic symbols in Figure 4.2(a) are more evocative and attractive than the letters of the alphabet, and are usually used in chess books and newspapers. However, people rather than machines read books and newspapers, so the widespread use of ASCII codes to represent characters in digital computers probably makes an alphanumeric representation more appropriate for machines.

4.3 From analogue to symbolic representation

One of the (difficult) problems a chess-playing robot has to overcome is to 'see' the pieces on the board, in order to know the current position and to move the pieces. After every move the robot needs to be able to make the same kind of transformation as there is between the image of Figure 4.1 and the symbolic representation of Figure 4.2(b). What might be involved in this?

Before discussing this for the robot's vision system, consider your own vision system as you read this book. Humans are capable of remarkable feats of visual perception, even those with poor sight. As an example, take another look at Figure 4.1 and point to the black queen's knight (the second piece from the right closest to you). Even though it is not all there in the picture, it is likely that you can 'see it in your mind's eye', especially if you are a chess player.

Now look at the fragment of the image of a chessboard in Figure 4.3. Do you think the piece on the right square closest to you is a knight or a rook? Spend a

Figure 4.3
A fragment of an image of a chessboard.

few seconds trying to decide. To me it looks like a knight, but things are not always what they seem.

To show what I mean, take a look at Figure 4.4, which shows a man bending over backwards in a most unlikely pose to watch an equally unlikely fish swim past. The remarkable thing about this picture is that the fish's head is visually the same as the heel of the man's foot. It is even the same size, although to me it looks bigger.

Human beings and many animals have astonishingly good vision. We can abstract subtle information from the least clue and in a remarkable range of illuminations, angles and sizes. How do we do it? The honest answer is that we do not know. We do know from Levine (1985) that the human eye has 120×10^6 rods (receptors that are achromatic, with slow response adapted to shades of dim light) and 6.5×10^6 cones (receptors that have rapid response to colour and bright light, concentrated directly behind the lens at the fovea). Roughly speaking, this array of rods and cones corresponds to a computer resolution of about 4000×4000 pixels. We also know that a large part of the human brain is concerned with vision, and since the human brain is one of the most complex things known to man, this suggests that biological vision is a very complex process indeed.

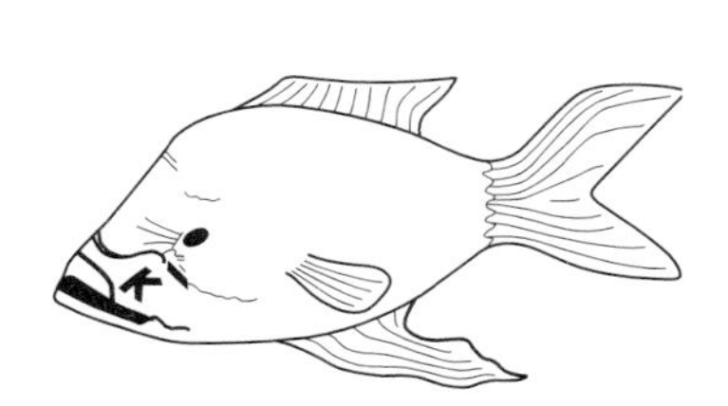

Figure 4.4
Image of a sportsman watching a fish swim past.

In the early days of computer graphics it was thought that the problem of abstracting explicit information from images was one of many technical problems that would soon be solved. Some thirty years later, after tremendous research effort, we have learned that it is not going to be that easy to solve the problem of *computer vision*.

A large and unsolved problem in computer vision is knowing what we are looking for and how to represent what we find. Should our symbolic universe contain phrases like 'fish's head' and 'sportsman's heel'. How would we represent the rest of the sportsman? And the fish? There are countless numbers of ways we could pick out coherent bits of each. Does this mean that our biological vision system has a 'pattern recognizer' for each possible part? Probably not.

Our fabulous and effortless human vision is the product of millions of years of evolution and specialization. Although it is the legitimate subject of research for those interested in human vision, it can be misleading for the mechatronics engineer to treat the relationship between human vision and machine vision as anything more than a metaphor. However, a comparison can be made, as shown in Table 4.1.

TABLE 4.1 COMPARISON BETWEEN HUMAN AND MACHINE VISION

Human vision	Machine vision
Eyes have receptors (rods and cones) in 2-D arrays.	TV camera produces a 1-D scan (camera technology may change).
Information reaches the brain down a 2-D analogue channel.	Information reaches the computers as a 1-D signal.
There is evidence of signal processing in the optic nerve.	Some cameras perform signal processing.
Billions of processing cells are involved in human vision.	Machine vision usually involves few processors ($<10^4$) and a few megabytes of memory ($<10^9$ bytes).
The processors are slow (milliseconds).	The processors are fast (microseconds).

It is unlikely that we will build machines with anything like the information-processing abilities of human vision in the near future. Nonetheless, machine vision can be developed in its own right as a form of artificial sensing which can be used as a powerful part of the perception process in mechatronics systems.

Computer vision illustrates the distinction between perception and cognition. A vision system able to begin with an image like that in Figure 4.1, recognizing the pieces, and able to convert it to a symbolic representation such as that in Figure 4.2(b) could be said to have *perceived* the chessboard. Performing this *pattern recognition* and correctly assigning collections of dark and light areas to symbols is part of the perception process. Knowing that the position is legal, for example, and that it is time to make a move, requires *cognition*.

4.4 The dimensions of sensory information space

4.4.1 One-dimensional information spaces

Consider a temperature sensor delivering an analogue signal: the hotter the environment the larger the voltage. Suppose this signal is put through an analogue-to-digital-converter to produce a stream of digital numbers. Finally, suppose we have a symbolic set of temperature states called *very cold*, *cold*, *warm*, *hot*, and *very hot*. In some cases knowledge is represented in these symbolic forms. For example, it makes sense to say that one of the factors in the *Challenger* space mission disaster was that it had been very cold the previous night, and that this was abnormal. So the perception information processing which transforms analogue signals representing temperature to symbolic form might be a program such as:

```
reading = READ_TEMPERATURE( sensor_5 )
if ( reading < -10 ) the_temperature = 'very cold';
else
if ( reading < 0 )    the_temperature = 'cold'; else
if ( reading < 25 )   the_temperature = 'warm'; else
if ( reading < 99 )   the_temperature = 'hot'; else
                      the_temperature = 'very hot';
```

So, the one-dimensional ***sensor information space*** has been ***classified***, with each point being assigned to one of the classes *very cold, cold, warm, hot, very hot*.

Where do such classes come from? Why is there not a class called *tepid* between *cold* and *warm?* The answer lies at the heart of the difficulty of representation: the word 'tepid' could indeed belong to the class *if it were useful for it to do so*. And what does this mean? It means that a symbolic representation may be perfectly suited to some purposes but not others. For example, the Inuit people of North America are said to have many names for snow of different types. There is less variation in English (e.g. snow, slush, sleet, hail) because detailed knowledge of snow has not been so important in our cognition.

Of course it is sometimes difficult to decide if a nuance of discrimination is worth while, and designing languages may be an important task for the mechatronics engineer when trying to represent information in an appropriate way. In this respect we should note that many of our theories as to how things work are expressed in relational terms rather than numerical formulae.

4.4.2 Two-dimensional information spaces

Consider the perception system of a robot which has to survey a set of coins and sort them into stacks according to type. In this case it is fairly easy to take a video image of the coins and recognize them as circular objects of different colours (Figure 4.5).

Suppose then that the robot's camera-based vision system detects sixteen objects and makes the diameter measurements in centimetres shown in Table 4.2. We can show how each coin appears on the diameter dimension in Figure 4.6, where the type of coin is shown, followed by its number. Although the robot could measure the diameter, which would allow it to recognize and classify some of the coins correctly, it will, within experimental error, confuse the British two pence coin (coin 5) with the old French ten franc coin (coin 3), and the French five franc coin (coin 4) with the old British ten pence coin (coin 12). (The 10p piece used here is the size of the historic florin or two shilling piece and was replaced by a smaller coin as decimalization was completed in 1992.)

Figure 4.5
The coins to be sorted by the robot (actual size; for key see Table 4.2).

TABLE 4.2 DIAMETERS OF THE COINS

Coin number	Denomination	Diameter/cm
1	5p	1.76
2	1p	2.02
3	10 FF	2.63
4	5 FF	2.86
5	2p	2.60
6	5p	1.78
7	£1	2.25
8	1p	2.03
9	1p	2.03
10	2p	2.60
11	1p	2.03
12	10p	2.83
13	5p	1.77
14	£1	2.23
15	1p	2.05
16	1p	2.05

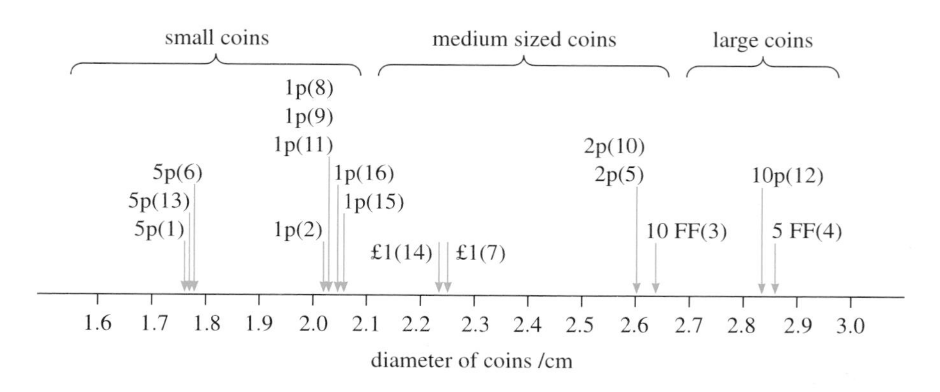

◀ ***Figure 4.6***
The coins arranged along the diameter dimension.

We can introduce another dimension to make the necessary discrimination, namely *colour*. The digitized image from which the diameters were measured can also be used to give values of, for example, the intensity of red light sensed. The data are as shown in Table 4.3, where the numbers are measured on a 100-point scale between 0% (no red at all, grey level = 0) and 100% (maximum red response from the camera, grey level = 255). The clustering of the coins according to their 'redness' is shown in Figure 4.7.

TABLE 4.3 'REDNESS' OF THE COINS

Coin number	Denomination	% red
1	5p	46
2	1p	60
3	10 FF	63
4	5 FF	39
5	2p	60
6	5p	38
7	£1	64
8	1p	60
9	1p	59
10	2p	55
11	1p	49
12	10p	48
13	5p	38
14	£1	70
15	1p	53
16	1p	54

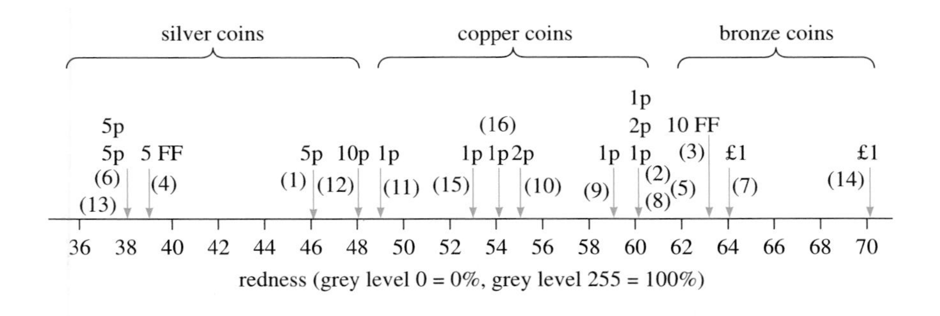

Figure 4.7
The coins clustered according to their 'redness'.

At first sight the clustering seems to be sensible since we can group the coins into 'silver', 'copper' and 'bronze'. However, the sceptic might ask where these groupings came from. If we clustered the coins by their objective similarity we would get the groups:

(5p, 5p, 5 FF)	silver coins
(5p, 10p, 1p)	?
(1p, 1p, 2p)	copper coins
(1p, 1p, 2p, 1p)	copper coins
(10 FF, £1)	bronze coins
(£1)	?

and these do not respect the neat division we have made between 'silver' and 'copper' coins.

In both of these one-dimensional classifications there has been undesirable ambiguity. But suppose we put the two together to make a two-dimensional classification. In other words each coin is to be represented by two numbers in two-dimension space, its size and its redness. We can plot these as points (size, redness) on a graph. Then we can see that it is possible to put each coin type into a 'classification box', as shown in Figure 4.8. For example, any coin smaller than 1.9 cm with redness less than 48% is a 5p coin. With these two-dimensional data our robot could distinguish the coins in order to perform its stacking task.

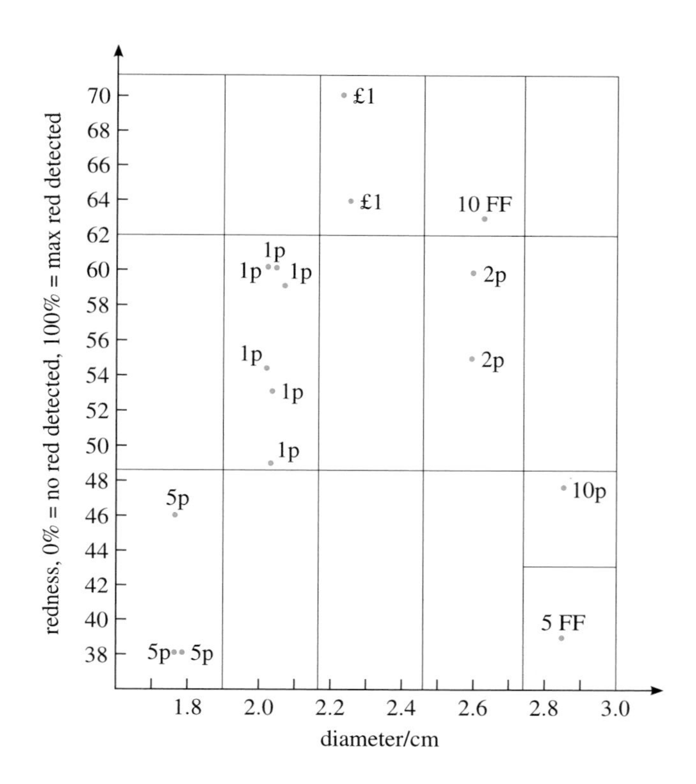

Figure 4.8 ***The coins in 'classification boxes' determined by two-dimensional data.***

4.5 Measurement scales and classification problems

In practice, the classification boxes shown in Figure 4.8 may not give robust results. The variation on the vertical redness scale might lead us to expect, for example, that the discrimination between the 2p pieces and the old-style 10 FF piece could not be made perfectly in all cases.

For example, the top 2p coin is separated from the other 2p coin by a difference of 5% redness, and it is separated from the 10 FF by a difference of 4%. How can the robot resolve this conundrum? The answer may be that with these data it cannot, and another dimension is necessary.

Before discussing the case of multidimensional data, let us consider the argument that a difference of 4% is less than a difference of 5%. Is this always the case? The answer is an emphatic *no*. In a large number of measurements the *scale* is *non-linear* so that equal distances on the scale may mean different things. Many car petrol gauges have non-linear scales so that the interval between empty and one-quarter full may be almost twice the size as that between three-quarters full and full. In such a case 5 mm on one part of the scale means something very different from 5 mm elsewhere.

A common cause of non-linearity comes from scaling two-dimensional objects. Figure 4.9 shows our coins reduced from the actual size shown in Figure 4.5. Even if you are familiar with British coins you will probably find it difficult to tell at a glance which coin is which. The reason is that our notions of 'small' and 'large' coins seem to be related to the *area* of the coin. It may surprise you to know that the 10p coin (12) is about 1.6 times the diameter of the 1p coin (8), although it easily looks more than twice as big. Of course the area of the 10p coin is more than 2.5 times bigger than the area of the 5p coin, so it is indeed more than twice as big.

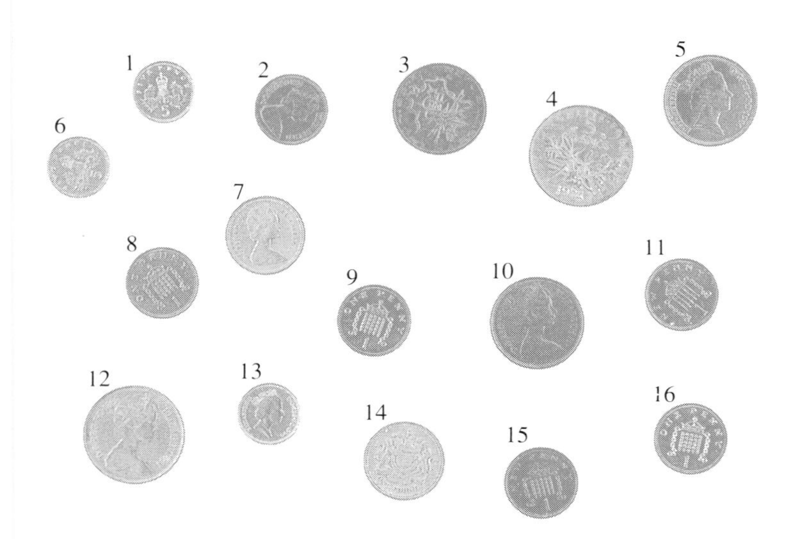

Figure 4.9
It is difficult to recognize the coins by their sizes when the scale has changed (for key see Table 4.2).

Many measurements result from statistical averaging, and sometimes these averages have peculiar arithmetic properties. Most algorithms for clustering make assumptions about the scales being used. It is important that any algorithm used for clustering is compatible with the arithmetic properties of the scales used to make measurements. This will be discussed in greater detail in Volume 2, Chapter 2 on Pattern recognition.

4.6 Feature and pattern recognition on the basis of multidimensional sensor data

If we wanted our robot to be more certain in its classification of the coins we could use another dimension such as the *weight* of the coins. This would give us a *three-dimensional* classification space: (1) diameter, (2) redness, and (3) weight.

When trying to classify on the basis of multidimensional data, we aim for the classes to be separated by clear gaps, as shown in Figure 4.10(a). In reality this is rarely possible, and frequently one or more classes intersect at the edges of their clusters, as shown in Figure 4.10(b).

How do we define these classes and what is their purpose? In general, the classes are determined by collecting data on what are supposed to be typical examples. The purpose of these classes is to allow us to classify new objects in terms of those

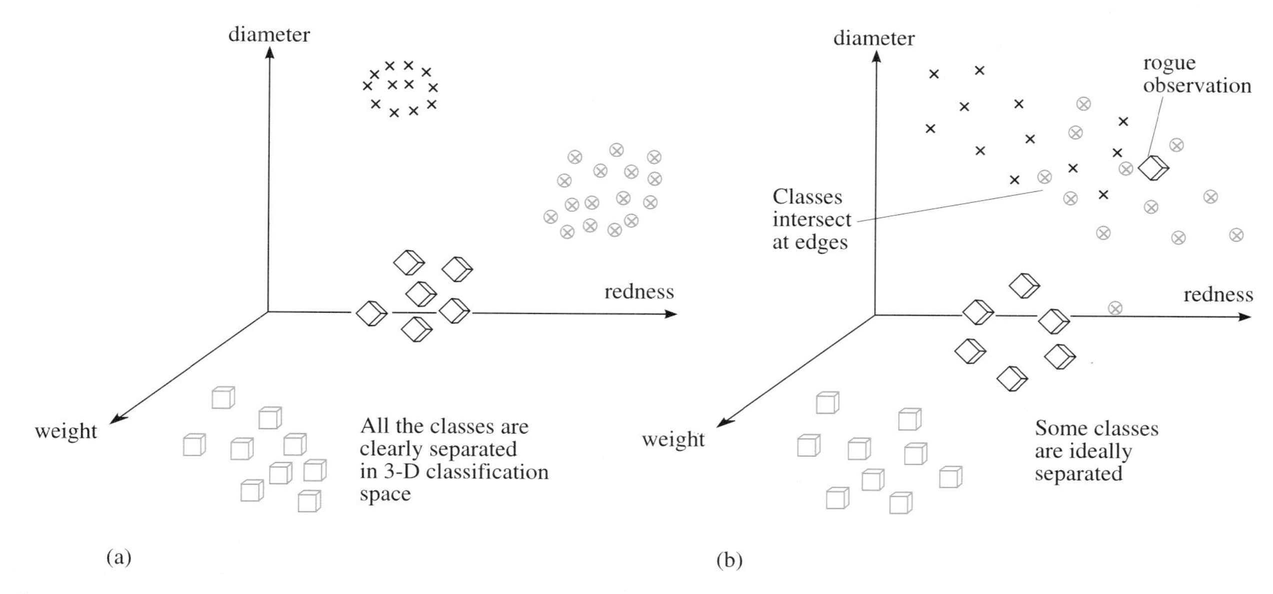

Figure 4.10
A three-dimensional classification space for pattern recognition.

that we know and have classified already. In principle, any observation that is similar to one of the observations on which the system was *trained* is said to belong to the same class as that observation. This approach is clearly problematic when the classes intersect, and it is especially problematic when there is a 'rogue' observation in the *training set*, as shown in Figure 4.10(b).

4.7 An example of sixteen-dimensional classification

To illustrate the collection of sixteen-dimensional data, consider the set of 2×2 configurations of black and white pixels shown in Figure 4.11

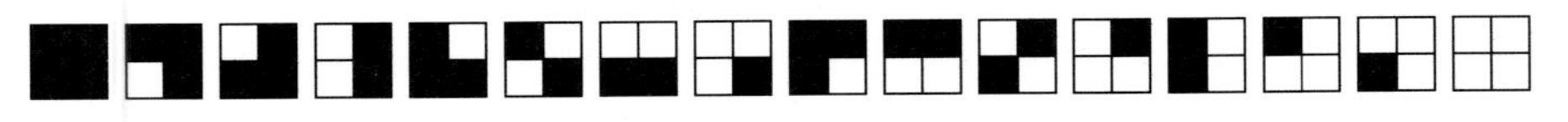

Figure 4.11
A set of pixel configurations used to classify the images of simple shapes in Figure 4.12.

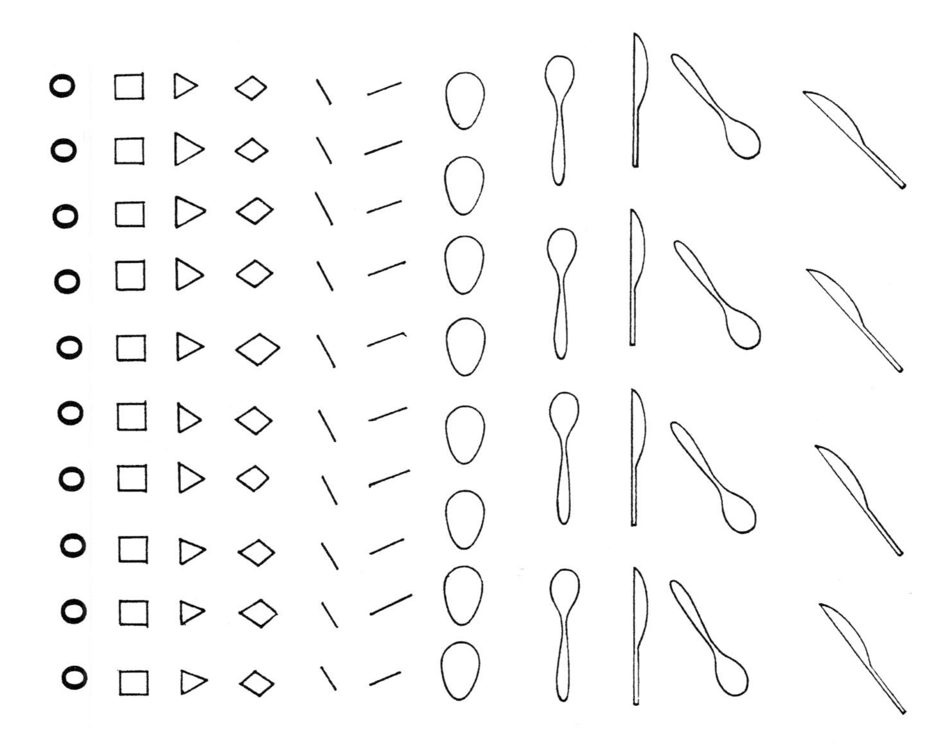

Figure 4.12
A set of objects to be classified by 2×2 pixel configurations.

Suppose each of the objects in Figure 4.12 were digitized. Then we can go through the image and count the number of times each of the configurations in Figure 4.11 occurs. In this way we can obtain sixteen numbers, one for each configuration. The computer program that finds these numbers works as if a box is put around the symbol being analysed, to isolate it.

Figure 4.13 lists the pixel configuration counts for just the first two examples of the circles, squares, diamonds and triangles. If you inspect the list of numbers in Figure 4.13 you will find that there are some similar configuration counts for some pairs of objects. However, in no case are all the configuration counts similar, except when the two objects are similar. These data belong to a sixteen-dimensional space, and within this space they can be classified on the basis of similarity. The details of this and other classification methods will be discussed in Volume 2.

Intuitively the reason this classification works is simple. We would expect the box shape to have many vertical and horizontal configurations, and so it does (despite imperfections in the digitizations). Similarly we would expect the diamond shape to have many diagonal configurations and few horizontal or vertical configurations, and this is indeed the case. We would expect the circle to have almost equal mixtures of horizontal, vertical and diagonal configurations, and again this is the case. In some sense these shapes *must* have the numbers they do: they have no choice. Thus they will tend to inhabit disconnected pieces of multidimensional space, and it will be possible to classify them.

Figure 4.13
Pixel configuration counts for two circles, two squares, two diamonds, and two triangles.

4.8 Template matching pattern recognition

The simplest form of ***template matching*** occurs when we have a structured set of sensed responses and compare them with templates of anticipated responses.

Consider a simple scanning device made of eight pieces of wire, each of which closes a circuit when it touches a common earth rail, as illustrated in Figure 4.14 (next page). For the purpose of illustration, six objects approximately 6 cm × 9 cm were made from soft modelling clay (Plasticine). When the scanner was drawn over these objects those wires that were raised by the Plasticine had their circuit closed. Thus the wires *sensed* the presence of the Plasticine as they moved over it. The scanner was connected to a computer through an interface which allowed the computer to 'read' the status of the wires through time.

When the scanner was drawn horizontally from left to right across the object, the data for the first scan were as shown in Figure 4.15(a). A '1' means the wire was raised and its circuit closed at the time interval. Each vertical list of 1s and 0s represents an instantaneous 'reading' of the states of all the eight wires. If we plot these data on a computer screen or a printer with '1' meaning 'black pixel' and '0' meaning 'white pixel' we find our scanner has produced a crude ***digital image*** of the original object (Figure 4.15b).

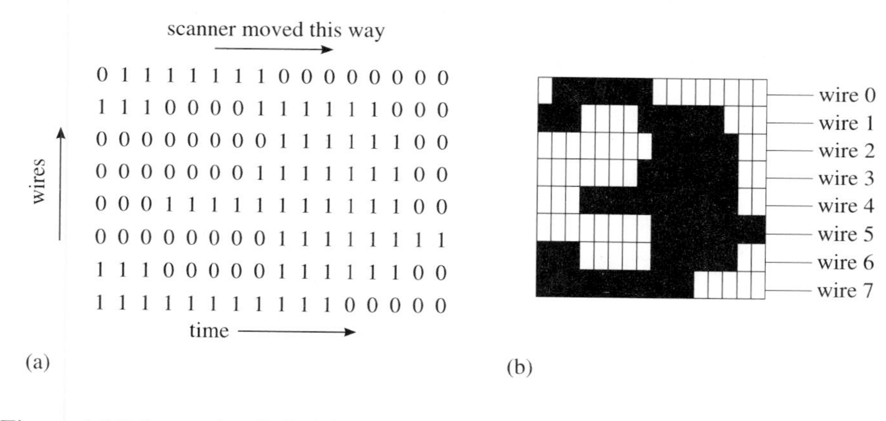

Figure 4.15 ***Binary data from the eight-wire scanner and its representation as a digital image: (a) the data from the eight-wire scanner sensor for sixteen consecutive time intervals of one-eighth of a second; (b) a digital image plotted from the scan data.***

Figure 4.16 shows the digital images of the six objects arranged vertically. These were the training set of templates, and the computer remembered their data.

The objects were then all scanned again and their digital images compared with each of the remembered patterns by template matching. If a pixel of a trained image and a test image are the same with both being black or both white, this is defined to be a match. In this way we could count the number of matching pixels for each of the test objects against each of the training objects.

You will see that the entries in the table in Figure 4.16 show the percentage of matching pixels. If you look at the numbers down the diagonal of the array you will see that each test digitization of an object matches the training digitization of

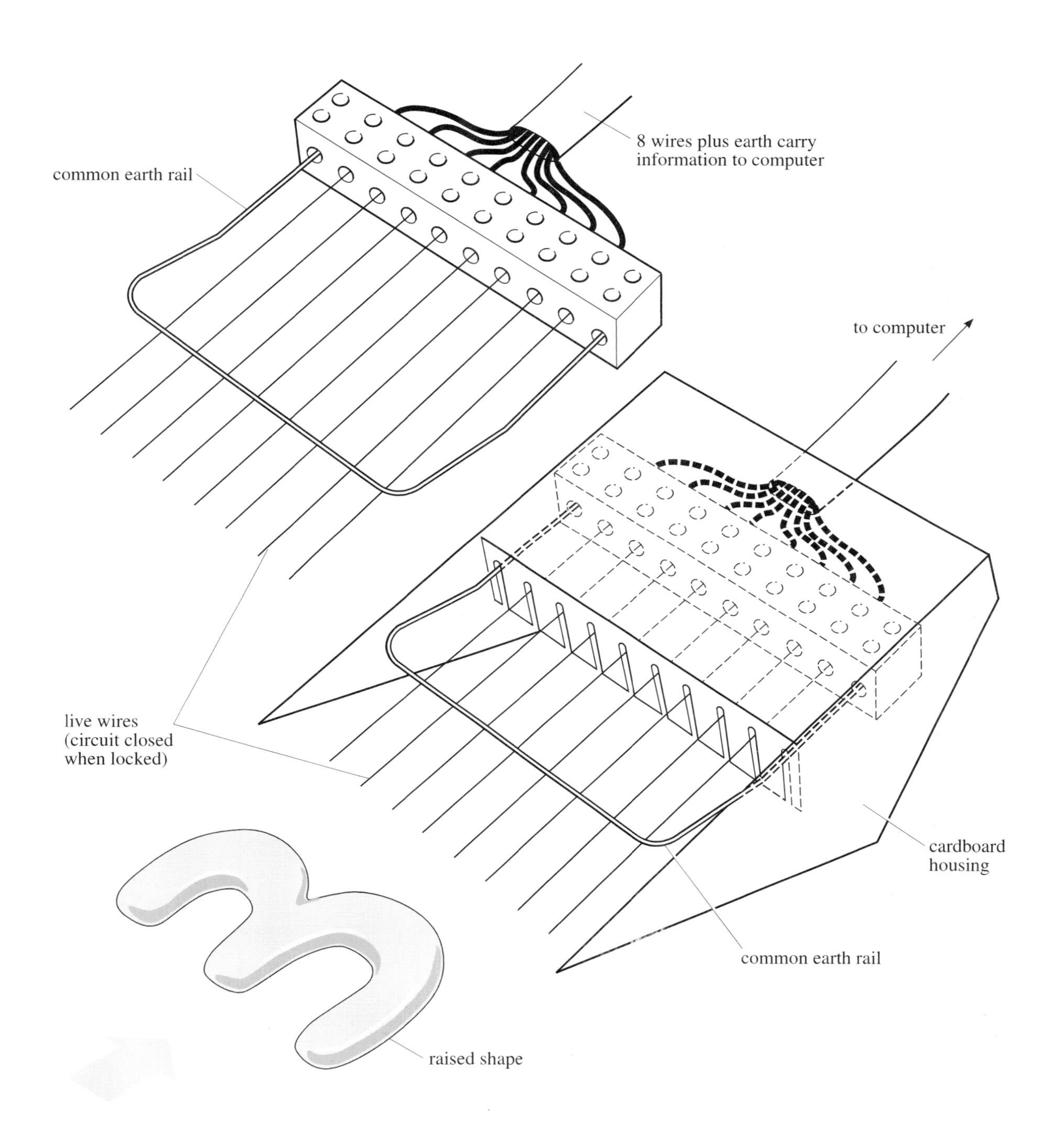

Figure 4.14
A simple scanner made of eight pieces of wire and one of the objects sensed.

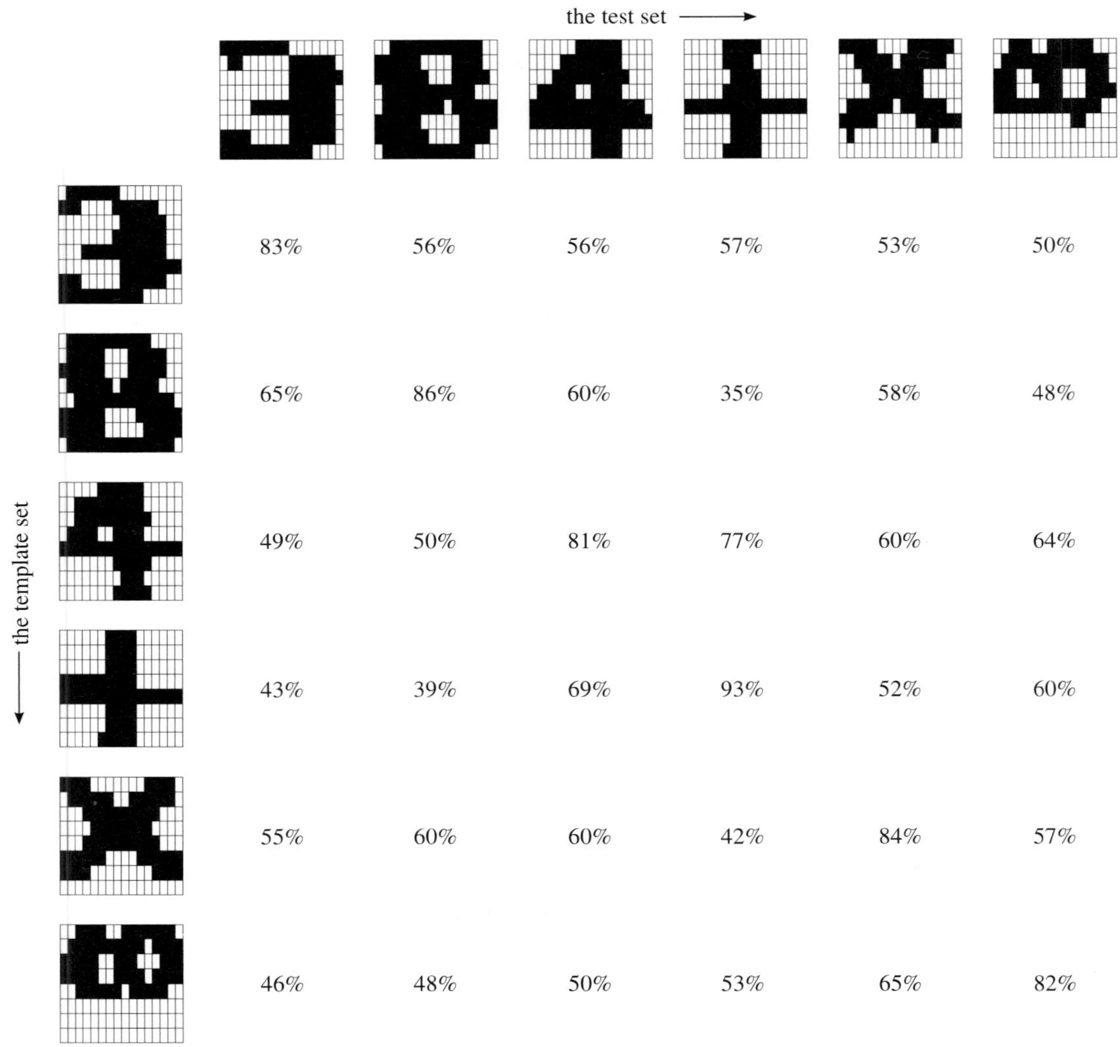

the test set →

the template set ↓

83%	56%	56%	57%	53%	50%
65%	86%	60%	35%	58%	48%
49%	50%	81%	77%	60%	64%
43%	39%	69%	93%	52%	60%
55%	60%	60%	42%	84%	57%
46%	48%	50%	53%	65%	82%

Figure 4.16
A set of test digitizations compared to a training set of templates.

that object best. They are all better than 80%. Most of the other matches are 60% or worse, showing that there is little confusion between these objects.

As one might expect, the most 'incorrect' matches are between different objects which are quite similar. For example, the test '+' matches the '4' template on 77% of the pixels, since both have vertical and horizontal strokes. However, this match is less than the match the test '+' has with the training '+' template, which is 93%. So there is little danger of confusion between the '+' and the '4'.

4.9 Invariance to position, orientation and size in perception

The template-matching approach of the previous section works well, even on those crude and inaccurate images. However, that approach has severe limitations.

As it happens, the same object was used to obtain the '+' and the '×' images. The Plasticine cross was simply rotated by 45 degrees. However, the matches between the 'plus' and the 'multiplication' signs are quite low. We say that the method is not ***invariant*** to rotation. Sometimes this is an advantage, but sometimes it is not. In this case we want to *discriminate* the plus from the multiplication configurations. However, a system trying to read addresses on envelopes might have to cope with numbers and letters written at all kinds of angles.

The pixel template-matching approach completely fails when similarly shaped but larger or smaller objects are used, for obvious reasons. Thus the method is not *scale invariant.*

It does not matter if the test object is moved right or left, or backward and forwards, and within reason the method is *translation invariant.*

Invariance to translation can be important in some applications. For instance, an institution that we will not name delivered a prototype vision system to perform some quality tests on motor cars. The method used was not translation invariant and the cars had to be positioned to within a few millimetres. In this industrial context such a constraint was totally impractical, and the system was scrapped.

Systems which are not translation invariant are rarely useful. Even in systems such as speed detectors we want to be able to detect the vehicle anywhere in the image and calculate the speed by calculating the difference in position between two sample times.

Apart from the geometric invariances we have discussed, there are others. In many applications it is desirable that perception is tolerant to changes in illumination so that ambient light doesn't have to be so carefully controlled. In other applications the perception should be temperature invariant. Some perception systems installed on spacecraft have to be gravity invariant.

4.10 An example of the perception–cognition interface

As we have seen, a simple sensor made of eight bits of sprung steel wire can be used to perceive objects in the shapes: 3, 8, 4, +, ×, and an 'infinity' symbol ∞.

At any stage the scanner-sensor subsystem can 'deliver' to the rest of a mechatronics system (say a robot) the data that 'I have been activated by my wires touching something, I have performed a template match on the objects I have been trained on, and I recognize the object as a 3.'

Of course the sensor would not communicate in English like this. More likely its electronic interface to the rest of the system can be 'read' at any time by the computer, showing zero if it has not sensed anything, and a non-zero numerical code if it has sensed something.

Suppose our robot's internal system keeps inspecting (polls) the scanner and finds it has perceived a 3. Then a second or so later it polls the scanner again and finds it has perceived a +. The next time it polls the scanner it finds an 8 has been perceived. Now suppose inside the robot we have a computer program which logs data from the scanner, and *parses* these data (i.e. looks along the string) to see if it is meaningful to add the numbers. The parser may contain the simple rule 'A number followed by a plus sign followed by another number should be added and the result spoken.' So the robot will respond to the string 3 + 8 it received from the sensor, calculate the result as 11, and send the data to its voice synthesizer to say 'eleven'.

In this case the perception occurs in the scanner and in the information-processing subsystem that does the template matching and classifies the sensed image. In so doing it has transformed the data from analogue form to an encoded symbolic form (the numerical codes). Anything that happens after this, such as the robot being made to speak, belongs to the cognition and execution subsystems.

4.11 Neural networks in perception

Neural networks usually perform well in classifying multidimensional data spaces. Typically, the network is presented with pairs of known input values and known output values. The input values could be diameter and redness, and the output values could represent the class to which a coin belongs. Alternatively, the input values could be a sequence of 1s and 0s corresponding to black and white dots in an image (arranged in a line with one row following another), and the output could be the character recognized. The details of neural networks are outside the scope of this book. They will be covered in much greater depth in Volume 2.

4.12 Model-based pattern recognition

Let's return to the problem of perceiving chess pieces. Figure 4.17 shows two different views of a set of chess pieces. In part (a) the knights are facing you, while in part (b) they are in profile. Unless you *knew* that the objects in Figure 4.17(a) were knights, you would have little reason for identifying them with the horse-shaped profiles in part (b). Similarly the king's crown appears differently in the two images. In this case we could train the system by telling it that anything that looked like the knights in Figure 4.17(a) is a knight, and anything that looked like a knight in Figure 4.17(b) is also a knight. However, sometimes the problem is more complex than this.

***Figure 4.17* Two images of the black chess pieces (compare the knights and the kings).**

Figure 4.18(a) shows a view looking down on the chessboard in which some of the chess pieces are invisible. Figure 4.18(b) shows a processed version of the digital image in which edges of various contrasts have been abstracted. In some cases absolutely nothing has been detected for the simple reason that (in the image) there is nothing there.

If we want to abstract more information from such poor images we will need knowledge encoded in a ***model***. The model tells us what to expect, and allows us to infer from partial information that something is 'probably' present.

The concept of 'model' can get very subtle and there is no universal agreement on exactly what constitutes a model. Without trying to be too technical we will use the following definition.

(a)

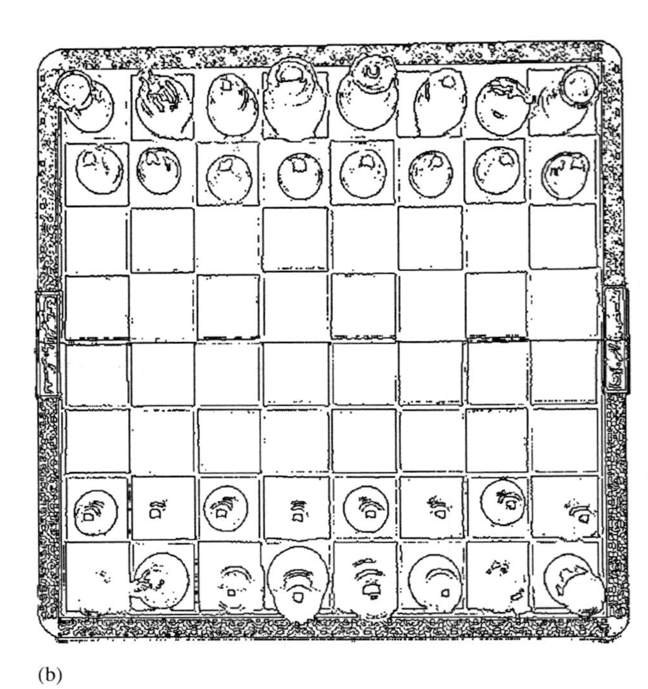

(b)

Figure 4.18
Some pieces are 'invisible' in this view of the chessboard.

A formal model of a system consists of:

1. a language in which to represent the system;
2. ways of identifying 'relevant' parts of the real system with named objects in the language;
3. ways of identifying: 'relevant' relationships between parts of the real system, explicit relationships between named objects in the language, and numbers attached to objects;
4. a logic for reasoning about the system and its behaviour in the language objects.

To find where the chess pieces are in the scene of Figure 4.1 (Section 4.2) we need more information than exists in the image. This could be provided by more or better digital images. Or it could exist in a model which has been abstracted from many other observations. We might reason as follows:

This game has probably not begun or has just begun (otherwise we would see more pieces in the centre of the board and less visual clutter in the foreground).
At the beginning of the game, if there is a piece next to the king's rook, then that piece is usually the king's knight.
But the visual clutter on black's king's side suggests there is a piece next to the king's rook.
Therefore we conclude that the king's knight is in the visual clutter on the king's side.
Therefore we can 'imagine' the king's knight in position, and draw what would be its silhouette.
And if we do, it gives rise to no inconsistencies (with this model), which reinforces our ability to 'see' the king's knight.

We do not know whether our brains engage in this kind of logical reasoning without us knowing it, or whether they appeal to a subliminal model of the chessboard. However, in order to draw a line between perception and cognition, we will say that any explicit use of models and reasoning in abstracting useful symbolic information from sensed data lies in the realms of cognition. This is treated in the next chapter.

4.13 Edge detection in digital images

Although cameras are a major source of digital images, there are many other sensors which can give a 'picture' of the state of a system or its environments. For example, magnetic resonance scanners give arrays of numbers which can be processed to form images. As we have seen, simple devices made of contact wires can also scan the scene to provide digital images. For this reason the problem of abstracting useful information from digital images remains an important research area.

It is often thought that it would be useful to detect edges in digital images. The early approach to this involved passing *masks* over the grey scale of the image to *filter* out the edge pixels. Most of the methods proposed for ***edge detection*** adopt this approach (which effectively classifies pixels as 'edge' and 'non-edge'), and most of them suffer from the drawback illustrated in Figure 4.19. Here the pixels have been classified according to the grey scale difference between the pixels on their left and right. As we can see, the method depends on setting a threshold. If the contrast threshold is too high (here a difference of 40 grey scale units) the edge is incomplete with gaps in it. If the contrast threshold is set too low, the edge becomes very thick (as shown here for a contrast difference of 20 grey scale units). But even in between these values there is no perfect threshold and we see both blobs and gaps simultaneously (as shown here for a contrast difference of 30 grey scales). In general, engineering techniques which require thresholds lead to machines which have to be carefully tuned, and which can be unreliable.

A further problem with edge-detection techniques which form edges greater than one pixel wide is that the 'edge' has to be *thinned* to obtain a line one pixel wide. Apart from adding another operation, this creates difficulties when edges run together to form larger blobs.

Of many other edge-detecting filters, one of the most commonly used today was devised by Canny at the Massachusetts Institute of Technology. The edges shown in the figures of this chapter have been found by a new proprietary method which produces edges with an accuracy of a fraction of a pixel (subpixel accurate). It is based on the notion of gradient polygons discussed in the next section. Edge detection and its applications will be discussed in greater detail in Volume 2.

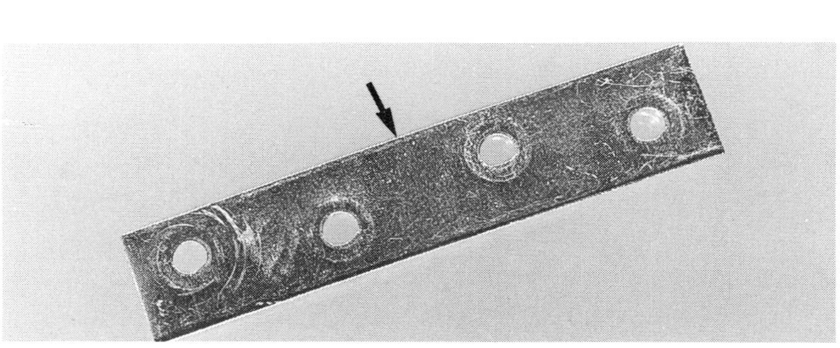

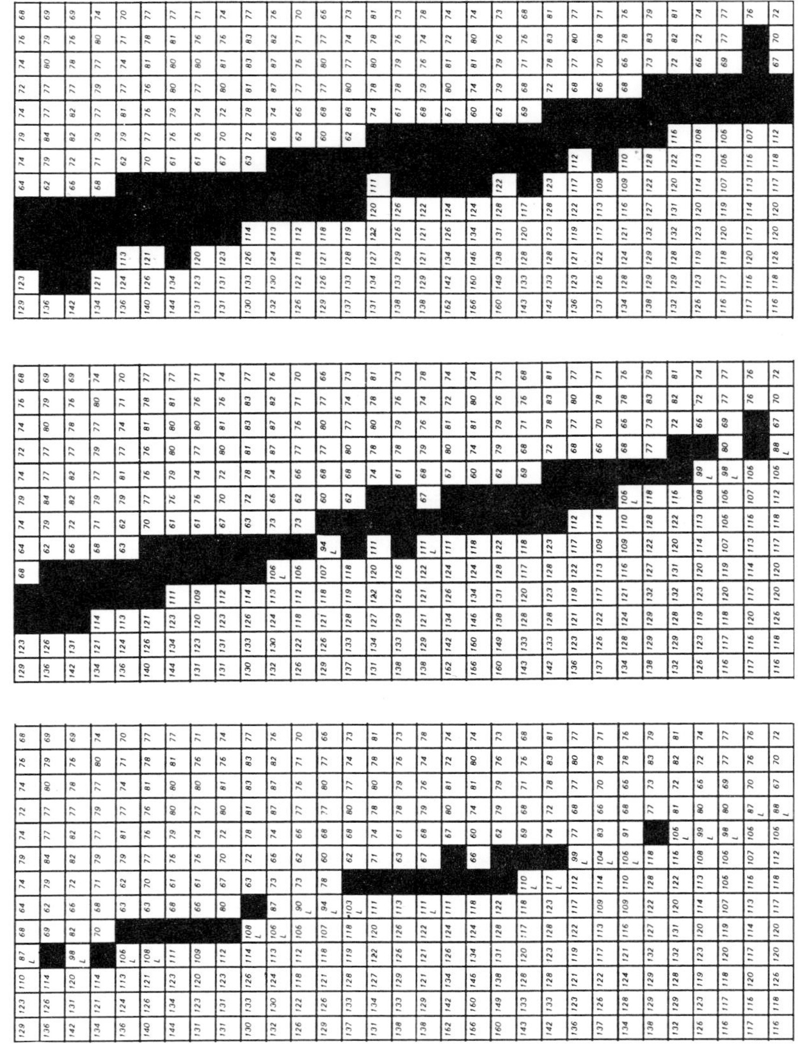

Figure 4.19

Finding the edge of an object using absolute contrast: (from left to right) edge of object; contrast >40; contrast >30; contrast >20.

4.14 An architecture for computer vision

The starting point in ***computer vision*** is the digital image which, for simplicity, I will assume is monochrome, with each pixel having a grey level between 0 (black) and 255 (white). When humans see digital images they can process them effortlessly to abstract all kinds of subtle information, and it is tempting to believe that it will be easy to program computers to do the same. Experience shows this is not the case.

The 'raw' digital image has no structure beyond the implicit knowledge of the adjacency geometry of the pixels, and the explicit knowledge of the grey scales. Therefore, one of the main approaches to abstracting useful information from digital images involves seeking *assemblies* of pixels that obey some rule or have some common property. This enables the pixels to be classified. Then, according to their classification, the assemblies may themselves be assembled to form higher level objects.

At some stage it is best to move away from configurations of pixels to objects which can be represented in other ways. As an example, the result of edge detection may be straight lines represented by four numbers such as <a, b, c, d>, where the line goes between the points (a, b) and (c, d). This kind of edge detection transforms the image from its original *raster* data structure to a *vector* data structure. In other words, information implicit in the rows of pixels (rasters) can be transformed into Cartesian coordinates such as the points (a, b) and (c, d) at the end points of a line (a vector). Thus the detection of edges in computer vision can be thought of as a stage in the perception process.

The configurations determined early in the perception process are called ***image primitives***. One lesson learnt from many years of research in computer vision is that it is essential to have *robust primitives*. By this is meant that it should be possible to recognize and abstract these primitives with a high degree of certainty. Often it is best to abstract relatively simple primitives with a relatively high degree of certainty.

Consider the problem of recognizing an eye in a digitized image of a face. Human beings find it easy to see eyes, even when in reality they are not there.

The upper part of Figure 4.20 shows a digitized face and the lower part shows the pixels of the left eye at a larger scale. It is very hard to see the eye in the enlarged version, unless you hold the book a long way from your eye and squint a little. The problem is to find some way of recognizing the eye in what appears to be a mosaic of pixels.

Grey-scale template matching does not work for this problem because the grey scales may never be quite the same for any given pixel. Also, a solution based on direct template matching would be size and contrast dependent. This problem is further discussed in the next section.

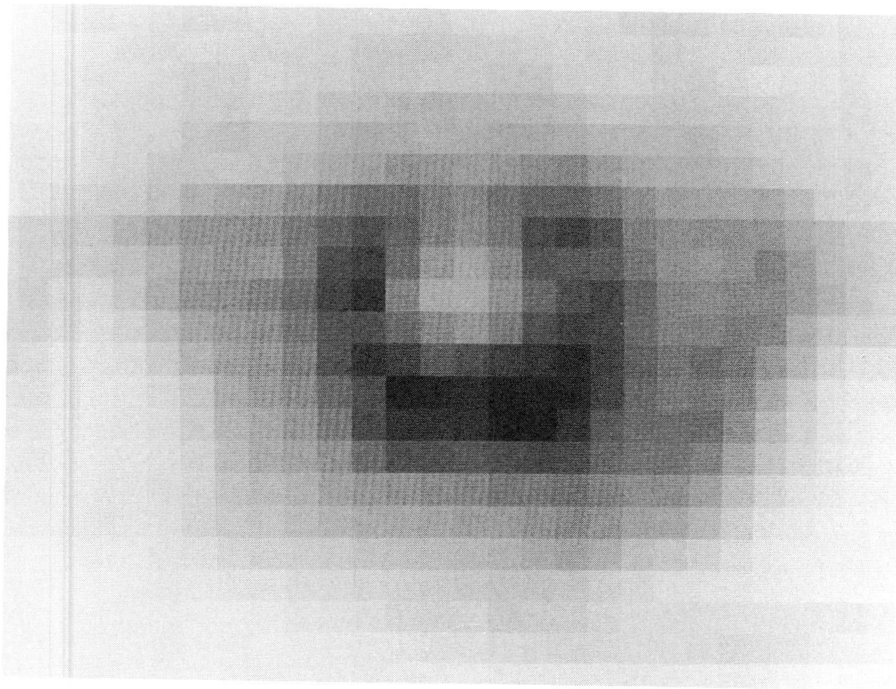

Figure 4.20
The mosaic of pixel grey scales in the eyes of a digitized face: (above) digitized image of a child's face; (left) enlargement of the left eye showing the mosaic of pixels of non-homogeneous grey scales.

One approach to this problem, which illustrates a particular architecture for computer vision, begins by looking at the pixels as the lowest level in a hierarchical structure, denoted level N. We can assume that the information in digital images is conveyed by the grey levels of the pixels, and the relationship between a pixel's grey level and those of its horizontal and vertical neighbours. So let us define some *relations* between the pixels to try to capture this information.

In general, a pixel will be lighter than some of its neighbours and darker than others. To make this precise, let's define the relations R_0, R_1, R_2, R_3, R_4, R_5, R_6 and R_7 between a pixel and its four neighbours as in Table 4.4.

TABLE 4.4

R_0	A pixel is R_0 related to its lower neighbour if it has the larger or same grey scale	**7** 3
R_1	A pixel is R_1 related to its right neighbour if it has the larger or same grey scale	**7** 2
R_2	A pixel is R_2 related to its upper neighbour if it has the larger or same grey scale	1 **7**
R_3	A pixel is R_3 related to its left neighbour if it has the larger or same grey scale	2 **7**
R_4	A pixel is R_4 related to its lower neighbour if it has the smaller or same grey scale	**2** 3
R_5	A pixel is R_5 related to its right neighbour if it has the smaller or same grey scale	**2** 5
R_6	A pixel is R_6 related to its upper neighbour if it has the smaller or same grey scale	6 **2**
R_7	A pixel is R_7 related to its left neighbour if it has the smaller or same grey scale	6 **2**

For simplicity it will be assumed that neighbouring pixels never have exactly the same grey scale. (This is reasonable since such uniformity is quite rare in practice, even for parts of the scene which appear to have homogeneous colour.)

Apart from these relations defined for the closest neighbours, similar definitions can be used for the 'next-but-one' neighbours. This can be useful when the grey scale gradients are relatively small compared with the *noise* levels (i.e. random variations or errors in the signal).

There are sixteen different ways a pixel can be brighter or darker than its four neighbours (Table 4.5).

TABLE 4.5

Relation	Lower pixel	Right pixel	Upper pixel	Left pixel
R_{0123}	brighter	brighter	brighter	brighter
R_{0127}	brighter	brighter	brighter	darker
R_{0163}	brighter	brighter	darker	brighter
R_{0167}	brighter	brighter	darker	darker
R_{0523}	brighter	darker	brighter	brighter
R_{0527}	brighter	darker	brighter	darker
R_{0563}	brighter	darker	darker	brighter
R_{0567}	brighter	darker	darker	darker
R_{4123}	darker	brighter	brighter	brighter
R_{4127}	darker	brighter	brighter	darker
R_{4163}	darker	brighter	darker	brighter
R_{4167}	darker	brighter	darker	darker
R_{4523}	darker	darker	brighter	brighter
R_{4527}	darker	darker	brighter	darker
R_{4563}	darker	darker	darker	brighter
R_{4567}	darker	darker	darker	darker

Each pixel in an image can be characterized by at least one of the ways of being related to its neighbours. A remarkable property of digital images is that pixels tend to form polygons when classified by these sixteen neighbourhood configurations. Furthermore, these polygons correspond to the kinds of polygons we see in digital images. They have been called *gradient polygons* because they are defined by grey scale differences. It has been argued that these polygons are fundamental features of digital images.

To see how these polygons can be used, consider the pixels in the pupils of the digitized face (Figure 4.21). This shows the gradient polygons for the centre of an eye (the next-but-one neighbours were used in this case).

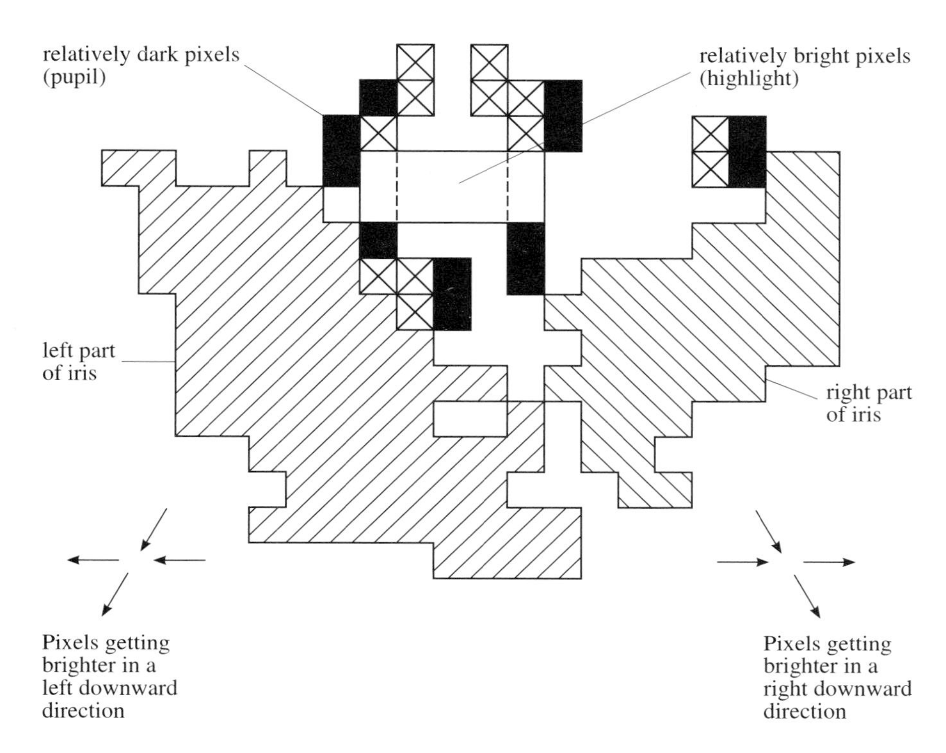

Figure 4.21
Polygon configurations abstracted from an image of an eye: these are inevitable given that grey scale gradients naturally occur in images.

Apart from the pixels being aggregated into polygons, we can see from Figure 4.21 that the polygons *must* have their relative positions: the dark 'pupil' polygons must be the centre of the eye, while the lower polygons must obey the R_4 and the R_2 relations as the pixels get brighter from the dark pupil to the whites and eyelids.

So the gradient polygons obey certain configurational rules, and they can be assembled to form the 'iris/pupil' part of the eye. In turn this can be assembled with other parts to form the 'eye'. This might then be assembled to form a 'face'.

What I have described is a *hierarchical architecture* for computer vision (as outlined in Chapter 1), and this is illustrated in Figure 4.22. It is in the first instance a *bottom-up* and *feedforward* approach to computer vision (these terms are defined in the next section and Volume 2).

The great advantage of this hierarchical approach comes in its tolerance to errors and omissions. Figure 4.23 shows four configurations of lemon shapes. We can see a face if we add a nose to the first and the second. However, we can make the pair of lemons into the wheel cowls of an aeroplane, as in the third case. Finally, the addition of a nose to the three lemon shapes makes us see two of them as eyes, while the one on the left is ignored. Thus in the first case we can still see the face in the absence of the right eye, while in the last we can see the face despite the presence of a spurious eye.

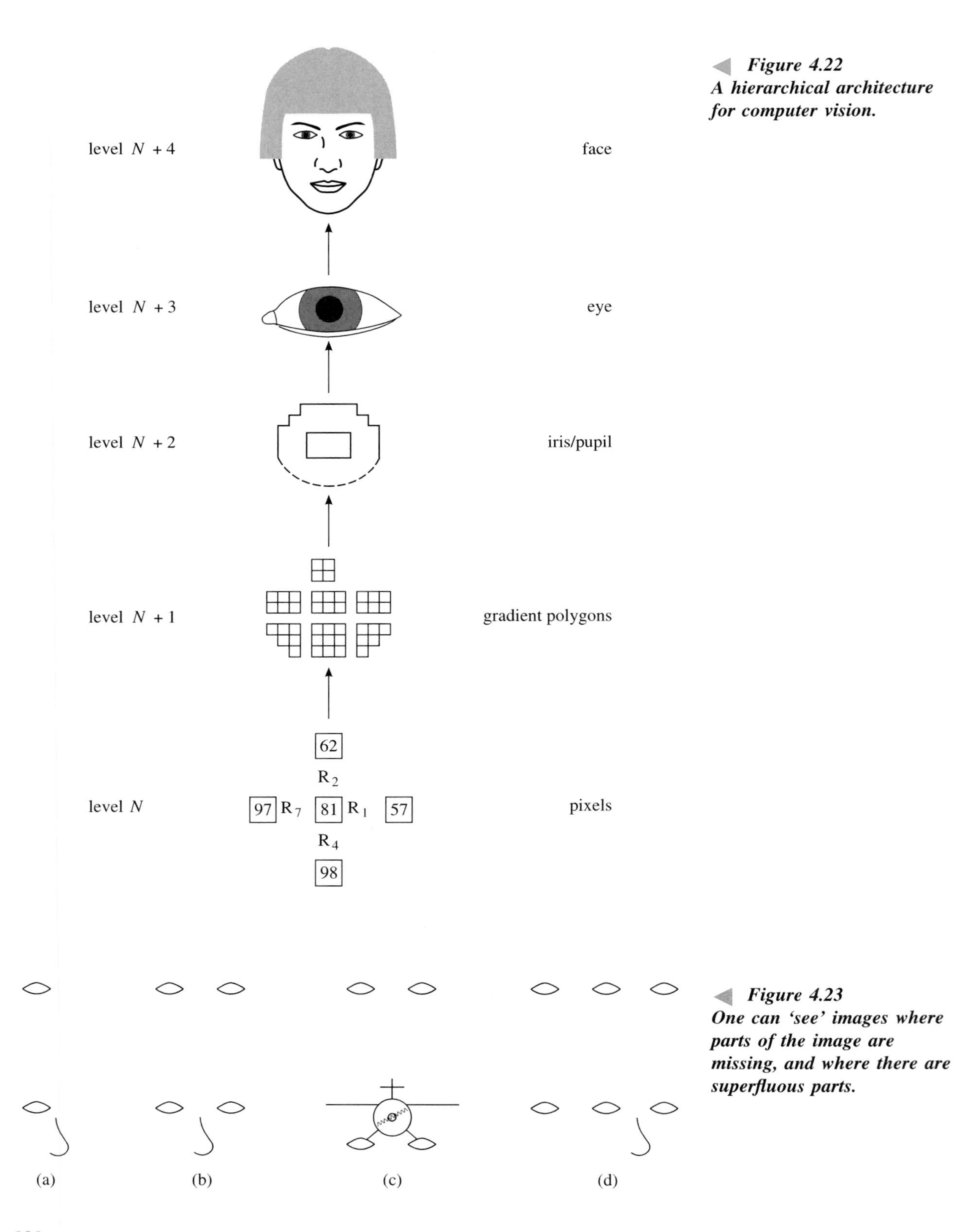

Figure 4.22
A hierarchical architecture for computer vision.

Figure 4.23
One can 'see' images where parts of the image are missing, and where there are superfluous parts.

The first and last cases of Figure 4.23 show the following assemblies:

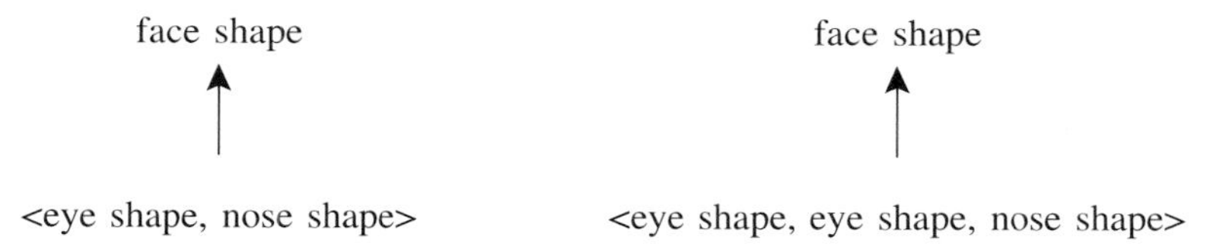

so the rules of assembly allow one eye shape or two eye shapes. Furthermore some configurations are shared by very different assemblies. In the second and third cases the assemblies are shown in Figure 4.24.

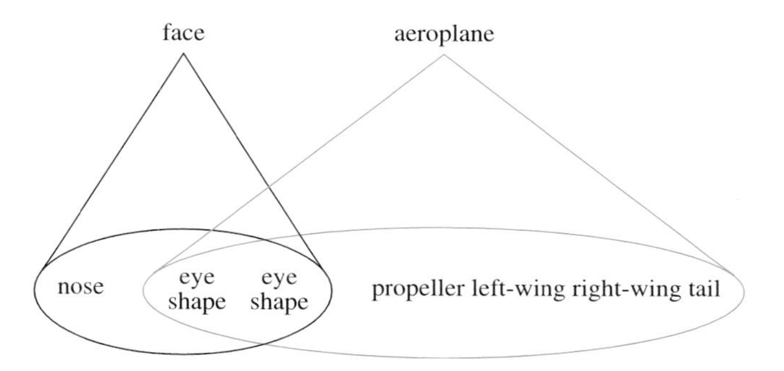

Figure 4.24 ***Higher level visual objects may share lower level features.***

This discussion emphasizes the importance of robust primitives and robust intermediate constructs within the bottom-up hierarchy.

4.15 Model-based computer vision

The hierarchical vision architecture described in the previous sections is *bottom-up* in the first instance: it starts with pixels at the lowest level, assembles them into low-level features, then assembles these into higher level features, and so on. It assumes that rules for assembly have been abstracted and entered into the system in some way. However, this process can only see what is there. Figure 4.18 showed an image of the chessboard in which there is litle or no information on the existence of some pieces; we would have to appeal to a model of the chessboard and to top-down reasoning in order to decide if the pieces exist.

Also it will be necessary for the system to have a geometric model of individual pieces if we propose to correct the image by the computer drawing them onto the digitized image.

4.16 Conclusion

In this chapter we have considered how the 'raw' data from sensors can enter a mechatronics system and begin to be processed into useful information. This process is called *perception*, and is distinguished from *cognition*, which processes the information provided by the perception subsystem in a much more goal-oriented way.

We have seen that *pattern recognition* is important in perception, and considered how this can be achieved by simple template matching and by trying to partition *n-dimensional* data spaces to create pattern classes.

We have discussed computer vision and have seen a hierarchical architecture for perception. Once we begin to reason about images, to infer that things are present that are not immediately signalled by the grey scale data, we are being brought close to cognition.

In summary, this chapter discussed the following ideas:

- perception as the means for information to enter a mechatronics system and be converted to a useful form;
- the perception–cognition–execution cycle;
- perception as low-level information processing;
- cognition as goal-directed information processing;
- analogue and symbolic representations in perception;
- machine vision contrasted with biological vision;
- classifying multidimensional data spaces;
- the importance of respecting the measurement scale;
- template matching;
- invariance to translation, rotation, scaling, etc.;
- models and model-based pattern recognition;
- digital images from different sensor types;
- edge detection, filters and feature extraction;
- computer vision architectures;
- primitives in perceptions.

Many of these concepts will be considered in more detail in Chapters 2 and 10 of Volume 2.

Reference

Levine, M. D. (1985) *Vision in Man and Machine*, Series in Electrical Engineering, McGraw Hill, New York.

CHAPTER 5
COGNITION

Chris Earl

5.1 Introduction

Mechatronics systems act using information obtained from sensors. They also use information given to them by their users or designers. Knowledge about how the system behaves and about the environment in which the system acts is also used. We will examine the ***link between perception and action***, making use of information on goals, system behaviour and environment.

This link may consist of ***planning actions*** to implement goals. On the other hand, the link may be direct, as when actions are the immediate consequence of sensing information. For example, an automated vehicle may take avoiding action when encountering an unexpected obstacle without any planning of future actions.

In contrast, a robot moving through a spatially constrained environment to reach a destination needs to plan its actions both to avoid collisions and reach its destination. These plans may not be complete descriptions of the robot's actions from the start to the end of its task. They may be local plans for reaching critical points on the route or strategic plans to inform and guide individual actions.

This chapter will examine the nature of this link between perception and action. The next chapter in this volume will put into place some of the basic 'machinery' for creating the link. More advanced tools will be developed in Volume 2.

5.2 Examples of cognition in mechatronics systems

The following examples expose the potential complexity of ***cognition*** in a mechatronic product.

5.2.1 A robot system

Let's start with a simple robot. It is an example of a mechatronic machine that physically acts on the world in a direct and obvious way when executing its tasks.

Consider a task where the robot has to lift a small box from a table and place it in a large box, as shown in Figure 5.1. Side and plan views are also shown.

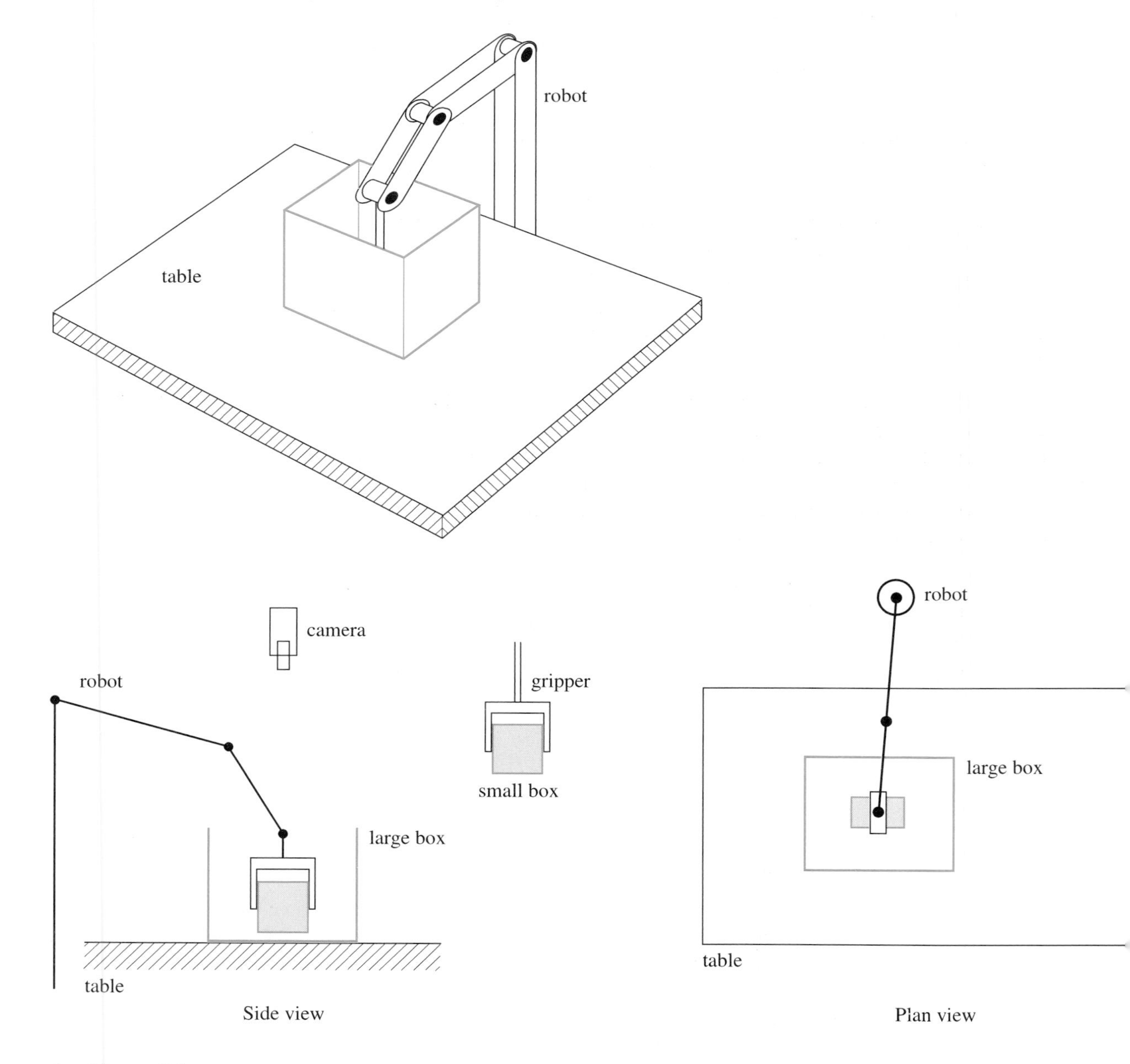

Figure 5.1
Robot placing a box within a box.

Suppose that a camera placed above the table has determined the position of the boxes (and distinguished large from small). With a bit of information processing on the camera's digital image it may be possible to specify a location where the robot's gripper hand should pick up the small box. This may be quite a complex problem of interpreting perception information. A valid pick-up location for the gripper may depend on factors such as whether there is enough space around the box to allow access to the gripper. Suppose further that a put-down location for the small box has been derived from the camera using information about where to place it in the large box (perhaps adjacent to one of the walls).

The robot system now has a start and finish position. The robot needs to move to the start. It needs an instruction to act on. The drives or actuators of the robot each move a single joint. They are individually controlled so that they can move to a specified position: an angle for a rotary joint and linear displacement for a sliding joint. The robot can implement instructions to move individual joints to specified positions. The robot thus needs a set of actuator instructions, but the perception subsystem has not provided these. The initial interpretation of perception information has given pick-up and put-down positions, but can the robot implement the task? In the worst case these positions may be out of reach.

It is not difficult to obtain these actuator instructions, starting from the required positions of the gripper. The calculations are based on the geometry and dimensions of the robot and its gripper. This is a significant step in cognition. On the basis of information about what we want the robot arm to do and the means available (the geometry and capabilities of the robot arm), a plan or set of instructions is provided to the machine to execute the task.

The gap between perception and action is easily bridged if the robot geometry is known. The geometrical configuration and parameters are encoded in a program that generates actuator instructions. Information on the location of the boxes, knowledge of the machine itself and the goal of the task are all required to link perception and action. A schematic of this process is given in Figure 5.2.

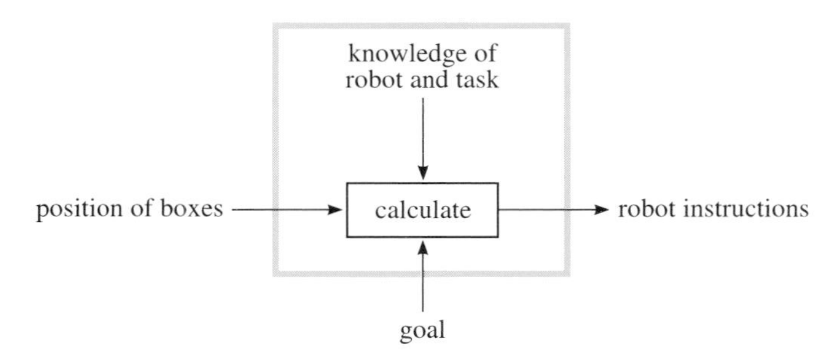

Figure 5.2
Schematic of cognition for robot task.

However, will the system work? Generally not, because the cognition system has only generated a set of instructions to reach the start and finish positions but has not taken into account how the arm will move between them. There could be a collision between the two boxes as one is moved inside the other. It is necessary to break the path into parts and generate intermediate positions of *depart* from *pick up* to a safe height and *approach* to *put down* over the large box (Figure 5.3). Corresponding arm positions for depart and approach are computed similarly to the pick-up and put-down positions.

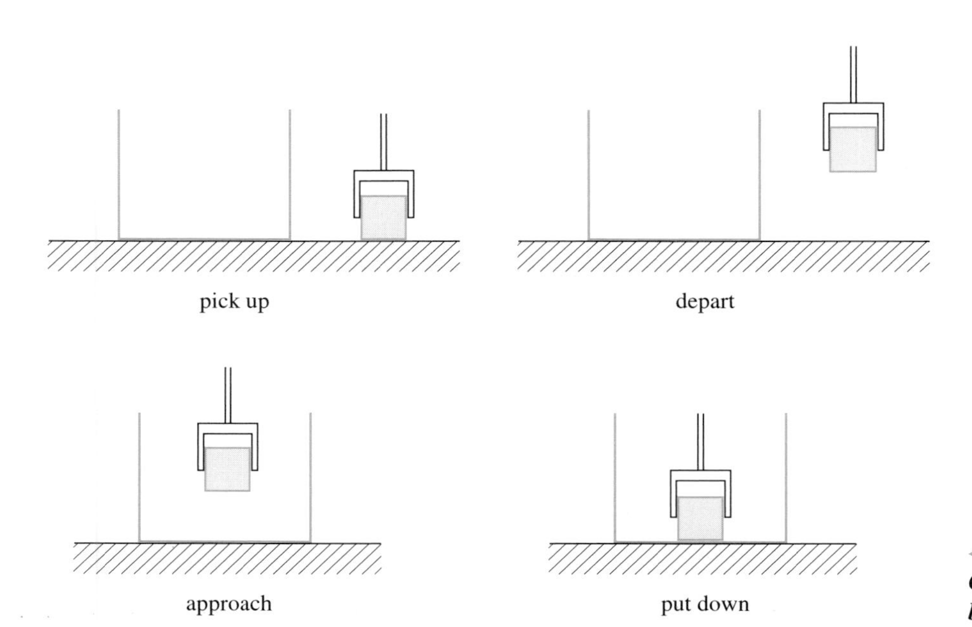

Figure 5.3
Critical positions for the box-in-box task.

The cognition subsystem needs more information. It might be able to call for more information, but how does it know that it needs more? It doesn't, unless there is knowledge of the task already available within the cognition subsystem.

The cognition subsystem must extract relevant information from the output of the perception subsystem. It looks for patterns in the data indicating the size and height of the edges of the box. It may not be able to find these. (Indeed they are difficult to extract from a single camera image.) If it cannot, cognition works on partial information. It is important to note that the concept of partial information is relative to the information demand and requirements of the cognition system. In such conditions, assumptions may be made about maximum likely sizes and dimensions of the cell within which a robot operates. However, for simple tasks of this kind it is likely that the dimensions of the boxes will be known and the information embedded in the program for the task.

Generally, the cognition subsystem interprets the output patterns of the perception subsystem but may also be able to initiate further perception functions – perhaps, for example, to instruct the execution subsystem to change camera position in order to supplement its knowledge and resolve ambiguous interpretations.

The seemingly elementary task is thus quite complex from a cognition viewpoint. The task needs planning and it is the cognition system that performs this function. The plan may be very simple and just respond to particular patterns in perception data in a unique way. Alternatively, there may be many possible responses, and choices must be made.

The plan for the robot task will be generated in the following steps:

1. Extract edges from the camera image (this is properly a perception function).
2. Transform from camera coordinates to world reference coordinates.
3. Compare clear space around small box with gripping requirements. If there is not sufficient space then either the task cannot be completed or cognition uses knowledge of task, machine and environment to plan actions to create sufficient space (see Volume 1, Chapter 6).
4. Determine gripping position (there are potentially many choices).
5. Determine *depart* position above box (offset from pick-up position by known distance and direction).
6. From large box location determine position and orientation of *put down*.
7. Determine *approach* to *put down*.
8. Transform *pick up*, *depart*, *approach* and *put down* positions to robot joint instructions.
9. Generate path from *depart* to *approach* and check for collisions.
10. Generate *pick up* to *depart* and *approach* to *put down* moves.

This represents a sequence of planning stages. The outcome is a set of instructions to the robot execution subsystem.

We will return to this example in the course of the chapter as we expose other features of the cognition subsystem. To summarize, the simple robot system has a cognition subsystem that coordinates individual joint actions to achieve goals, such as picking up the small box. Cognition also creates sequences of such goals that help in avoiding collisions. Note that there is a two-level hierarchy of cognition functions in this example: an overall plan of constituent moves is created, followed by the detailed planning of each of these moves.

5.2.2 Game playing: chess

Let's imagine that we are playing chess with a computer system. The board and piece positions are shown on a graphics screen. We could include a physical chess piece mover, but this is an unnecessary complication and many of the relevant issues in this area of cognition have already been considered in the example of the robot system. Without any physical actions on the world we do not strictly have a mechatronics system but it serves to illustrate features of planning relevant to cognition.

The pattern of the pieces on the board may be considered as a perception model in which each piece has been identified and labelled with its type. For our purposes this labelling of piece type and its position on the board (namely the label of the square it occupies) is sufficient.

The overall goal of the computer chess player is to win the game by forcing a checkmate. At each position there are many possible actions. One of these is chosen, usually by looking ahead (or simulating) the possible effects of different valid moves. There may be a strategy at each stage of the game that will help to determine appropriate moves, but this depends on the machine being able to recognize particular states of play in order to invoke the strategy.

In this example the chess player does not plan a single course of action to achieve the goal but plans locally at each move. After each move the opponent generates an 'unknown' move. In a sense the chess player is acting in an uncertain environment. Although the space of uncertainties (the opponent's possible moves) is known, at least in principle, the particular choice made cannot be foreseen.

The chess player requires a representation of the board and knowledge of the rules of the game in order to plan future actions. There are explicitly defined possibilities for actions. It is the purpose of the cognition system to examine possible states that arise from these actions and devise a move to help in achieving the overall goal or any local goals that are formed as part of the strategy of play. As I have noted above, the computer chess player is not a mechatronics system, unless it also physically moves the pieces, because its actions are within a symbolic world representing the chessboard, pieces and moves. However, it does illustrate a basic cognition process of searching possibilities and selecting a course of action.

The sequence of steps for planning chess moves may be described as follows:

1. Examine possible moves from the current state.
2. Evaluate each according to the state created and order with best first.
3. Examine possible responses to the best move.
4. Estimate most likely response.
5. If position improves then make move, else backtrack to next best move.

This is a shallow search with a one-step look ahead. Generally we need to search possible moves and their consequences a few moves ahead to determine the best move from the current state. These search methods will be developed in Volume 1, Chapter 6.

5.2.3 Environmental management

As a third example of a mechatronics system, consider the management of the environmental conditions of temperature, humidity and ventilation in a commercial greenhouse.

The inputs are measurements of air change, temperature and humidity as well as temperatures of the thermal masses within the greenhouse. Further inputs are provided from prevailing weather conditions and forecasts.

The perception inputs require processing to determine the actions of ventilation openings, shading blinds and artificial heating elements. These actions need planning: they are not just a matter of immediate response to particular conditions. Possible courses of action require comparison and evaluation for maintaining a pattern of environmental conditions as well as for minimizing costs of operation (such as fuel for artificial heating or service plant maintenance costs). The ability to reason is essential to judge what the effect of these actions would be in terms of the current and future environmental conditions inside and outside the greenhouse. This example illustrates the relations between perception, cognition and execution, as shown in Figure 5.4. We are particularly interested in the relation between perception inputs and instructions for action.

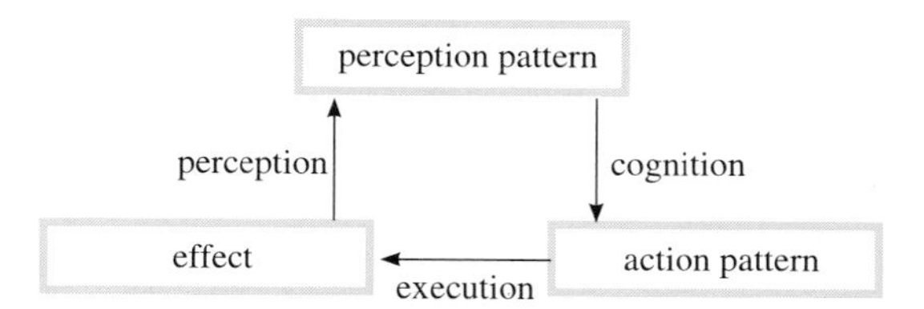

Figure 5.4 ***Perception, action and effect.***

One way to do this is for the cognition subsystem to have a physical model of the greenhouse and its responses. The system needs to relate to its state expressed in terms of the controllable elements (ventilation, shading and heating) and to its state in terms of internal environment (temperature, humidity). This is not just a problem of associating patterns in each space for each set of current weather conditions (taking into account lags in response) but should take into account future patterns of weather. The way that cognition links these states is shown diagrammatically in Figure 5.5.

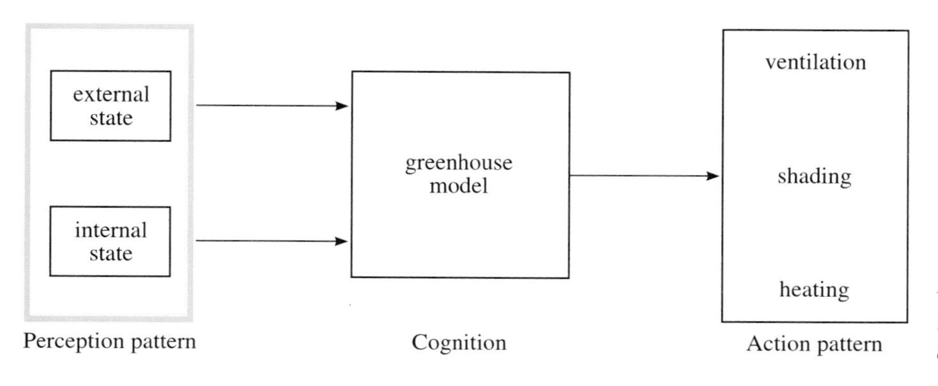

Figure 5.5 ***Links between perception and action patterns.***

5.2.4 Features of cognition

Consider a safety system for a machine tool in which there is a light curtain across the area of machine access. A light curtain (Figure 5.6) consists of a linear strip light source and a corresponding linear receptor. We can imagine a sheet or curtain of light between the light source and the receptor. If the light is interrupted at any part of the curtain this will be detected at the receptor. Usually the light is in the non-visible, infra-red part of the spectrum. The machine tool has two states: set up (during which the machine is prepared for process and the light curtain would need to be broken) and active (during which the process is executed). The light curtain has two states: intact and interrupt. The model of the environment in which the system acts, the *world model*, has two states: normal and emergency. These are associated with machine and light curtain states as in Table 5.1. All planned activities can continue in the normal state. However, when the emergency state is registered, all activities cease, such as machine processes and peripheral operations (e.g. loading). Each state has a set of specific actions that are triggered or blocked when the machine is in a particular state.

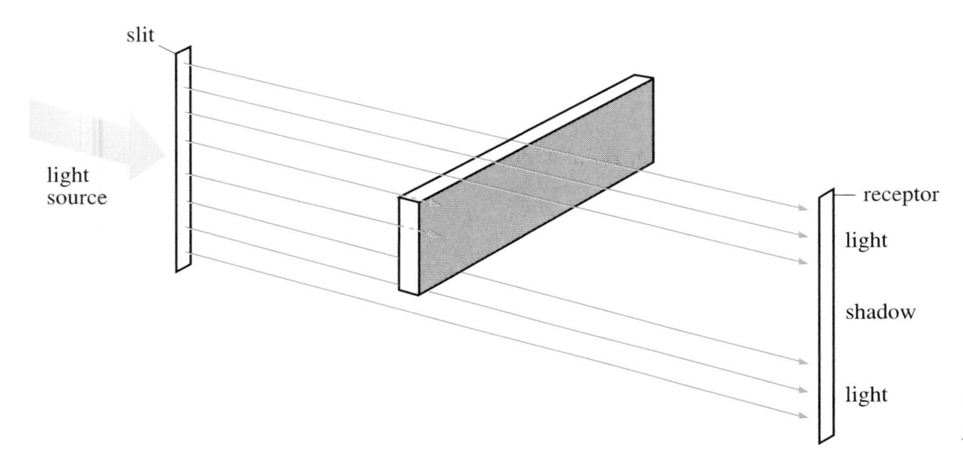

Figure 5.6 Light curtain.

TABLE 5.1 MACHINE AND LIGHT CURTAIN STATES

Machine/ light curtain	**Set up**	**Active**
intact	normal	normal
interrupt	normal	emergency

In contrast, the examples earlier in this chapter do a little more than this. They examine possible actions and choose those appropriate to a specified goal. The cognition subsystem is *'designing' a plan of action.* The incorporation of a cognition subsystem within the mechatronics system gives it the ability to design its actions and behaviour. In designing the mechatronic product, we are therefore developing a kind of creativity and intelligence associated with design within the system itself.

In designing a course of action the cognition subsystem will deal with different levels of detail. In the chess example the cognition system in the chess-playing computer should be able to interpret the patterns on the board in a way that is meaningful for developing strategy. The computer chess player should exhibit not only a knowledge of the rules of the game, together with the ability to assess the strength of a given position and to use the rules to search for a good move, but also a knowledge of the strategic rules for playing the game successfully. The strategic rules encapsulate knowledge of patterns on the board and appropriate responses. Further, they provide guidance on significant types of pattern and the strategies for achieving them. This knowledge can help in two ways: first, it reduces the search of unrewarding moves, and secondly it provides intermediate goals against which moves can be generated and evaluated. There is a distinction between cognition at a level of individual actions and at a level of strategic planning.

So far, we have considered three cases: the robot example illustrates the coordination of actions to achieve goals, the chess example illustrates the concept of search, and the greenhouse example indicates the importance of models to determine actions. In each case the interface to the execution system may be considered as a straightforward issuing of instructions that are then executed. But this is not sufficient, as shown when the robot can cause collisions between the boxes. A method of avoiding the collision was suggested, which involved interpreting both the perception information and the goal. It is often difficult to cover all cases. For example, the box might be assumed to be of uniform height, but how would the system respond when this condition was not satisfied?

A temptation for the mechatronics system designer is to try and cover all cases. The target is a machine cognition subsystem that will create as accurate and as complete a model of objects as possible in the world of interest and use this complete description to plan actions. However, a complete model may contain much information that is of little or no use to the task in hand, and the system cannot know that it has a good model, except by performing 'experiments' and examining the consequences of actions in the world. This suggests that *learning* about actions and their consequences is an important part of the cognition function.

Models are often used which are incomplete or uncertain. The effects of actions may be uncertain, as in the case of the chess player described above where the response from the opposing player is by definition uncertain. In the robot system the model assumes that the task environment remains constant between sensing and execution, and that planned actions will have the desired effect. In more uncertain environments estimations are used.

In other cognition subsystems a *local model* for immediate action may be used. Plans of action based on these models may be generated incrementally and locally, actions may be executed, the effects observed and further actions generated. For example, in the case of the robot arm placing the box within a box, it may be appropriate to take a local view of the task. Rather than attempt to plan all details before execution, suppose that the decisions are to be made during the sequence of actions. A range-finding sensor may be attached to the end of the robot arm to detect proximity to an obstacle and to take avoiding action (consistent with achieving the final goal). The perception data now need to be interpreted for local action. The data from the proximity sensor have meaning locally. There may be an overall framework for the task at a strategic level of cognition but the details of moves are considered at a local or incremental level of cognition.

An important aspect of cognition is the *relation between levels*. In the robot example we want the system to exhibit a local collision avoidance behaviour but in the context of 'higher' level behaviour we want it to move to place the small box inside the larger box. This strategic planning can simplify cognition by decomposing the task into a series of 'simple' tasks for which cognition can be short-circuited, giving direct perception–action links at the local level.

A conceptual model of cognition may be useful in providing a reference for the issues and problems discussed. Figure 5.7 shows this diagrammatically.

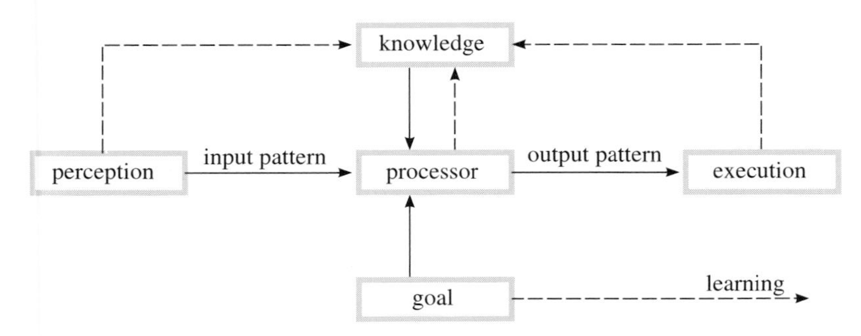

Figure 5.7 Elements of cognition.

The *computation* (for example, search) in the processor is guided and controlled by knowledge about the world in which the actions take place and the physical capabilities of the machines performing the actions. This knowledge may be available as models of the world and the mechatronics system. Further knowledge may also be available about how to use the knowledge of world and machine. It may be possible to improve these models during action. The models can be expressions of current best estimates, and associated plans of action can be constructed. Subsequent actions are based on new estimates and evaluation of past actions. Knowledge is often embedded in the processor that transforms an input pattern to an output pattern. The case of the machine safety system is an example of this.

The action of the intelligent machine is usually characterized by an intent or *goal*. For example, the simple robot system goal is to move one box inside the other. The goal in chess playing is to develop a checkmate, and in the greenhouse the aim is to maintain an environment suitable for plant growth at minimum cost. Goals, and corresponding intentional behaviour to meet the goals, are a feature of advanced mechatronic products. We may consider the goals as the specification for a course of action. The cognition subsystem designs the plan of actions against this specification.

Note that in Figure 5.7 there are *links* by which knowledge is acquired or learned. This is a powerful capability of some mechatronic products as it offers the potential to improve performance and learn new tasks within the capabilities of the machine. Developments in neural networks offer one method for incorporating learning in a mechatronics system. These appear to combine the knowledge and the processing elements. The processor is configured by learning to act intelligently. In this sense the schematic in Figure 5.7 may not be adequate, but for the purposes of this introduction it captures many of the characteristic features of cognition subsystems as currently implemented in mechatronic products.

We will next look in more detail at the terms and concepts introduced so far in considering the cognition function.

5.3 Basic elements of cognition

5.3.1 Perception pattern and response

The links between perception and action can be stored as reflex responses or reactions to particular perception patterns. In the latter, the patterns and actions are linked in pairs. Although a single action may be linked to many perception patterns, each pattern is linked to only one action. Cognition is like invoking a rule when a specific pattern is found.

The set of perception–response pairs may be stored in memory, but are usually engineered so that input patterns are associated with ***action instructions*** through a program. The implicit type of perception–response program will look for particular features of the input pattern, which are then associated with actions through the procedures of the program.

A machine tool safety system may have circuits with switches activated by light curtains and pressure mats. If the circuit is interrupted, voltage changes in the circuit stop the machine. A single input is examined. For a machining system with many parts only one part may need to be shut down to ensure safety. There are many interrelated circuits and there should be a unique response to each pattern of danger signals.

5.3.2 Formula-based planning

Suppose that the simple robot introduced in Section 5.2.1 is required to move along a straight-line path, perhaps to pick the box from the table or place it inside the larger box. The coordinated action of a number of joints is needed. If the path is known mathematically by the line it follows, a formula can be used, based on the geometry of the robot arm, which gives the required motions of the joints at any point on the path. Thus, embedded within plan production is the generation of a straight-line path. This may be regarded as a cognitive function whose result is a stream of instructions to the execution system based on the repeated application of the formula. The overall effect is to coordinate the actions of the execution system to achieve the specified goal.

We should note that the formula requires information on the positions of the joints at any time along the path. This is because the appropriate action at each joint according to the formula changes as the path is traversed. Perception information is thus required from the sensors on the joint actuators. We shall see later that these measurements of joint position are central to the control of execution.

Formula-based planning uses a program to determine an output response from sensing inputs. Goals set parameters in the program. The formulae and associated programs are models of system behaviour and are commonly used to coordinate elementary operations. This is the case in control of robot motion, in which joint motions are coordinated to move the gripper along a specified path.

5.3.3 Multidimensional patterns and response

The above discussion on pattern and response has concentrated on the situation in which given goals and perception patterns give rise to specific and often unique responses. A direct link between pattern and response is established within cognition or by cognition. These can be referred to as deterministic systems in the sense that output actions are determined from the perception inputs. The range of possible inputs is set down and corresponding output actions are associated with each. An example of such a system is an autofocus camera where the image and light conditions are used to determine focus and exposure. The functions of setting focus, exposure and shutter speed are formulated as programs whose output is uniquely determined by sensor input.

However, the link between perception and action can be more complex when it includes:

multiple interpretation of perception patterns;

multiple actions associated with each interpretation.

This flexibility is important for mechatronics systems. Multiple interpretations allow perception patterns to mean different things in different contexts. Multiple actions provide the means of exploring possible courses of action.

We can observe elementary examples of this flexibility of response in the autofocus camera that allows manual setting of some of the system parameters. For example, metering only the central part of the image corresponds to an alternative interpretation of the perception pattern (by ignoring segments of the input data). Further, manual override on shutter speed, aperture and depth of field requirements will create multiple actions corresponding to each perception pattern. However, the choice of mode is entirely manual and under the control of the operator.

These properties of multiple interpretations and actions allow mechatronics systems to be combined into more complex systems. Interpretation of perception patterns by one system will depend on the context of other patterns, and the actions of one system will depend on the context of the actions of other system components towards the goal.

Note that the multidimensional flexibility is also the basis of learning. New interpretations of perception patterns and associated actions can be examined and evaluated with respect to the system goals. Appropriate interpretations and actions can be learnt so that goals can be achieved more effectively.

Consider the example of a submersible vehicle used to explore the sea bed and to investigate deposited minerals. The submersible will have a navigation mechatronics system enabling some degree of autonomous movement under sensor guidance to seek out specific types of mineral sample. There is also a mechatronics subsystem that acts while the submersible is stationary, handling specimens and performing analysis. The output of the handling and analysis subsystem may instigate movements to find other samples of the same minerals, perhaps to establish frequency or consistency among the samples. The movement system receives local instructions from the handling and analysis that alters its behaviour. It will now interpret its perception patterns differently by recognizing different specimens of interest and it will act on these interpretations in ways influenced by the results of previous analysis. Conversely, the handling and analysis system will act according to the types of sample encountered by the movement system.

5.3.4 Links between perception and action

To link perception and action, a pattern of signals from the perception system (associated with identified objects or features and properties of objects) must be interpreted. For example, automobile sensors monitoring the operation of a car's controls are used to indicate the driver's drowsiness. The pattern of sensor readings is compared with known patterns indicating a drowsy driver. The resulting association gives the sensed patterns a meaning in terms of state of awareness to the perception pattern. If the interpretation is provided by matching patterns (or possibly the degree of mismatch) then this is usually called recognition. The system has found a pattern that it knows about and to which it can associate an action. In this case the driver is given a warning.

Interpreting a perception pattern lies at the boundary between perception and cognition. Information from sensors has a structure that is used to give it meaning. In the case of a vision system camera, the recorded information at each pixel is composed into the image. Relevant features such as edges represent structure in the image and are used as the basis of recognition and subsequent action. The role of perception has traditionally included the initial interpretation of sensing information, as seen in earlier chapters. The processes of cognition act upon these interpretations.

Meaning can be given to a perception pattern by association with knowledge about the task. In the box-in-box robot example a configuration of edges and dimensions is to be associated with the idea of the small box. Perception identifies a box and cognition associates it with knowledge about task and goals in order to identify actions. A suitable representation of the box may be in terms of outline edges.

The system must distinguish the small box. The robotic system should associate the configuration of edges in the perception pattern (if edge representation is used in perception) with knowledge about the box to be picked up. A preliminary association or matching may occur, but may be discarded because of size, for example. All general conditions for a box may be satisfied, such as the outline square shape, but not the particular conditions relating to the one to be picked up. This constitutes a simple example of reasoning about perception information. The perception pattern of the box acquires meaning in the sense that it is associated with the box to be picked up. This association must be done to achieve the goals of the task, unless, of course, the small box is already in the large box, in which case no action is required.

If outline edges are used to represent boxes, the processing will search for configurations of edges representing boxes, and will extract dimensions (to distinguish large and small boxes) and compare their coordinates (to determine their relative positions). In three dimensions this can be a tricky problem as not all edges are visible. For simplicity I take a plan view and deal with the outlines of the tops of the boxes, reducing the problem to two dimensions.

Any knowledge about tasks of this type in the cognition system can be used once the two boxes have been distinguished and their locations determined. The knowledge about the task is often expressed in terms of states and available transitions between states. The states in this case are identified by the spatial relations between the boxes. Translations are possible moves of the boxes. This knowledge is closely related to the possible actions to achieve goals. Knowledge about states and transitions between states can be used in planning.

For example, Figure 5.8 shows a simple set of four states based on the relative positions of the boxes. The purpose of the task is to move from state A to state D (while maintaining a hold of the box). The cognition system examines possible states and transitions in making its plans. The state sequence <A,B,C,D> represents an appropriate plan.

The plan of action can be formulated in terms of transitions between states. Knowledge about the feasibility of transitions can be attached to the state transitions or may be deduced from knowledge about task and machine. For example, geometrical conditions applied to transitions B→D or A→C identify possible collisions.

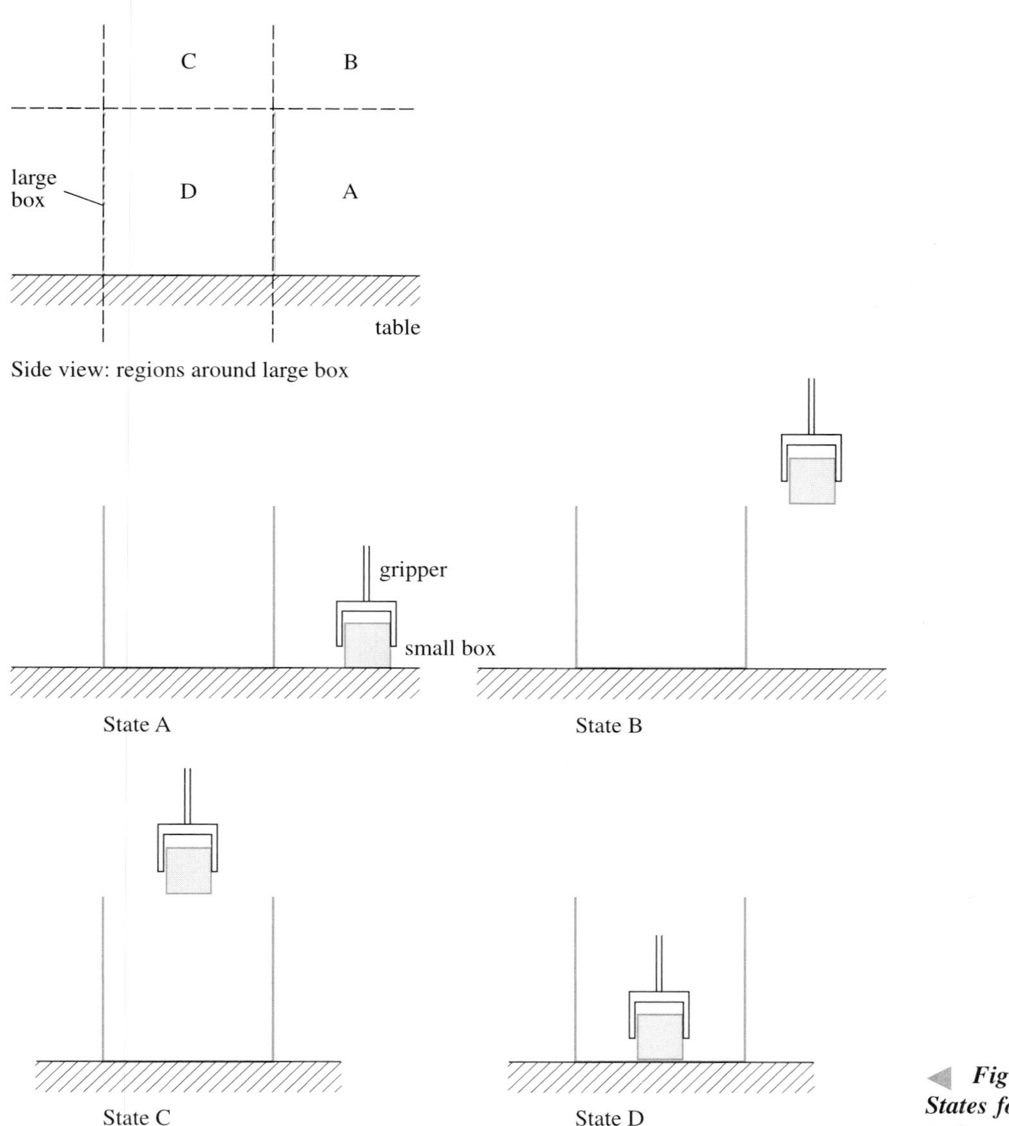

Figure 5.8 **States for the box-in-box task.**

Note two aspects of this formulation:

1 The state space representation used is appropriate for a particular task. In different tasks other relations, and thus other states and transitions, are used.

2 This representation needs further elaboration before execution.

The sequence of state space transitions is in itself not sufficient to generate actions. They form the basis or framework for the plan of actions. Note that there is a natural hierarchy of planning actions, starting with a rough state space representation allowing strategic choices to be made. The plans are then elaborated within the context of the high-level decisions.

5.3.5 Goals and tasks

A ***goal*** is an essential element in cognition and represents the task specification. The goal is to be achieved under the constraints, limitations and capabilities of the perception, cognition and execution subsystems as well as the environment in which they operate. In the example of placing the box in a box, the task is quite distinct if the two boxes are moving, perhaps on different conveyors. The goal remains the same, but the task is different. The term 'goal' is often used in a broad sense, whereas the task description includes the conditions under which the goal is to be achieved.

There can be many levels of task description and the goal is generally a high-level description. The output of cognition is a set of instructions, which can be considered as the lowest level task description. From this point of view, the whole process of cognition may be seen as taking place in a space of task descriptions, moving from high-level goals and then by a process of decomposing the task into lower level tasks that can be readily executed.

The simplest types of task description are the specification of parameters in a formula-based response. For example, a change in the destination point for the robot's straight-line path alters the path and thus all the motion instructions have to change.

In more complex cognition, different goals may change the structure and type of the response. They may control the interpretation of perception pattern and therefore the choice of response.

Actions with a purpose are a hallmark of intelligent machines. Purposeful behaviour may emerge and may not be explicit in any specific function of the machine. The coordination of complex patterns of elementary actions may provide purposeful or goal-seeking behaviour. For example, an automatic vehicle may have elementary actions to sense and avoid obstacles, while staying as close as possible to the line of a distant beacon. The elementary collision-avoiding actions, will, when accumulated, achieve the purpose of reaching the goal position. The purposeful behaviour arises from the way that the elementary actions are composed.

The systems in which there is a fixed link between pattern and response are not formulating goals or exploring hierarchies of task description. Nevertheless these systems may appear to possess purposeful behaviour. They work in this way under a restricted set of circumstances but are unable to adapt to new or changing conditions, which could lead to drastic failure. The appearance of purposeful behaviour may be the result of restricting the conditions under which the machine is observed to act, rather than in the inherent properties of the machine.

The overall goals may be decomposed into local goals by the cognition system. This decomposition of function may continue further so that a hierarchical structure of goals will be exhibited. Figure 5.9 shows a partial hierarchical structure of goals for the box-in-box task. The goal structure is divided into levels. The goals at level N are connected to their subgoal at level $N \pm 1$. Thus to achieve an overall goal 'box-in-box', we must successfully execute the subgoals 'pick up', 'transfer' and 'put down'. Note that these subgoals are given in time order from left to right (this is not always possible in the tree representation).

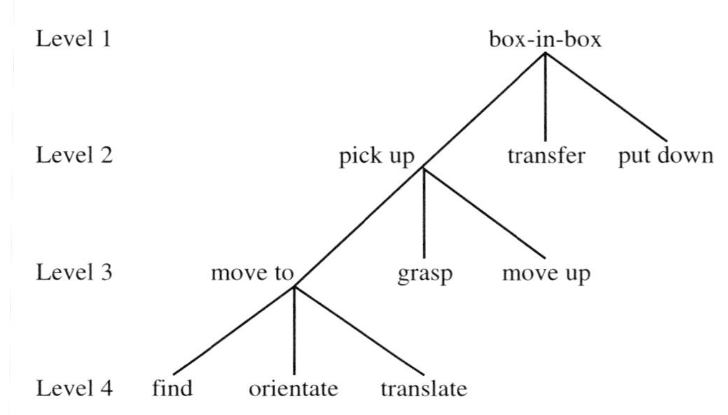

Figure 5.9 Box-in-box task: hierarchical goal structure.

Goals guide choices within cognition. Many other factors, including constraints and available resources, will also govern choices. For example, the path through the state space (see Figure 5.8) for the box-in-box task <A,B,C,D>, to place a box within a box, is guided by the goal. The path <A,C,D> is inappropriate because of collision constraints and the path <A,B,C,B,C,D> is inappropriate as the transition <C,D> at the first occurrence of C would achieve the goal.

5.4 Information processing and representation

Information in cognition has four main elements:

input perception patterns,
knowledge about the action domain,
knowledge about the machine itself,
specification of the goals.

These elements are organized so that relevant items can be found, relations between data items established, operations performed and new data derived. We will consider some of the basic information processing and representation tasks in the cognition component of a mechatronics system. The ways that information can be structured and organized are considered first.

5.4.1 Structuring information

Consider the box-in-box task. Data from the vision system may include positions of the corners of the two boxes. Suppose these positions are represented as coordinates after appropriate interpretation of the primary image. The corners will be represented by coordinates placed in order around the box.

The straight line edges between corners can be inserted by the cognition system, provided of course that the perception system has recognized a box (with straight edges). The grasping positions can then be calculated. Note that information is extracted from the perception pattern.

The information about the box is organized as an ordered list of coordinates. Now consider a situation where there are four identical small boxes to be placed in the corners of the larger box (Figure 5.10). The set of small boxes can be represented as a set of the lists of corner coordinates. There is no obvious order for the small boxes, so they may be listed according to an arbitrary order derived from processing the perception pattern. (For example, this regime may organize them according to the position of their bottom or left-hand corners.) In Figure 5.10 the four boxes are ordered 1, 2, 3, 4.

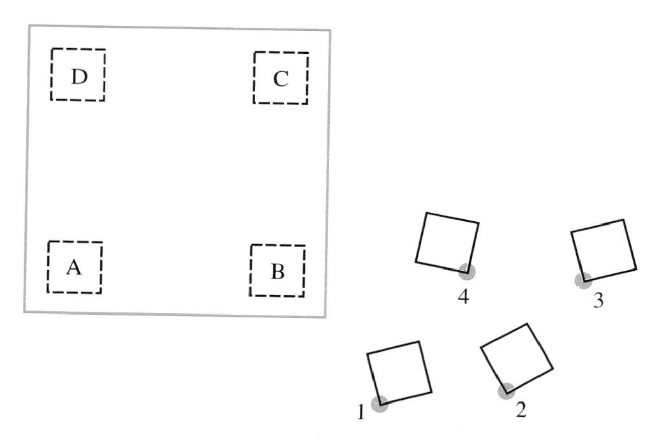

Figure 5.10 Box-in-box task: multiple boxes.

Suppose that in this assembly it is required to minimize the assembly time. We try and do this by minimizing the total distance moved in placing the small boxes (1, 2, 3, 4) in the four corners labelled A, B, C and D. To do this a search is conducted for each corner in turn, examining the boxes not already placed. The list of boxes to be placed changes as the plan is generated. This is conveniently arranged by including with each item in the list (the corner coordinates of a box) a pointer to the next item on the list. It is easy to search through the list of the remaining unplaced boxes following the pointers and to delete an item by changing the pointer to skip the 'deleted' item.

The plan created depends on the order in which the corners are considered. I will show how the information might be organized to allow a search of all possibilities and evaluate total distance in each case. The structure is more complex than the simple list and takes the form of a tree.

The critical data for planning is the distance between small boxes and the corners of the large box. This information is derived from perception (note that the means of calculating this distance, perhaps by using centres of area, is required). The appropriate structure for searching all possible placements of boxes in the corners can be represented as the tree shown in Figure 5.11.

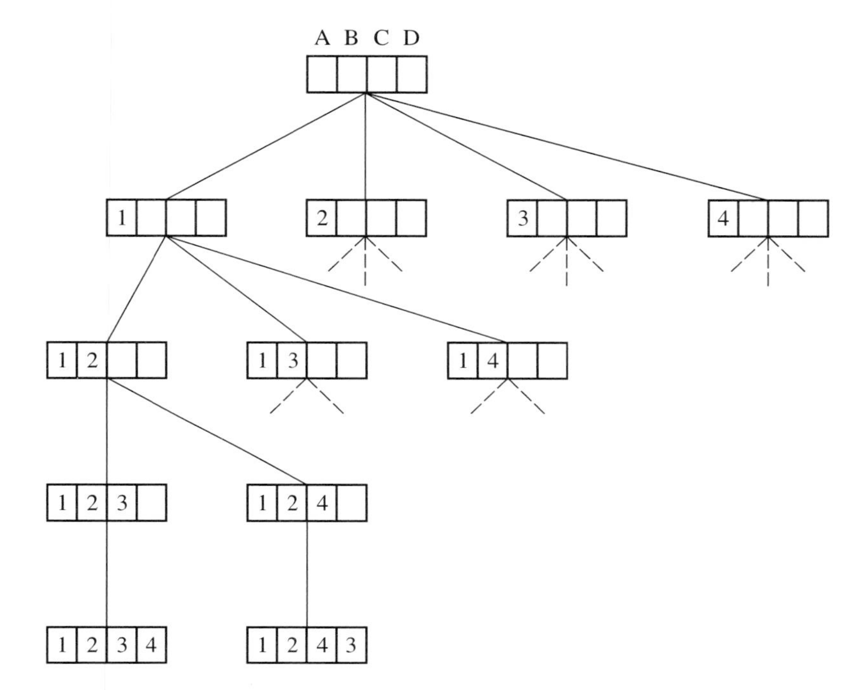

Figure 5.11
Boxes-in-box: structure for exhaustive search.

Each node represents the four positions A, B, C, D in the corners of the large box and the entries in each position correspond to the number of the small box positioned at that corner. Each entry at a node corresponds to a particular corner of the large box. For example, the first position corresponds to the bottom-left corner A and succeeding entries to corners B, C and D taken in anticlockwise order. The top level of the tree represents the empty large box. As a path is traced through the tree, small boxes are progressively allocated. At the leaf nodes, a complete allocation has been made.

Search may be initiated

- by considering all successor nodes at a given node and performing distance calculations of the total distance at the point in the task before moving down the tree structure to the next level; or
- by following paths from the top (root) of the tree to the bottom (leaf nodes).

Paths may be considered in an organized way by moving back along the path until another path downwards (not previously explored) is found. In this way the leaf nodes and the paths that lead to them are traversed in order. Distance calculations can be made as the search proceeds. Leaf nodes and the corresponding allocations are given total distances that are used to find a minimum distance allocation.

The first method is called *breadth-first search* and the second *depth-first search*. In general, exhaustive search is not possible in large problems that have many possible courses of action. Some guidance or direction is needed to limit search. This is possible with both types of search but does not necessarily produce a best result. This is considered further in Volume 2.

5.4.2 Models

We have seen in the last subsection how information about the *task* can be organized to enable processing in cognition. We now examine information about the machine and its environment.

The information that describes machine capabilities and the environment in which the machine operates is usually represented in a structured and coherent way as models. The model of the environment is often called a ***world model***. Models form an essential part of the cognition system and are used to determine the interaction of the machine with the environment and predict the effects of possible actions.

The model may be partial, inaccurate or possess uncertainties in its formulation. A partial model will only contain information on selected characteristics of the world. For example, an automatic vehicle may use a two-dimensional map of the world rather than a full 3-D model of the space through which it navigates. Alternatively, the model used may only contain information on the space in the immediate vicinity of the vehicle's current position, being updated as the vehicle moves through the workspace. In the latter case the partial range of the model will restrict planning to local decisions.

A model may be inaccurate for a number of reasons. The perception system may make mistakes in recognizing features of the world, or the resolution of perception may not be sufficient to discriminate between cases or to make accurate measurements. Further, the construction of the model may be based on assumptions rather than observation. The assumptions may be inaccurate. For example, an automatic vehicle navigator may assume that distant objects remain static when planning moves under a known and observed set of local conditions.

A model may be uncertain in the sense that the measurements and observations recorded by perception have associated probabilities or likelihoods. Thus in the example of the automatic vehicle, distant objects may be assumed to move but the precise extent of movement is uncertain. Estimates of likely movement such as average speeds can be used to plan paths.

The models allow:

exploration of the space of possible actions,
prediction of the effects of actions,
coordination of machine resources.

Exploration of possible actions is exemplified by state space search mentioned in Section 5.4.1. Models useful for this usually possess representations of the results of possible actions. This contrasts with the models for predicting the effects of actions. These are usually physical models of a system. Actions cause the system to behave in a particular way. Coordination of machine resources requires models of the capabilities and relations among the component elements in the machine. Thus for coordinating the actions of a set of machine tools forming a flexible manufacturing system, we require models of machine production times, production sequences and machine capabilities.

It may be tempting to think of a cognition model as being a detailed, centralized description. This may not be the case. A model can be: outline or schematic, distributed among parts of the mechatronics system, partial, or hierarchical.

- An *outline* model will often be an abstraction of the structure or relations among elements. For example, a model required to schedule the production in a flexible manufacturing system may use distances and times between the various machines but not a detailed geometric description of the layout.
- A *schematic* model is one that deals with general cases and represents how types of element are related. It is not an abstraction of a specific case but rather covers a range of cases under its scheme.
- A *partial* model was discussed in Section 5.2.4. A further example of such a model is that used in a machine safety system in which only selected information on the state of the machine is required to determine the actions of the system.
- A *hierarchical* model works with many levels of detail. For example, a model of a flexible manufacturing system able to produce a range of products may possess a high-level model for planning the schedule of actions based on machining times and outline capabilities. Detailed models of machine capabilities and tooling are then used to create process plans at each machine.
- The *distributed*, non-centralized type of model is noteworthy. Parts of a model may be created and maintained locally, possibly physically close to the sensors and actuators of a component of the system. The local model may be sufficient for the actions of that component. However, to function effectively in coordination with other parts there must be connections to other local models. This is because (a) local action choices impose constraints on other models, and (b) information from one site may be needed at another to complete that model or make a choice between alternative actions. The essential element in this distributed representation is the structure of connection in terms of what information (and at what level of detail) is available to the local subsystems.

5.4.3 Models in control

The essence of traditional control lies in the relation between information received from instrumentation and a model of the process or machine to be controlled. The model is used to estimate how to correct deviations in machine behaviour from instructions. In general, control uses cognition models that predict the effects of action. Developments in control have moved control and cognition closer in the sense that machines learn appropriate action responses without the need for an explicit physical model.

A good example of such a scheme involves a well-established control problem in which a pole or rod is hinged to a cart (Figure 5.12). The pole is hinged to rotate about a horizontal axis that is perpendicular to the direction of travel of the cart. The cart can be driven backwards and forwards along a single direction. The control problem is to drive the cart backwards and forwards keeping the pole balanced vertically upright.

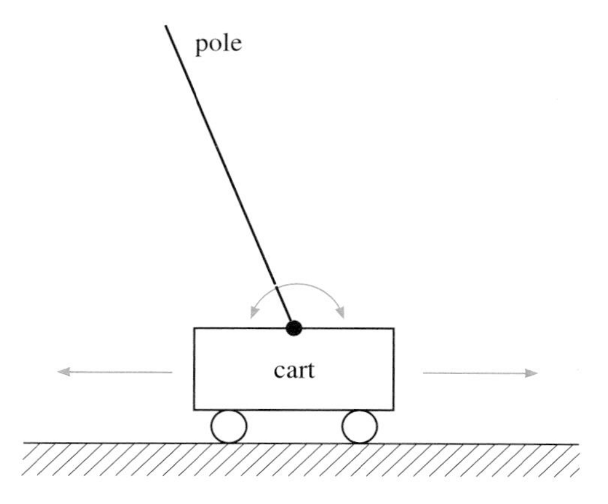

***Figure 5.12* Driven cart and freely hinged pole.**

It is possible to model the dynamics of this system such that measurements of deflection and speed of the pole invoke a correcting action in the movement of the cart. Continual monitoring of the pole enables deviations of the pole from the vertical to be corrected via the model of the system.

Alternative approaches have been examined which take two actions of the cart (*move right* and *move left*) and associate them with states of the pole as measured by its angular position and velocity. When the system starts, it has no knowledge of this association and the pole quickly falls over. However, the system gains some empirical information on appropriate (and inappropriate) actions so that when it tries again the pole will remain upright for longer. As the trials continue, the association between states of the pole (angular position and velocity) and the actions of the cart (left or right movement) is refined. A pattern of responses to the pole positions is created which, after a number of learning sessions lasting a few minutes, allows the pole to be balanced in an oscillating mode for long periods.

The system has learnt by constructing a complex pattern of perception–action pairs that achieve the goal of obeying the instruction to keep the pole upright. Essentially, a set of rules of action has been learnt by experiment.

The information processing in this learning control system is more like that of a cognition system building a model from perception. However, the basis of the model is not perception of the world (the pole) or the machine (the cart) separately, but is the interaction between world and machine.

Systems such as this may not need explicit models of state space in order to plan actions; instead they may search the state space physically or experimentally. Actions are modified according to their effectiveness in achieving goals.

This type of control starts to assume some of the functions I have assigned to the cognition system. Outline instructions are provided by the planner, but the execution takes care of the details. In effect, a mechatronic product can resemble a set of cooperating (and possibly hierarchically organized) execution systems with very little overall planning and consequent requirement for explicit representation of task, machine and environment. The state of actions of each element in the cooperating set is known to others, but there is no centralized representation of the interaction of the intelligent machine with its environment.

5.4.4 Computation and cognition

Computation provides the means to represent and structure information in models. Further it allows these models to be used to examine transitions in the states of these models. Association of transitions with actions performed by machines enables possible actions to be explored and sequences of actions to be planned to achieve goals.

Computation can provide models for cognition processes. Research in this area generally concentrates on the human cognitive processes and their understanding in terms of computation. It acts as a guide to the implementation of cognition functions by digital information processing. However, it is not of immediate interest to the mechatronics designer, who wants to exploit computational processes to transform information from perception, knowledge and goals into machine instructions. The mechatronics designer wants a machine to achieve a specification and meet a perceived need rather than one that mimics aspects of human cognitive behaviour.

Computation and its capabilities are being extended through developments in areas such as parallel computation and neural computing. The latter, which will be covered in Volume 2, seems to offer the prospect of machines that learn. These machines build their models by direct interaction with the world. The link between perception and action becomes dynamic. The simple cognition systems we have examined are static. Better data or improved interpretation of perception will improve their performance at restricted tasks but it is likely that flexibility in response and a capability to learn new tasks will characterize future intelligent machines.

5.5 Conclusion

To conclude the chapter we return to the mechatronic product as an intelligent machine. The essence of these machines is that they exhibit a flexibility in action. They can respond to many situations. The physical actions of the machine are driven by goals. The range of physical actions available imposes fundamental limits on machine capabilities. Within the range of actions possible, the task of the cognition system is to design a set of actions that meet a specification or goal.

The design of the mechatronics system is concerned with assembling the information needed to design and plan the set of actions. From the specification of the task and from information on:

the characteristics of the execution system,

the current state of the world, and

the knowledge about relations and properties of objects,

the cognition system creates a set of action instructions.

The cognition system may be an algorithm that translates input perception directly into instructions. Embedded in the algorithm is knowledge about appropriate responses for the given data and goal inputs. Cognition may be a combination of knowledge about objects and properties and mechanisms for its manipulation.

The next chapter of this volume will address the issues of representing and using knowledge to plan actions. In terms of the schematic of the elements of cognition in Figure 5.4, this chapter has considered the horizontal dimension of input from perception, processing and output to execution. Chapter 6 will consider the vertical dimension of knowledge, processing and goals.

Cognition requires a knowledge base to interpret perception information. Actions change states in the world. They can change the objects and their properties and change their relations. In order to reason about these changes the intelligent machines must be able to work out consequences of planned actions from the knowledge base.

The intelligent machine requires the flexibility to respond but it also requires a flexibility in using information and knowledge about its possible actions to make effective use of its capabilities in perception and execution in meeting its goals. The flexibility to respond is the business of the execution system and control. The flexibility in planning *how* to respond is the central function of cognition.

CHAPTER 6
PLANNING

Chris Earl

6.1 Introduction

Chapter 5 of this volume covered the main features of *cognition* as the link between perception and action. A major feature of this link is ***planning***, in which knowledge of the intelligent machine and its environment is used to determine courses of action to achieve goals and complete tasks.

The previous chapter concentrated on the ways that *perception* information is processed to provide *action* instructions to an *execution* subsystem. Processing can be regarded as a program initiated by perception patterns. The program may encapsulate knowledge about the task, the machine's capabilities, or the world in which the task takes place. The components of the cognition subsystem were outlined in Figure 5.4 and we found that the knowledge and goal elements of cognition were embedded in the process of linking perception and action.

This chapter will examine in more detail the ***knowledge*** in a mechatronic system, its ***representation***, and how it is used in conjunction with the *goals* to create *plans* (***goal-oriented behaviour***). Note before starting that the match between knowledge and goals may not initially be in a useful form. New (or composite) items of knowledge may have to be derived. On the other hand the goal may need to be decomposed into subgoals before the knowledge becomes applicable. In the first case we need a search of the knowledge base of state transitions to determine sequences appropriate for the goal. In the second case we need an examination of the overall task and a search of the ways that the goal can be decomposed into subgoals.

The theme of this chapter will be the relationship between knowledge and goals for planning. I start by examining the representation of knowledge and then move in later sections to consider how knowledge and goals are used to create plans.

6.2 Knowledge representation

A mechatronic system plans its actions on the basis of its knowledge of:

(a) the capabilities of the machine executing the task,
(b) the world in which the task is executed, and
(c) the task itself – this knowledge may be specific knowledge about the current task or may be in the form of general methods and techniques for creating plans.

The next sections review three methods for representing knowledge as part of cognition in intelligent machines. We will use the example of the box-in-box task from the previous chapter to illustrate the ideas presented.

6.2.1 Procedural knowledge

Perception identifies elements in the world. These might be objects and their locations. For example, in the box-in-box task used in the previous chapter, boxes are identified, distinguished and their locations determined.

Consider a specific goal, such as to place the small box inside the larger. If the small box is already in the large box, no action is required. If the small box is outside the large box then a *procedure* for a sequence of actions is initiated. This procedure relies on knowing the precise parameters of where the object is located and will produce a *sequence of instructions* so that the robot arm can complete the task.

The sequence of motions of the machine is determined by the position of the object. The sequence held in the procedure is in terms of the types of motion, such as approach, pick-up, depart and put-down moves. There is an *algorithm* specifying the steps in the procedure. The algorithm may consist of a number of stages. For example, the first stage determines a sequence of discrete positions for the gripper. The second stage creates paths between the discrete positions. In each stage the algorithm is designed to ensure that collisions are avoided.

The input from perception is interpreted directly in terms of actions. This example shows an ***action space*** interpretation of perception. In contrast, perception may be interpreted initially in a ***state space*** model for subsequent interpretation in action space.

In the box-in-box task, perception gives details of start and finish relations between the boxes. A state space interpretation gives a sequence of intermediate relations (or relative positions) of the boxes. This provides a sequence of subtasks expressed as transitions between states. The state space description is then further interpreted to produce the robot actions to achieve the transitions.

These systems use information in perception models to initiate algorithmic planning procedures which result in machine actions. The algorithms encapsulate knowledge about the world and the execution of tasks. This may be called *procedural knowledge*.

Consider again the box-in-box task. Assume that algorithms are available for generating action sequences to pick up the blocks. To do this a choice has to be made as to which pair of sides of the boxes is to be used for grasping (Figure 6.1a). The procedure would have to specify which choice was to be made. Perhaps those sides closest to a preferred direction are chosen.

However, suppose the two blocks are touching one another (Figure 6.1b). The choice presented above is not appropriate. The cognition model is inadequate. Perception has identified the two blocks, but their spatial relation is significant in completing the task. Cognition cannot interpret this configuration of boxes in terms of its knowledge about available actions.

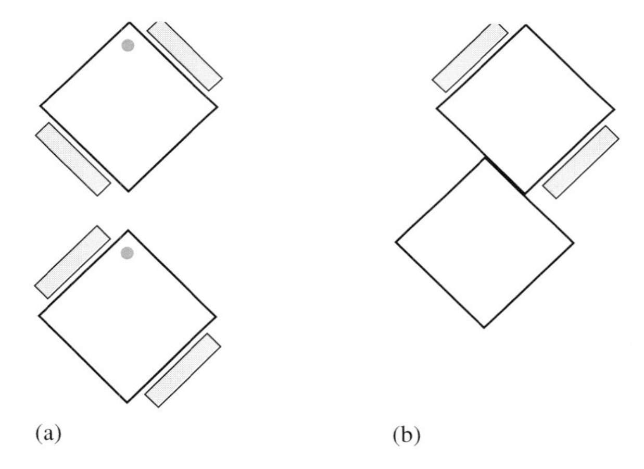

Figure 6.1
Gripping orientation on boxes: (a) if boxes are isolated then there is a choice of gripping orientation, but (b) choice is limited when boxes are adjacent.

The cognition system is required to recognize this configuration of blocks and initiate appropriate actions. The problem here is that the number of cases all needing special consideration can multiply out of control (Figure 6.2a and b) if we adopt a procedural knowledge representation.

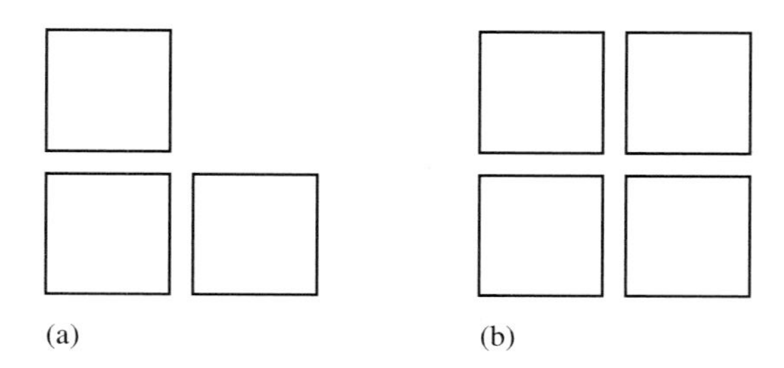

Figure 6.2
Tricky box configurations.

Look again at the configuration in Figure 6.2(b). Suppose that the boxes are resting on a flat table and can slide. Knowledge of the possibility of this relative motion can be used to generate the plan indicated in Figure 6.3. This can be incorporated into the algorithm for the task. But, we might ask, where does this consideration of special cases end? This seems to be a potential drawback to procedural representation.

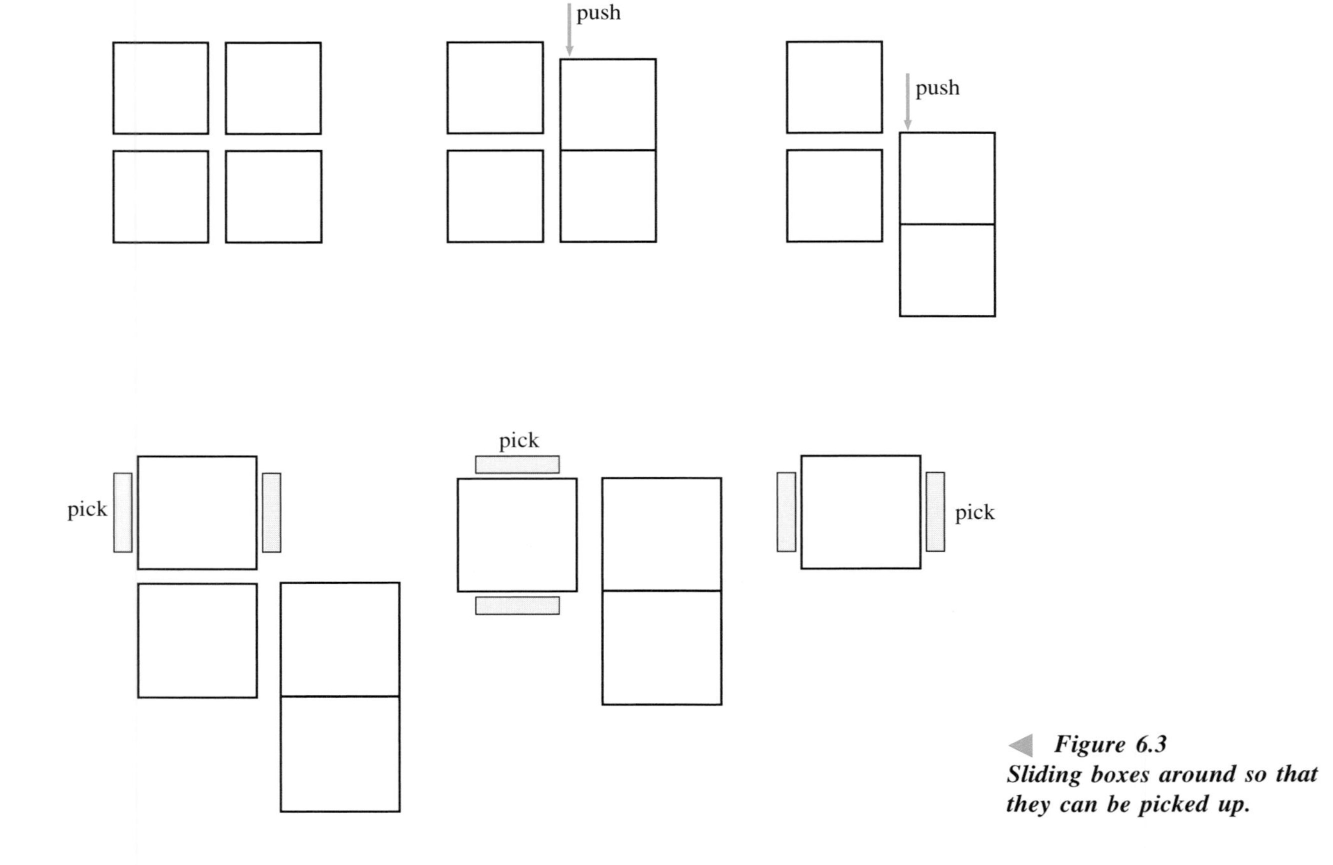

Figure 6.3
Sliding boxes around so that they can be picked up.

6.2.2 General aspects of knowledge representation

It would be an advantage to be able to reason from known facts. In the previous example, if we knew about the relative sliding freedom between boxes and a flat table we could conclude that pushing them offers the possibility of freeing the gripping surfaces. The configuration of the boxes then needs to be analysed to determine the location and direction of the pushing action by the gripper. Cognition should be able to plan such actions on the basis of general knowledge of the world rather than have such actions resident in procedures that are 'called' in specific circumstances.

Knowledge needs a suitable representation to allow:

manipulation and restructuring knowledge,
modification and updating knowledge,
creation of new knowledge,
support for reasoning,
creation of plans of action.

An intelligent machine is likely to contain a knowledge base. The nature of this knowledge base is characterized by:

representation of knowledge,
the types of knowledge represented,
the way the knowledge base is used.

We have seen how knowledge can be represented procedurally. Knowledge in a procedural scheme usually represents a sequence of actions to achieve a goal. One of the difficulties with procedural schemes is that knowledge about the world (facts) and knowledge about actions in the world are closely related. Facts are associated with actions that need the facts. For example, when the robot system picks up a box from the table it explicitly accesses the information on the box's position and generates a grasping position within the procedure.

Suppose, however, that it is necessary to examine this procedural knowledge, perhaps to assess the adequacy or efficiency of the procedure for executing the task. The knowledge about the world (such as that boxes lying on the level table can slide on the table) is deeply embedded in the procedures for picking up the boxes. If this knowledge is to be used to modify actions (and thus procedures) it needs to be extracted from existing procedures.

The knowledge about the world (fact: boxes slide on table) and procedural knowledge about actions (action: boxes can be pushed along table) are equivalent. But it is often difficult to extract knowledge for general use from the procedural representation. On the other hand the factual knowledge is more easily used in generating action. This knowledge is a general property of the relation between the objects 'box' and 'table'.

There are thus advantages in considering explicit ways of representing factual knowledge. Two main methods are used: logic-based representation schemes and network-based schemes.

6.2.3 Logic-based representation

Logic-based schemes declare two types of knowledge: facts about particular objects, and rules that assert that specified patterns of properties imply other properties. Thus for the boxes on the table we have the facts: (*a* is a box) and (*a* is on the table); and the rules: if (x is a box) and (x is on the table) then (x can slide on the table). From these facts and rules we infer the fact (a can slide on the table).

The type of knowledge represented in terms of facts and rules is usually referred to as declarative or logic based, in contrast to the procedural schemes. The knowledge base can be constructed in declarative or logic programming languages, such as Prolog, which provide the mechanisms for inferring new facts, deduced from the declared facts by using the rules.

These systems contain particular knowledge about objects (facts), general knowledge about classes of objects (rules) and the mechanisms for deducing further facts. New facts can be generated by rule application and added to the knowledge base. This is known as *forward chaining*. In using the knowledge base, it is often necessary to determine whether an object has specified properties. If this is an established fact in the knowledge base, a positive answer is given. If not, the rules are examined to discover how this might be established. The rules are considered in reverse (*backward chaining*).

The drawback to declarative representations is that there is no natural organization of the knowledge base. The structure of the knowledge, including associations, classes and hierarchies, is not easily represented. The lack of structure means that all elements are potential candidates for use at any stage of reasoning. This can present a significant computation problem.

6.2.4 Network representation

An alternative form of declarative scheme focuses on this missing structure. These are the associative or semantic network representations. There are two basic elements: *objects* and *relations*. The objects may be object types. Relations may associate a specific instance of the object with the type, or more generally may specify a type within a class. In this case properties of the object type are inherited by the instance. In other cases, properties of the objects may form the associated objects. At higher levels of representation general types of object may be related to subclasses.

The *objects* are thus all-purpose and may represent types, instances or properties. The *relations* are of correspondingly different types, indicating instances, states of objects and properties (including spatial and physical properties).

Let's consider the network in Figure 6.4 which represents knowledge about the boxes on the table in the box-in-box task. Explanations of the annotations on the links of the network are as follows: 'on' is a spatial relation between objects of different type; 'is_a' is a relation indicating subclass or a type within a class; 'is' means that an object possesses a property; N means next to.

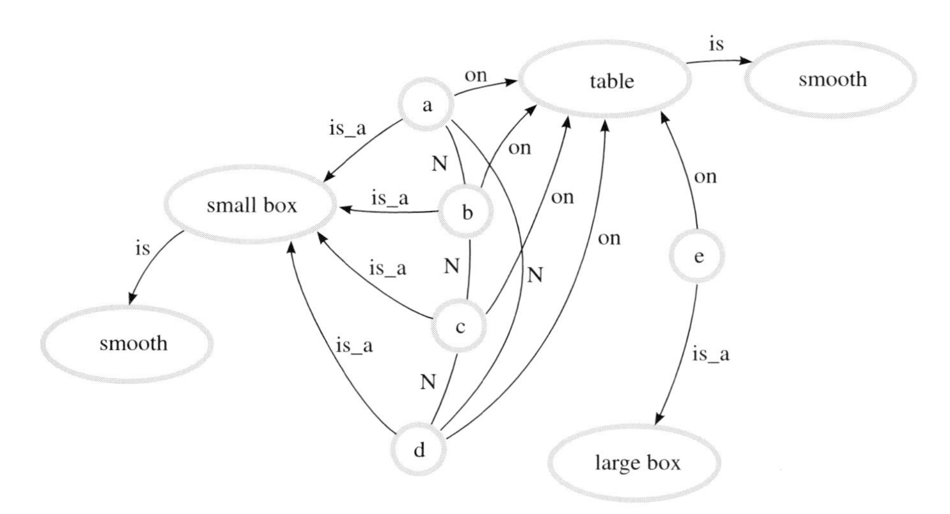

◀ ***Figure 6.4 Network representation of close-packed boxes on a smooth table.***

The knowledge contained in semantic networks is a description of objects, properties, and relations. Information about objects is defined in terms of relations to other objects. Some of the information about particular objects is stored as more general information about objects of that type. For example, the knowledge that small boxes are smooth is carried over into knowledge that boxes a, b, c and d are smooth. Other information is stored at the nodes representing subobjects and properties. The knowledge associated with a particular object may thus be widely distributed across the network. The meaning of the nodes and links may be unclear.

6.2.5 Summary

Each of the three representation schemes, procedural, logic and network, has shortcomings. Procedural knowledge hides facts and structure, but is focused on action. Logic schemes hide structure but explicitly state facts. Network schemes exhibit structure but do not always explicitly present facts as they are sometimes hidden in long chains of inherited information among the objects.

Knowledge bases for cognition in intelligent machines need elements of procedural knowledge to plan actions but they also need declarative knowledge to reason about the world. None of the above schemes is completely suitable. We would like to simplify the structural content of network representations to make knowledge easier to access, integrate logical operations and incorporate local procedural knowledge at the nodes in such a way that if information is required on a particular object then this can be generated as needed. If this requires information from other objects, these local procedures need to be able to 'call' procedures at other objects. This is the basis of frame-based approaches to knowledge representation and associated developments in object-oriented programming.

Cognition requires a knowledge base to interpret perception information. Actions change states in the world. They can change the objects and their properties and change their relations. In order to reason about these changes intelligent machines must be able to work out consequences of planned actions from the knowledge base.

The intelligent machine requires the flexibility to respond but it also requires a flexibility in thinking about its actions to make effective use of its capabilities in perception and execution.

The rest of this chapter examines how the cognition system plans actions.

6.3 Planning – tasks and actions

Before embarking on the discussion of planning methods, I reflect briefly on the nature of planning. There are two sides to any planning problem involving intelligent machines – planning the task itself and planning machine resources for execution.

6.3.1 Tasks and machines

First, there is the task itself. As we have seen, this may be described at many levels of detail. The planning system is often required to break down the task into subtasks or activities which the mechatronic machine is to execute. These activities are not described as actions of the machine but as changes of state or transitions in the world on which the machine is targeting its actions.

Second, there are the actions of the machine. These actions involve changes of state of the machine, in much the same ways as changes of state in the world. However, these actions are the causes and the tasks are their effects.

The two sides must be brought together in planning. The actions available at the machines (the action resources of the mechatronic system) must be matched against the required tasks. The deployment of resources across tasks is a central aspect of planning. Note that in some planning systems these two aspects are integrated by explicit limitation of state transitions in the task to correspond to machine capabilities. In these cases knowledge of machine capabilities is embedded in the task-planning aspect of the planning system. In more general cases state transitions are not explicitly limited but must be chosen to correspond to available resources.

It is important to be aware that in any mechatronic system the plan is not just an abstract entity consisting of a sequence of state changes to reach a goal, but must conform to the machine capabilities. To illustrate this point we will develop an example of a planning system in which planning for machine capabilities requires integration with task planning but where it is difficult to embed one in the other. This example will also serve to illustrate the relations between planning and execution, particularly the role of local planning during execution.

6.3.2 Planning a manufacturing task

Consider a manufacturing cell consisting of a bending press which is capable of producing linear bends in a sheet material. This is achieved by pressing the sheet into a V-shaped cavity as shown in Figure 6.5(a). The upper and lower press tools are moved in a vertical press stroke which determines the angle of the bend produced. By moving the sheet between the tools of the press a sequence of bends is produced. Complex components such as the enclosures shown in Figure 6.5(b) can thus be developed.

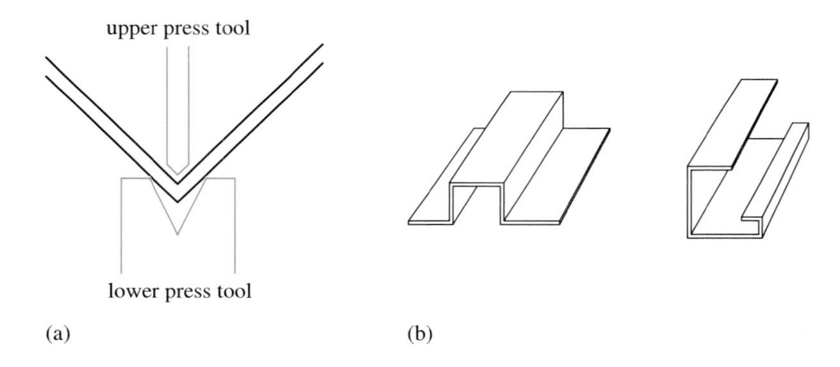

Figure 6.5
Pressing linear bends in a sheet.

Special tools may be required in the press to fit 'inside' the bends, and a means of changing tools may be needed for automatic operation. This aspect of the manufacturing system will not be covered here – we will concentrate on planning the sequence of moves of the sheet in the press.

The first planning task is to determine a sequence of bends to produce the component that is consistent with the geometric constraints of the press (Figure 6.6). This means that the production of each bend should avoid collision with the press. The problem here is that the two halves of the component (labelled A and B in Figure 6.6) rotate about the bend line. Each half may be quite a complex shape and a planner must ensure that collisions do not occur at any time during bending.

Feasible bend sequences satisfy these conditions for each bend in the sequence. The aim of the first planning task is to choose a particular sequence of bends that is optimal. In terms of manufacturing times this may be the sequence that requires the least manipulation of the sheet between bends. These manipulations include

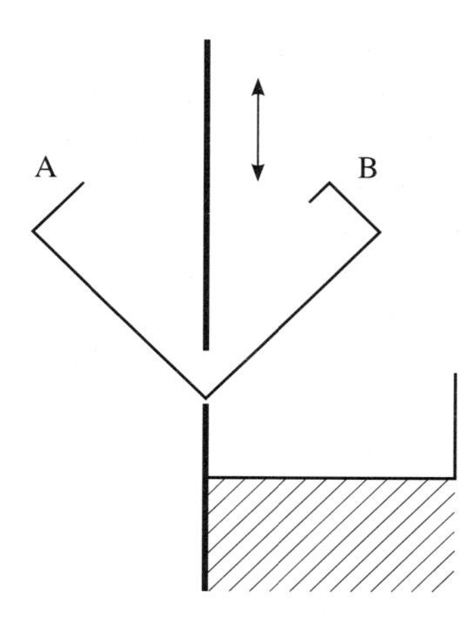

Figure 6.6
Bending press geometry.

moving the sheet to a new position in the press by threading it between the upper and lower press tools, removing the sheet from the press and turning it end to end or, perhaps, turning it over. For example, the four-bend component shown in Figure 6.7 requires a turnover between producing bends (1, 4) and (2, 3). It would be uneconomical to use the bend sequence (1, 2, 3, 4) as this would require two complete removals of the sheet from the press during the sequence.

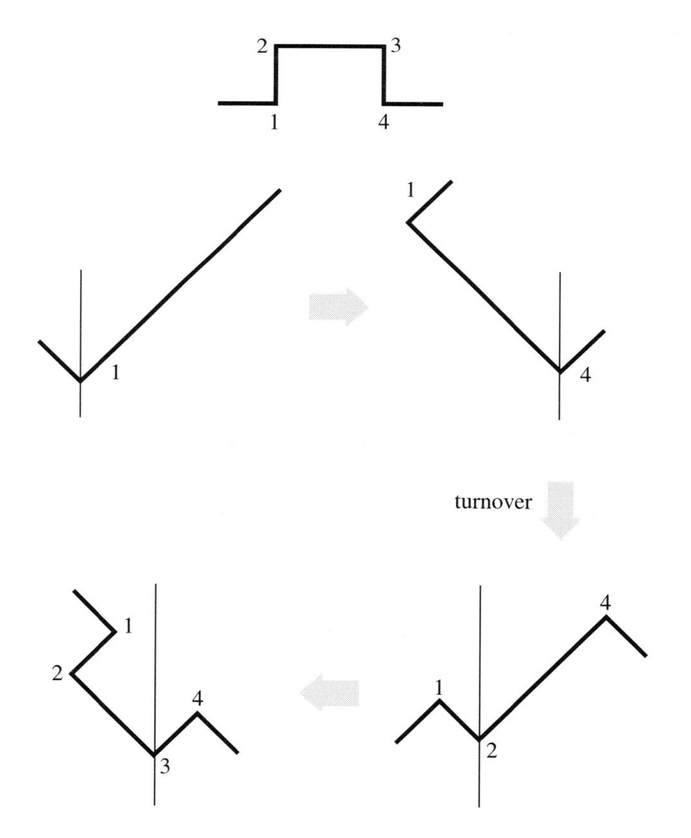

Figure 6.7
Component manipulations.

A planner trying to determine a sequence of bends needs to search possible sequences, checking that the bends they contain are possible and evaluating the sequences for manipulative complexity. This can be accomplished by examining the problem in reverse. Starting from the finished component, possible ways in which unfolding can take place are generated. This involves searching the space of state transitions corresponding to bend generation. The states are the shapes of the partially completed components and the transitions correspond to removing another bend from a partial component. The methods for such search will be considered later in this chapter. Note here that it is possible to associate a measure of manipulative complexity (based on sheet motions needed to thread, remove and turn the partially formed component) with each transition, and that these transitions accumulate to provide a means of evaluating the effectiveness of a particular bend sequence in achieving the goal. We will see in Section 6.4.2 how such evaluation of state transitions is used in guiding search of the possibilities.

The above describes the task. We now move to describe machine capabilities. Suppose that there is an automatic handling device, such as a free-standing robot arm positioned in front of the press. This is used to position the sheet in the press through the sequence of bends. The moves of the sheet have been created without reference to the handling device. The nature of the robot will affect the task plan because of the following characteristics:

- The robot arm occupies its own space and must not interfere with the sheet or the press. In this context we are referring not only to the space occupied by the gripping device at the end of the arm but also to the space occupied by the moving limbs of the arm.
- The robot arm has limitations on where it can reach. This workspace of the machine is characterized by both the positions and the orientations of the gripper at the end of the arm.
- The gripping device has requirements for grasping the sheet. For example, if a gripper consisting of an arrangement of vacuum pads is used to grip flat surfaces between bends, the size of the vacuum cups imposes limits on the surfaces to which they can be attached.

These characteristics of the robot arm and gripper impose constraints on the task plan. Note that different handling devices would impose different constraints.

The bend sequence plans provide a set of ordered positions of the sheet in the press before and after bending. These positions are descriptions of the partially formed component and its position relative to the press.

The capability of the machine is a critical factor in deciding which sequence of bends to choose. The initial search of the state space of bending sequences should include consideration of the validity of state space transitions with respect to the handling device. Different handling devices will provide different criteria for validity.

There are further planning operations to be undertaken. The moves of the sheet between bends have to be planned. Again, these moves need to be generated within the constraints of the handling device. Note that the choice of gripping position may be critical in the creation of a valid path between bending positions. Path planning for an intricate component will be a non-trivial task and is a planning task in itself. This particular type of planning will be considered in Section 6.6.2.

To review briefly, we have a planning problem in which the task can be decomposed into a sequence of bending operations. The particular decomposition chosen is based on a rough measure of manipulative difficulty. Further planning to realize the sequence of subtasks may yield difficulties that require the original bend sequence to be revised.

This example demonstrates a characteristic feature of planning the actions of a mechatronic system. The creation of a description of the task (in terms of bending sequences in this case) is relatively straightforward, but the decisions on physically valid routes through the state space are complex, requiring the development of detailed plans before evaluating transitions.

This may entail considerable wasted effort in generating details of moves just for the purposes of evaluation. Ideally what is required is a way of assessing whether a move is valid, leaving details of move planning to a later stage. In many cases a good guess is required, based on characteristics of the states. This is the role of *heuristics*, which are essentially the guiding principles of good guesses.

This economy may not require a provably correct answer but may provide an evaluation of the likelihood of the move being achievable. Comparison with other state transitions and their associated measures will provide a 'most promising' scenario for further analysis. This may not prove successful on detailed examination and another task decomposition may have to be chosen. Ideally we would like to include the reasons for failure in making the choice of a new sequence.

This example has demonstrated a number of planning issues in mechatronic systems:

- Planning task decomposition in terms of state changes to meet goals. This is *task planning*.
- Planning the machine actions to execute task plans. This is *action planning*.
- Planning to match resources of the system to the planned actions. This is *resource planning*.

The third item, *resource planning*, is not strongly illustrated in the bending example. The idea here is that there are multiple resources which can be used to complete the planned actions. Not all resources can be used for all actions and some actions may require multiple resources. Further, there is a time dimension in that resources are used for finite times on actions, and actions have precedence and concurrency requirements.

This situation could arise in the bending task if there are several bending machines and robots in a single cell. Suppose that removing a large component requires two robots to support it. The second robot may be involved in a task elsewhere and the unloading must wait for its completion. The scheduling of resources and matching them to the task involves complex planning procedures, including dealing with the precedence relations among activities, and ensuring optimal or efficient use of the resources in terms of completed tasks against time. An example of precedence relations will be considered later in the planning of assembly sequences (Section 6.5).

The relations between task, action and resource plans are complex. Choices made in each plan domain constrain and are constrained by what is feasible in the other domains.

6.3.3 Planning and execution

To conclude this introduction on the nature of planning, a further distinction is useful. In the previous chapter and the example above, the creation of a plan is presented as a complete set of instructions to the execution system. We have seen that there are different kinds of plan and have appreciated the necessity for their integration in producing a final plan of instructions. However, planning is not necessarily centralized or separate from execution. Some areas of planning may be closely interleaved with execution, such as when further information on current states is required before the details of a plan can be formulated. In the case of the sheet bending system the component will deflect under its own weight. Thus the exact position of grip cannot be determined beforehand unless detailed physical models of sheet properties are available. The approach move may be made in response to sensor measurements of sheet position or an approach move may start from a position clear of the sheet and continue until a vacuum is created and grip achieved.

Planning may take place in all areas of the machine's activity. Global plans of action may set tasks for local planning to be conducted during execution. This distinction between local and global plans is important and mirrors the distinctions between plan levels. It is the relation between the local and global plans that determines the nature of the relation between planning and execution.

As an example let's consider an autonomous vehicle (AV). Suppose that it is following a planned path through a sequence of materials-handling activities among workcentres in a factory. It may be loading and unloading components and assemblies from assembly stations, transferring machined parts between centres or providing a tool delivery and recall system. The patterns of activity and the paths to follow have been planned in advance. I will discuss how this is done later in the chapter.

Suppose that there are uncertainties and unexpected events which the AV needs to accommodate. For example, an activity at one workcentre may be delayed, leading to the possibility of collision with other AVs in the system on the route to its next destination. Two modes of resolution are possible:

1 The overall model of AV paths and schedules is referred to, which might allow rescheduling.
2 While the remainder of the system proceeds as planned, the 'disturbed' AV makes a series of local plans along its route to avoid potential collisions. It can do this in two ways:

    by reference to the planned activity patterns of other AVs,

    by reacting to immediate conditions in slowing down, stopping or taking local diversions where necessary – local modifications rather than global replanning of its moves are needed.

The ability to create these local plans is not confined to collision situations. The AVs serving the numerous workstations in the factory will often need to dock at centres to load and unload components, assemblies and tools. The details of these manoeuvres do not need to be planned before arrival at the workcentre. The particular mode of approach, docking and parts transfer may be determined by a local AV planner. In this way the actions to provide transitions between task states specified by the global planner are planned during execution. Such schemes have the following advantages:

- Unnecessary detail is avoided in the central planner.
- Uncertain aspects of the task environment do not have to be registered in the overall model. Small perturbations do not affect the majority of the task and can be factored out of consideration.

However, care must be exercised in constructing this distinction between local and global plans. The local plan should not affect the global plan. If decisions and choices made at local plan level imply that the global plan requires alteration then the advantages of the distinction are lost. Further, the global plan must present goals to the local planner that can be achieved. For example, the task planner for sheet bending must provide plans such that the action planner, which deals with sheet moves between bends, can achieve success or must return to the task planner for revised instructions. Ideally a guarantee based on the capabilities of the local planner and the action capabilities of the execution system must be available to the global planner, before handing over to the local planner.

A good example of a local planner can be drawn from the sheet bending example. The press used to create the bends has a controlled stroke which allows bends of different angles to be formed. The bend angle will be specified in the plan. The particular stroke may be left to local planning.

The calculation of the press stroke to achieve the bend may be quite complex, since most materials will spring back when the pressure on the sheet is released. A certain degree of overbend is required. The amount of springback is generally uncertain as it depends on the material properties, which may vary from sheet to sheet and from batch to batch. In manual operation a correct press stroke is determined by trial and error. The correction factor is then applied as the basis for correction to bends in the same direction on that batch of sheet (bends at right angles in the sheet will often need different corrections because manufacturing methods produce non-isotropy in the sheet properties). If bend angle can be measured, perhaps using probes or a vision system, correction can be made automatically in a cycle of bending corrections under the guidance of a local planner. This creates a sequence of instructions to release the sheet until springback ceases, to record the angle and compute a new press position. The task of the local planner is to create the press stroke to yield the correct bend.

6.4 Methods of planning

The last section gave an outline of what we want a planner to do in a mechatronic system. We now examine the ways that this can be achieved.

6.4.1 Goal and state space planning

Planning works with representations of task and machine to generate actions to meet goals. In order to do this the planner must examine possible actions that the machine can execute, and so select sequences that achieve the goals.

In simple terms, a planner searches possible actions – but not quite, for the planner is working with representations or models of the actions and their effects. The representations are critical. They are partial descriptions designed for the planning task. This raises difficult issues in the design and choice of representation schemes. We have touched on some of these in the previous chapter in this volume.

There are many approaches to representation. I will examine two.

The first considers *states* of the machine and environment together with state transitions. The actions of the machine change states and are represented by operations. Planning involves finding a sequence of state transitions and associated actions that yields the goal state. The state space is represented diagrammatically (Figure 6.8) with nodes representing states and edges joining the nodes representing operations. Different edges may have the same operator.

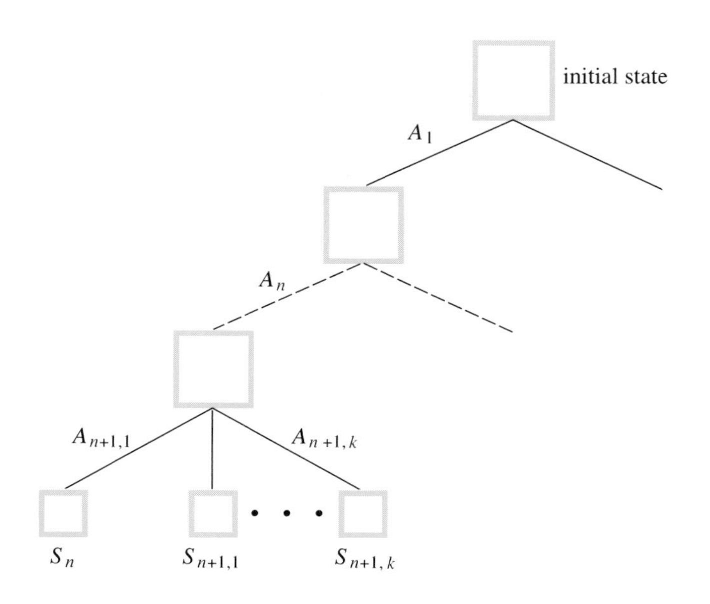

Figure 6.8
State space representation.

In Figure 6.8 the effects of actions $<A_1, A_2, \ldots A_n>$ have been predicted to yield state configuration S_n. From this state there is a finite number of possible changes of state $S_n \rightarrow S_{n+1,1}, S_{n+1,2}, \ldots S_{n+1,k}$ associated with actions $A_{n+1,1}, A_{n+1,2}, \ldots A_{n+1,k}$ and modelled as operators in the representation. The operators change states. The task of the planner is to find a sequence of states connected by operators leading to the goal state.

The second representation describes, not what the machine can do, but what the machine should try and do to complete the task. The *goals* of the task are decomposed into subgoals, leading to further subgoals, until they can be identified with primitive actions. Often many subgoals must be satisfied to achieve a higher level goal, and there may be many sets of subgoals that can satisfy a goal at the higher level. A diagrammatic representation is shown in Figure 6.9. A set of subgoals joined by an arc indicates that all of them must be satisfied if the parent goal is to be achieved. The task decomposition may not lead to a set of subgoals that guarantee the achievement of the original goal. The subgoals may represent a set of necessary conditions rather than sufficient conditions.

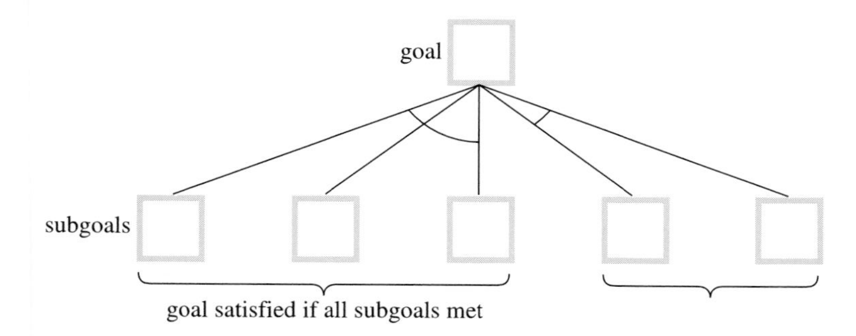

Figure 6.9
Goal space representation.

Planning in each case requires searching a solution space. In the state space representation, search takes place to identify paths that lead to the goal state. In the goal space representation, search identifies a particular task decomposition, with a set of primitive actions at the lowest level. They are just sets of actions without any idea of their order. This will come from conditions of precedence and protection on the subgoals generated during task decomposition. Precedence means that subtasks should be done in a specific order. Protection means that after a subgoal is achieved, other tasks are not allowed which destroy it. For example, a subgoal may be a required feature of the final solution, which once achieved should be protected.

The possibilities to be searched are composed of collections of actions or state changes. The search can be very large, with many possibilities at each stage. In general this leads to an exponential growth of eventual possibilities making exhaustive search infeasible. It is necessary to guide search using knowledge to focus on certain subgoal decompositions or state transitions. Note that the representation of the planning space uses knowledge about the task domain in its construction.

As an example of planning in state space consider the problem of planning the route of an AV through a sequence of activities. For the purpose of illustration let us take a snapshot of planning requirements. Consider the set-up in Figure 6.10 in which the AV carries three pallets. It is required to load pallets from the store and visit workcentres. At each workcentre it unloads an appropriate pallet and receives a pallet of completed parts from the workcentre. It returns to the store after serving three workcentres to obtain three more pallets. At the snapshot the workcentres (WC) must be served before the following times: WC1: 10 minutes, WC2: 5 minutes, WC3: 5 minutes, WC4: 10 minutes, WC5: 15 minutes, WC6: 15 minutes. The state of the system can be represented by the contents of the pallet. If it has a pallet to load on WC2 and has completed pallets from WC4 and WC5 then it would represent the state as (2,4′,5′). Further suppose that the AV takes two minutes to travel from area of the workcentres to the store and return, one minute to move to, unload and load at a workcentre in the same row, and two minutes if traversal between rows is required.

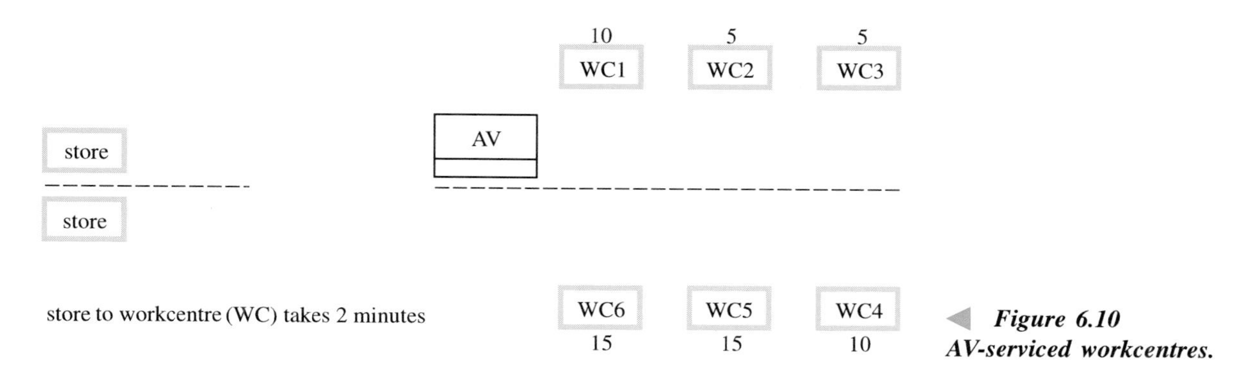

Figure 6.10
AV-serviced workcentres.

A portion of the solution space is shown in Figure 6.11. Beside each state is the current elapsed time from start.

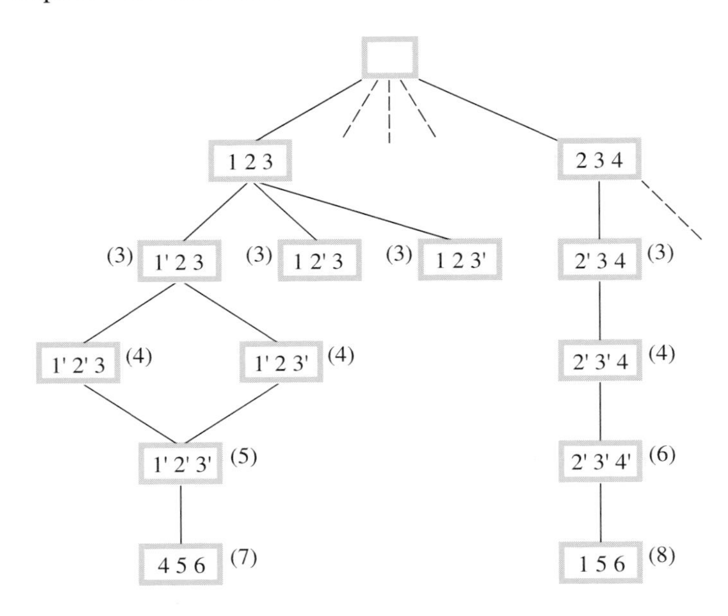

Figure 6.11
Portion of solution space for AV-serving workcentres.

Note the large number of possibilities, some of which come to an end as time limits are exceeded and some that look as if they will succeed. Indeed, there are some quite simple guidelines that could be used in the search, such as trying to service the time priorities first and reducing the travel between rows of workcentres. These are simple guides to the search which should be used to obtain a solution. There is no guarantee that such a solution exists or that it will be optimal, but in most cases we expect to obtain a solution after searching only a small part of the solution space. We will see how these guidelines or heuristics are formalized a little later.

For an example of goal space planning let us consider the example from Section 6.2.1 in which four small boxes are on the table and the task is to put a small box in each corner of the large box using the robot with a simple jaw gripper. Recall that we can slide the small boxes on the table if required. Further note that the configuration is only one possible configuration of the boxes with which the planner should deal. A portion of the goal space is shown in Figure 6.12.

The subtasks in the task decomposition have some clear precedence relations. These are shown by wavy arrows in Figure 6.12. A wavy arrow from subtask A to subtask B indicates that A should be completed before B can start. These will induce structure on the primitive actions at the bottom level of the task decomposition.

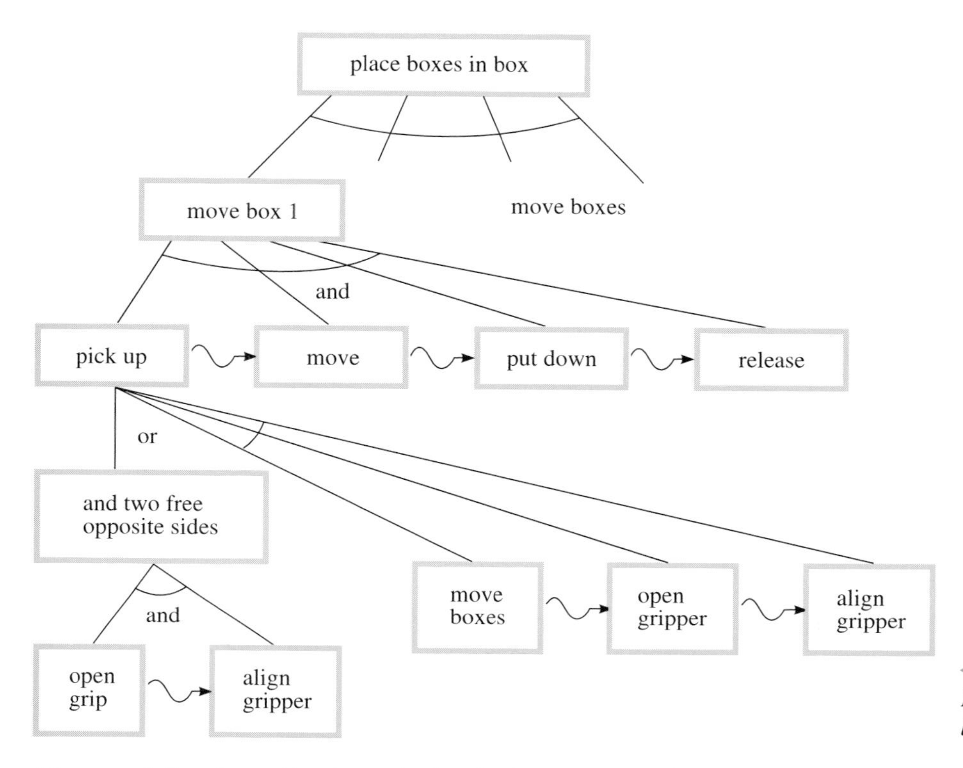

Figure 6.12
Portion of goal space for box-in-box task.

The search for plans of action, whether in the state or goal space, is central to the planning activity of intelligent machines. Search takes place in a space representing possibilities, as the next section explains.

6.4.2 Search

Setting up the representation of a problem to allow the search of possibilities in the state space requires three elements:

1 the initial state and knowledge about the task;
2 specification of goal state;
3 operators which when applied to a state, transform it to another state.

In fact, the search of a goal space representation can be formulated in the same way as a search in state space after a redescription of the goal space. Briefly, the goal space is redescribed as a space of partial plans through which search is made. This makes the formal details of the search equivalent to the search in the state space.

The operators will include information about the conditions under which they can be applied and their potential for execution by the system. There may be evaluation weights attached to the operators indicating the difficulty, likely problems or the time involved in applying that operator.

The state space and goal space representation for planning are essentially the converses of one another. In state space, planning actions are being matched to goals (sequences of available actions are built up and matched to required goals) and in goal space planning goals are being matched to actions (goals are broken down and matched with available actions).

The explosion of possibilities (sometimes referred to as the combinatorial explosion in that the possible combinations of actions to form plans can grow very large) means that it is usually necessary to guide search. This may be done in two ways:

1 ordering the operators available at a given state;
2 evaluating the results of applying operators in terms of distance from a goal state.

Operator ordering attempts to set values on the likely best operators to apply a rough method of guiding search, and it can cause problems if used without evaluation. This is because the search may pursue an unprofitable route with no means of returning to a better route. The planner will base its search on a priori attributions to transitions indicating which are best in creating a plan. This may not be the case in practice under conditions of uncertainty and in applying search to a range of similar but related tasks. The best transitions for one task may not be the best for others. This case arises when the conditions of the task change during execution and new plans need to be formulated. The major shortcoming of using just operator ordering is in assigning the ordering for each of the tasks. Unless ordering is valid across a range of tasks or the ordering can be modified to suit the task, this is not a viable route to task planning.

As an example, consider an AV which is searching for a particular workcentre but only has a set of actions (such as move left, right, forward, backward, or stop) to take at specified beacons in the factory. If these actions at each beacon are ordered independently of a particular task, the AV could easily cycle repeatedly through a set of states making no progress to the goal. Generally we need some means to evaluate the actions against goals.

The use of state evaluation allows the revision of decisions. If a particular operator is judged the best in getting closer to a goal, but subsequently gets further away, then another choice is made. If this is then found to be worse, a return to the initial choice is made. The guiding principle is to keep track of the most attractive states as they are generated. This kind of search is usually called *heuristic*, because the evaluation functions are often expressed as good guesses based on partial knowledge, rather than precisely evaluated functions.

In the AV example, when finding a path to its goal the AV may use a distance evaluation function. However, this is only one dimension in the search as other factors such as path obstruction could make distance irrelevant if the AV has to take circuitous detours.

The use of a state evaluation with distance as a measure may not lead to a good solution. For example, the AV may be able to get close in distance to a subgoal workcentre (Figure 6.13) but obstacles mean that it has to make a series of intricate moves to achieve the goals. Alternatively, with another choice of approach it may be able to make a direct line to a goal state. In this example we have assumed that the AV can service the workcentre from either side, as shown in Figure 6.13. Both positions are goal states. This provides a lower cost solution in terms of time and resources. In order to build this into the evaluation, the composition of distance from goal and depth of search (as a function of the number of operators applied) can be used. This modification will cause the search to return to examine alternative ways of reaching the goal after many state transitions have been encountered in negotiating obstacles. If the plan takes the AV into a space with many obstacles, it will look for ways around them.

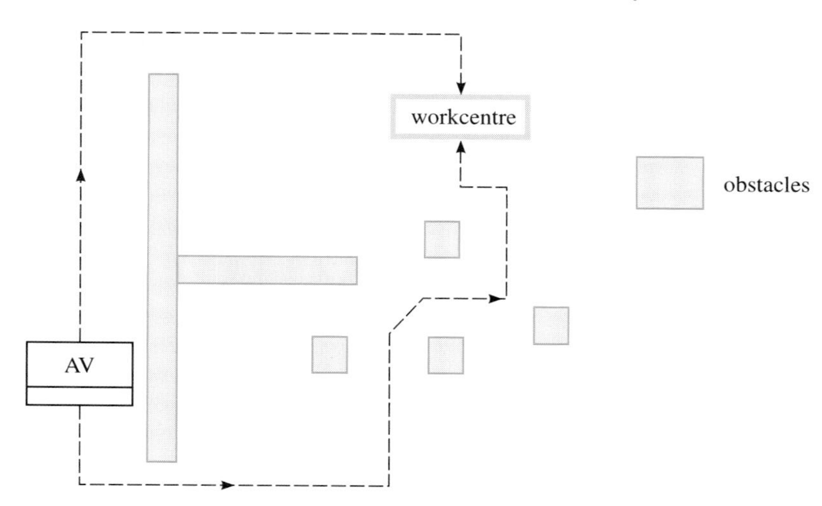

Figure 6.13
Alternative paths for the AV.

The algorithm for this kind of search, sometimes called ***best-first search***, involves a specification of the initial state and the following functions:

> `test` – to test a state to determine if it is a goal state;
>
> `successor` – to generate immediate successors to a state;
>
> `evaluate` – to calculate distance to nearest goal state.

The algorithm for best-first search executes the following loop until `test` is true:

1 `successor` to current state;
2 `evaluate` successors;
3 insert output of `successor`, ordered by `evaluate`, into `queue` of states to be examined (put best states at front of `queue`);
4 take first member of `queue` (remove it) as current state;
6 repeat until `test` is true.

The algorithm above is restricted in its application to search of a state space that forms a tree. This means that each state can be reached in a unique sequence of transitions from the initial state. But for many route-planning problems this condition of unique paths does not hold. Modifications to the algorithms can easily be made to accommodate more general state space structures such as graphs and networks. To handle these we need to keep a record of examined states and their evaluation in case they are encountered again. The two transition sequences are then compared for the minimum weight path. Also note that the algorithm above needs to be enhanced by the addition of pointers back from the current states to their parents (the states from which they were derived by state transitions) so that the sequence of transitions to the goal state can be recovered. It is these sequences that constitute the plan.

This algorithm maintains a `queue` of states (with associated evaluations) waiting for further examination. Note that if `successor` finds states which are worse than those already in `queue`, the search jumps across to the best state so far. If this turns out to be a blind alley then the search jumps back to the state which has come to the front of `queue`.

In some cases the search is made in a blind way with no evaluation function. This is quicker and, in circumstances where the search space is small or where many of the search paths lead to a solution, may be appropriate. This method is called ***depth-first search***. The algorithm works as follows:

`dfsearch`:

1 If `test` is true then current state is goal.
2 Else put output of `successor` in `queue`.
3 Loop through members of `queue` applying `dfsearch` to each member.

This algorithm exhibits a recursive structure in that it calls itself. The `queue` is particularly simple and elements are added to the front of the `queue`. The last addition to `queue` is pulled off for examination. A 'last in first out' policy is adopted and a stack structure is sufficient for `queue`. Figure 6.14 shows the order in which the states are examined.

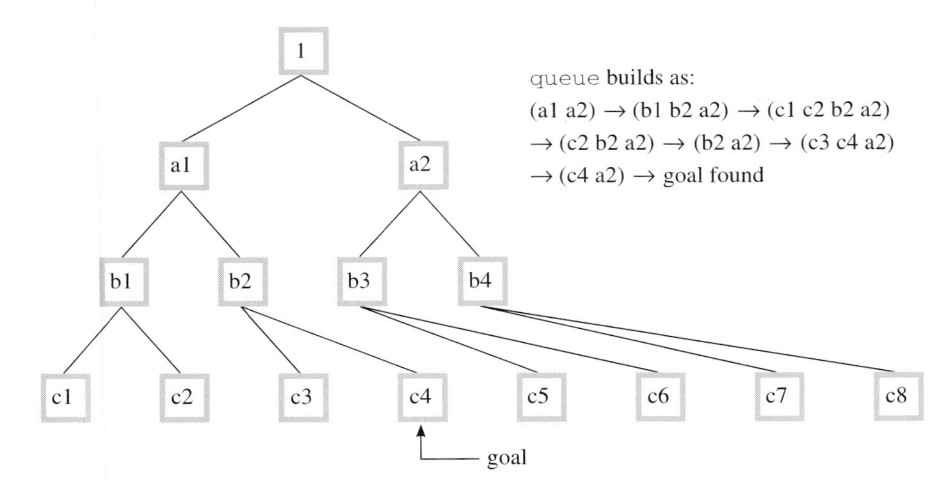

Figure 6.14 Depth-first search.

In practice, depth-first search works with a depth cut-off so that the search does not generate long and costly solution sequences and does not cycle around a sequence of states without getting closer to a goal. The depth-first method may be used as a look ahead during search to achieve an intermediate position. Suppose, for example, that we want to look ahead to find a path to a state with evaluation distance to the goal lower than some specified value. In other words, we want to get within striking distance of a goal. Depth-first may be used with depth cut-off and `test` replaced by a distance evaluation function.

Another blind search method is ***breadth-first search***. In this case all options from the current state are examined before proceeding to their successors. In breadth-first search the algorithm is similar to depth-first search but puts the output of `successor` at the back of the `queue`. Figure 6.15 illustrates the order in which the states are examined.

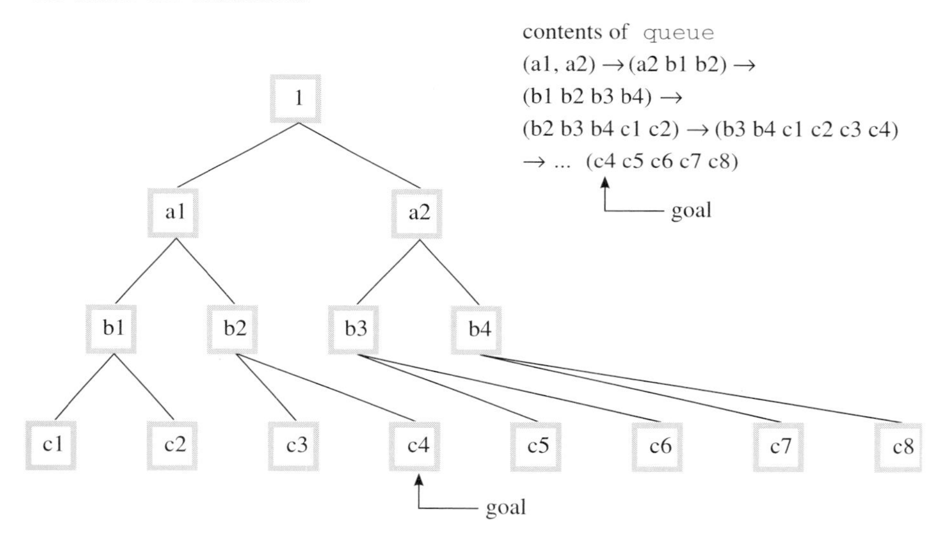

Figure 6.15
Breadth-first search.

The best-first method of search, using an evaluation function with a combination of distance to goal and distance from start, is effectively a combination of depth-first and breadth-first methods with additional guidance effected by ordering the elements in `queue`.

This section has given an overview of some methods of search used in planning. Greater detail and worked examples are provided in Volume 2.

6.5 Managing plans – the task perspective

We have described the ways that planning uses representations in state or goal space. The previous section examined search in the state space. We now concentrate on planning in the goal space. The discussion in Section 6.3 examined possible goal reductions and task decompositions in which the tasks are broken into sets of subtasks that lead to completion of the task. The subtasks are further decomposed until they can be identified with primitive actions. This section examines the process of task decomposition which leads to a complex structuring of the subtasks and thus of the primitive actions. This structure may include, for example, conditions of precedence and concurrency among the subtasks.

6.5.1 Plan structure

In creating a plan which involves a set of subtasks required to achieve a goal we must recognize the connections between the subtasks and the consequent links between the associated subplans.

- There may be precedence relations among the subtasks – certain tasks need to be completed before others can be satisfied. Precedence relations are shown in Figure 6.12 for the decomposition of the box-in-box task. Precedence may either require that one task be completed before another is begun, or that parts of one task should not be completed before another task is finished. In some cases conditions of concurrency may be specified in which a number of subtasks are to be undertaken simultaneously.
- The subtasks, once completed, provide conditions that should be maintained. The results of a particular task are being protected. They may represent conditions for subsequent subtasks or they may be part of the goal. The state achieved must be protected from change by other subtask plans. Plan failures can arise from protection violations because the conditions for the application of the subsequent plan have not been met.

The plans are created by a plan generation stage. This is a highly domain-specific activity tailored to particular kinds of tasks. Thus plans generated for the path of an AV will be of a different type from those generated for a sequence of assembly operations.

However, there are numerous decisions to be made in putting the plans into effect. The mechanisms for implementing these decisions may be common to many plan domains. These include protection of subgoals and the allocation of resources where a number of subtasks with separate plans draw on the same resources.

A general scheme for plans may be formulated with the following elements:

1 *objects* used to carry out plans (the resources which may contain both physical and information processing resources);
2 *steps* or *subgoals* which compose the plan;
3 *order* of the sequences of steps in the plan;
4 *protection* among plan steps.

These elements constitute a schema for the plan. They are required for managing task decomposition and organizing subtasks.

The subtasks may require further plans, possibly of the same type as the original plan (calling the plan recursively). In this way we have available plans that are brought into action to meet subgoals and subtasks. The component plans themselves are domain-specific but the methods of managing the use of the component subplans may be quite general. These methods specify how the subtasks are to be organized.

In managing planning decisions it may be important to look ahead for potential protection violations caused by the interactions between separate parts of the plan. Further, the use of the objects or resources may need managing. Suppose that two robots are being used in an assembly task. One robot could hold the assembly while the other mated the parts. If, however, there was a subassembly to complete for subsequent assembly, it would be preferable to use an assembly fixture (if available) to allow the two robots to assemble main and subassemblies concurrently.

6.5.2 Uncertainty

When making planning decisions we might consider it beneficial to be able to evaluate plans in terms of the likelihood of achieving overall goals. We have assumed that actions have predictable consequences. This assumption is based on the completeness of the representation used in planning and the confidence that execution will yield the expected results. Either or both of these conditions may not hold. The plans may be based on uncertain knowledge, or the outcomes of the execution of plans may be uncertain.

Decision theory allows us to estimate the probabilities of particular outcomes of the course of action. These probabilities may be affected by performing tests such as requesting specific measurements which can resolve uncertainty. The tests have potentially uncertain results and may not resolve uncertainty completely. However, their results change probabilities. The incorporation of probabilities into planning requires that plan options are available and that there is sufficient knowledge of the problem domain to assign probabilities.

Neither of these conditions may be met in practice. Planners may not offer comprehensive plan generation, and the assignment of probabilities is hard to make without extensive 'experience' in evaluating actions. This requires the planners to accumulate a body of knowledge which can often only be gained by evaluating execution. A process of learning is required. This will be covered in Volume 2.

In the absence of uncertainty in plans, the planner works in an ideal world of its representation. The execution may not be strictly possible. Conditions or states predicted by the planner may not arise because of execution errors. The execution system will be looking for the current conditions (as specified in the plan) before moving to the next planned action. Three ways of dealing with this situation are possible:

1. Make the execution system correct errors.
2. Accommodate the errors with local execution-level planning.
3. Replan actions in the case of large errors which would preclude the original plan proceeding.

Further, there are some situations in which the planner should not be trying to output plans in complete detail. Certain components in the plan should be left to the execution stage. In practice, planning and execution are closely interleaved. There is little point in planning every move in detail because of intervening events that will make plans for later tasks wrong. This is not to say that planning in advance is pointless but that appropriate levels of forward planning should be assessed for the task domain and associated execution system.

In a mechatronic system failures and errors may occur which render the immediate sequencing in the plan impossible. For example, a robot might drop a part or be unable to complete an assembly because of ill-fitting components. This requires understanding of a completely new situation, planning recovery, executing recovery and possibly returning to the original plan. This is an active research area in planning.

6.6 Sample planning domains

The treatment above presents a generalized view of the nature of the planning activity and an introduction to some of the methods and techniques. To conclude this chapter we will examine a number of mechatronic systems, indicating the characteristics and methods of planning in each case. These do not all make use of the general planning structures outlined above, but demonstrate the kinds of mechatronic systems for which planning is needed. They also illustrate methods of planning being used in mechatronic systems.

6.6.1 Robot motion planning

A task was introduced in Chapter 5 of coordinating the joint motions of a robot arm so that the end of the arm moves along a prescribed path between start and finish. This is a trivial planning problem as not only is the goal state specified, but also the sequence of intermediate states (of the end effector) is known. What is not known are the states of the machine itself (in terms of its joint motions and displacements).

In such a problem we need to make serious and far-reaching assumptions about the task, namely that there are no obstructions and other hazards. The prescribed path represents a valid plan of action as created, perhaps at a higher level within the robot planning system. (We will deal with path planning in the next subsection.)

How is this planning accomplished? There are several ways:

1 Divide the path into small segments.
2 Treat each segment as a subtask.
3 Start and complete each subtask at a prescribed time.

4 For each segment endpoint calculate values of the joint displacements from the geometry of the robot arm.
5 Issue these instructions to the joint controllers at prescribed intervals, regularly updating their inputs.

Alternatively, complete steps 1, 2 and 3 above and then:

4′ For each segment endpoint create a direction and speed of motion for the end effector.
5′ When the robot completes a segment use the joint displacements recorded by transducers to calculate corresponding joint velocities.

In practice we notice two things about the execution of this path:

(a) the path between segment endpoints approximates to the prescribed path,
(b) the robot may not reach any of the endpoints exactly before moving on to the next segment. Provided the segments are small and the controller is tuned correctly, a close approximation to the path will be obtained.

This low-level planning problem directly linking the planned path to machine motions may be regarded as a control or execution problem. However, if we link motion planning along a prescribed path with the planning of the path then a clear cognition or planning problem emerges. Three outline examples are given to illustrate this link and to show the role of motion planning.

1 If the path is planned in outline because it is subject to change during execution, there may be a requirement to refer back to the path planner.
2 The requested move is beyond the capability of the robot arm through limited joint motion or overall reach. In this case the planned path cannot be executed and reference back to the path planner is needed. If the path planner contains a model of the robot's workspace then these constraints can be introduced at the planning stage. In principle, it would therefore be possible to perform joint coordination at the planning stage.
3 The example in Section 6.3 of planning automated sheet bending provides another example. The bend sequence gives rise to paths for the component between bends. However, these paths must be generated in combination with knowledge about the motion capabilities of the robotic handling. Paths need testing and evaluation by the motion planner to avoid collisions between robot, sheet and press.

6.6.2 Robot path planning

Robot motion planning has concentrated on creating moves of the end effector along a pre-planned path. In its general form the problem of robot path planning is to move between specified positions without collisions of the robot (including all limbs) with other objects (including other robots in the workspace, machines, AVs or handling devices) in the workspace.

The problem is geometrically complex for a robot arm in three-dimensional space. In general, the planning problem is looking for a sequence of configurations of the robot arm that completes the task.

To simplify the problem let's consider the two-dimensional case with a geometrically simple robot. The task of planning the route of an AV across the factory floor is geometrically simple. The AV moves on a surface (possibly undulating but all points reachable). Other objects within the workspace are on the surface. The planning system is required to create paths between a series of locations (say, WC1, WC2, WC3, …) between workcentres. A simple path plan in the absence of obstacles may follow a straight-line path between workcentres (Figure 6.16) with the robot reorienting itself during the path or perhaps using a circular arc. It is useful to consider the move taking place in a state space described by two linear position variables and an orientation variable (in this case an angle). This is called the configuration space of the AV. Planning takes place in this three-dimensional configuration space.

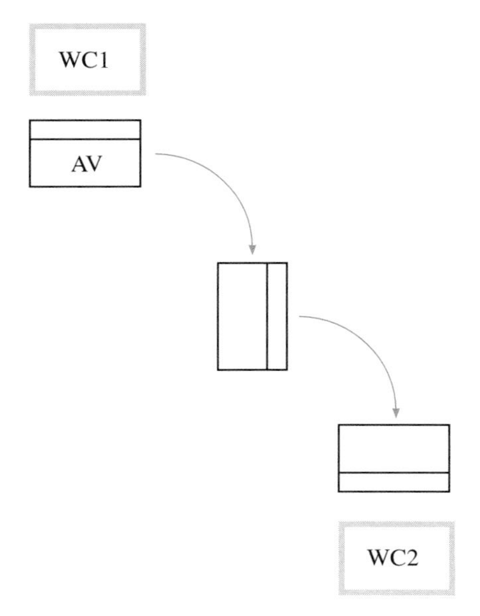

Figure 6.16 Moving between workcentres.

When there are obstacles in the path of the robot the configuration space becomes more complex. The planning task will need to modify the unobstructed path. In the task shown in Figure 6.17 the modified path through the gangway may be unsuitable, perhaps because it takes too long. Path planning needs to examine other paths: for example, the 'detour' path shown in Figure 6.17.

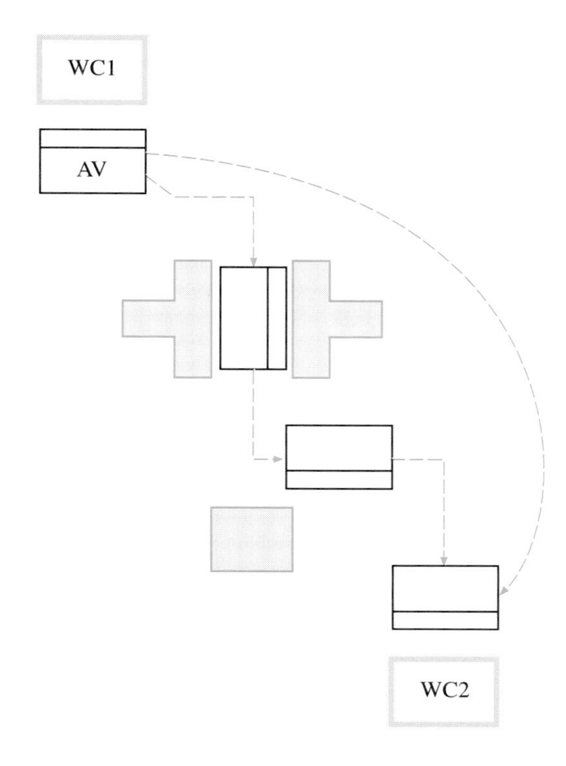

Figure 6.17
AV paths between workcentres.

I now consider how the problem of path planning through an obstructed workspace can be represented, and will indicate some corresponding solution methods.

One method commonly used is to consider a representative point on the AV, expand each obstacle (Figure 6.18a) by half the size of the AV, and find a path of the representative point through the expanded obstacles. However, we need to give closer attention to the orientation. If the object expansion corresponds to the maximum dimensions of the AV, there would be no path through the gangway (Figure 6.18b).

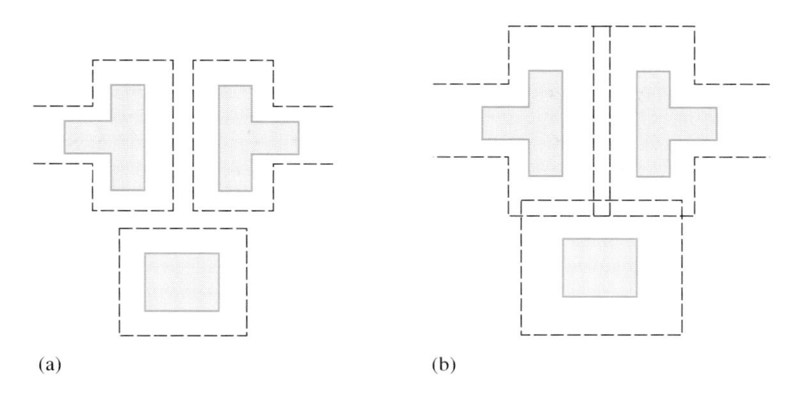

Figure 6.18
Expanded obstacles for task in Figure 6.17.

For this reason the path should be planned in the configuration space that models both position and orientation. If a slice (Figure 6.19) is taken through the 3-D configuration space representing a particular orientation of the AV, obstacles are expanded in that slice by an amount corresponding to that particular orientation. If there is free space in the configuration at some slice, the AV may be moved between the obstacles with that orientation. The general analysis of the configuration space is complex and we will not deal with it in detail here.

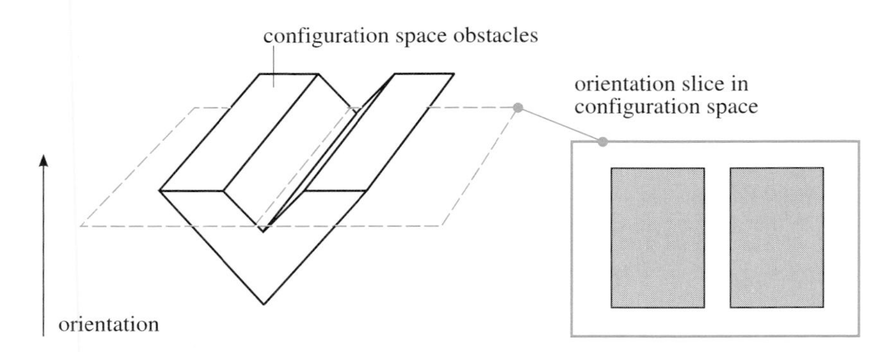

Figure 6.19 Slices of an AV configuration space.

The path is planned in the configuration space. In some cases we might be looking for the shortest distance whereas in others we might be looking for a minimum of combined distance plus direction changes. In order to plan collision-free paths we can focus either on the obstacles or on their complement – the free space. We will consider the obstacle space first.

The objects in the working space are expanded to take account of the dimensions of the AV. This is not a trivial exercise. For simplicity, suppose that for a particular orientation of the AV (i.e. a particular slice through the configuration space), expanded objects are considered. The problem is now to find a path in the slice of configuration space. Note that there are some cases in which the path may require reorientation as it moves through the configuration space and in these cases a single slice is inappropriate.

As an example, let's return to the problem earlier in the chapter of planning the path of a sheet through the bending press. The threading of the sheet through the tools of the bending press requires reorientation along the path and thus planning will involve the whole configuration space. In the obstacle space we take the expanded outlines of the obstacles and examine paths between their corners (Figure 6.20).

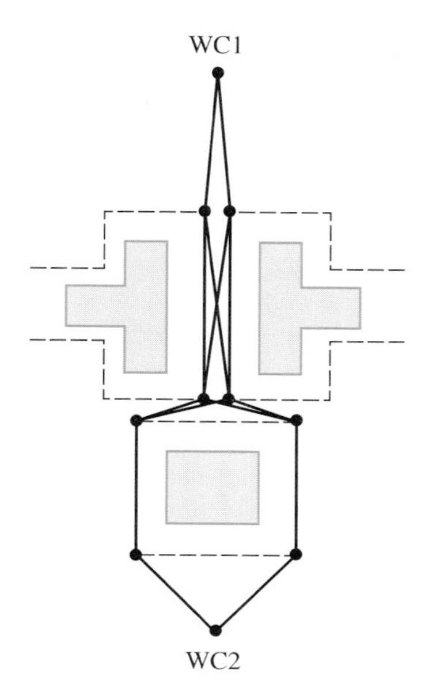

Figure 6.20
Possible path segments around expanded objects.

The first stage in path planning is to draw 'lines of sight' between the corners of the expanded obstacles. The network of these lines, including the obstacle edges, produces a line-of-sight graph. The application of an algorithm for the shortest path on this graph will result in the shortest path and the appropriate plan for the AV. Note that this simplified approach may not consider all paths and thus may not yield the shortest path. Neglecting this, we examine an algorithm for constructing the shortest path on a graph between specified nodes. The graph edges will have weights associated with physical distance and potential speed of the move related to orientation changes.

Consider a graph with a start vertex u_0 and a goal vertex v_0. Refer to the graph in Figure 6.21 corresponding to the path segments for the task similar to that in Figure 6.20. Weights are attached to the edges of the graph.

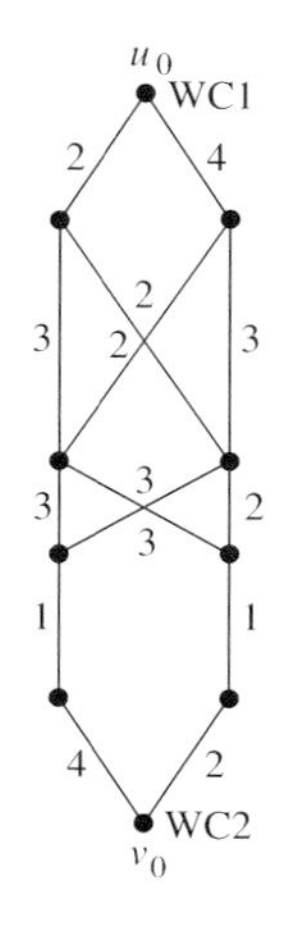

Figure 6.21
Graph of path segments.

Let weight (u,v) be the distance on the path segment from u to v. Let distance (u_0,v) be the shortest distance along a path from start vertex u_0 to a vertex v in the graph. Let S be the distance from u_0.

Step 1: Find vertex u_1 with shortest distance from u_0. That is, find u_1 such that weight (u_0,u_1) is minimum.

Step 2: Let $S_1 = (u_0,u_1)$ and P_1 be the path (u_0,u_1).

Step 3: Let $S_k = (u_0,u_1,u_2, \dots u_k)$ with shortest paths $P_1, P_2, \dots P_k$. (P_k is shortest path from u_0 to u_k).

Step 4: Find a vertex u_j in S_k and vertex u_{k+1} not in S_k such that (u_j,u_{k+1}) is an edge of the graph and distance (u_0,u_j) + weight (u_j,u_{k+1}) is minimum.

Step 5: Add u_{k+1} to S_k to form S_{k+1}.

Step 6: Add edge (u_j,u_{k+1}) to path P_j to form P_{k+1}.

Step 7: Repeat steps 4 to 6 until the goal vertex is reached.

The algorithm effectively searches for all shortest paths from a given node. It is essentially a kind of breadth-first search which eliminates certain paths that cannot lead to minimum weight paths.

As an example consider the graph of path segments in Figure 6.21. The history of the generation of the shortest path is shown in Figure 6.22.

Alternatively, the search for a path takes place in the free space. The spaces between obstacles are divided into polygons. This may be accomplished in many ways, including those which successively divide the whole space, including obstacles, into cells, searching for sequences of adjacent empty cells between initial and goal positions. Other methods divide the free space into largest convex polygons and examine the ways that the AV can pass through these regions and join paths in adjacent regions. Orientation is again critical, with allowable orientation attached to each region and reorientation required between adjacent regions. The business of path planning is again finding a minimum weight path between adjacent regions along the edges of a suitable graph of routes. The minimum path algorithms are used again.

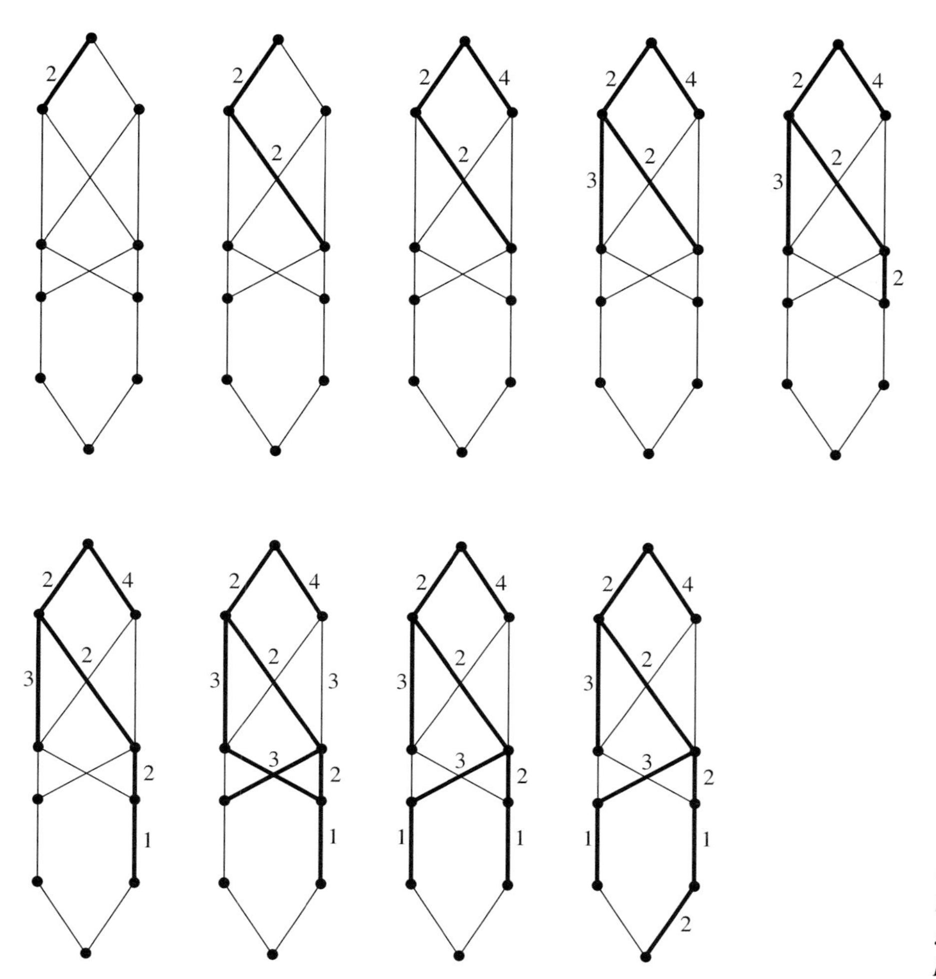

Figure 6.22
Successive steps in the generation of the shortest path.

6.6.3 Assembly planning

Planning for assembly requires the generation of an assembly sequence consistent with the constraints imposed by the components of the assembly, their relations in the final product, as well as the characteristics of the assembly machine. For example, the assembly may require to be turned over or otherwise reoriented during assembly to give access for successive components. This may be a consequence of the limitations in the motion dexterity of a robot machine or because of the fixtures which locate and retain subassemblies.

Constraints arising from the components and their relations include the case when a particular order of component assembly is physically impossible. A simple example is completing a motor assembly before fixing the top cover.

The first requirement in assembly planning is to represent knowledge about the assembly. This can be done as a set of connections or liaisons between the components which represent the basic spatial relations between components in the assembly. In Figure 6.23 the critical spatial relations are <F,E>, <B,C>, <A,C>, <A,E> and <E,D>. Object E is a cylindrical tube. D and F are endcaps. A, B and C fit inside the tube. These liaisons are represented as a graph or diagram. The problem is how to represent the possible sequences of liaisons.

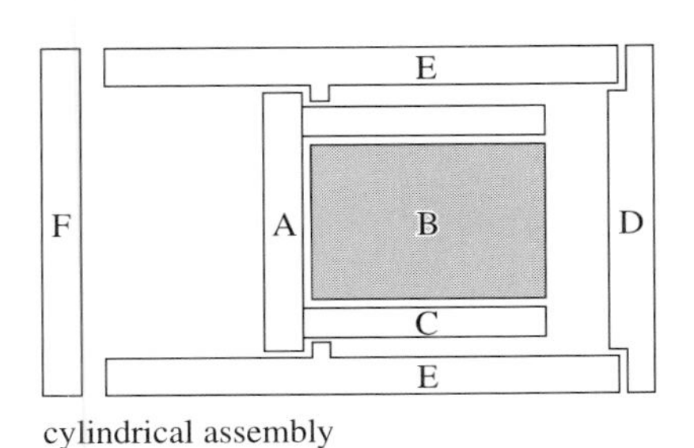

cylindrical assembly

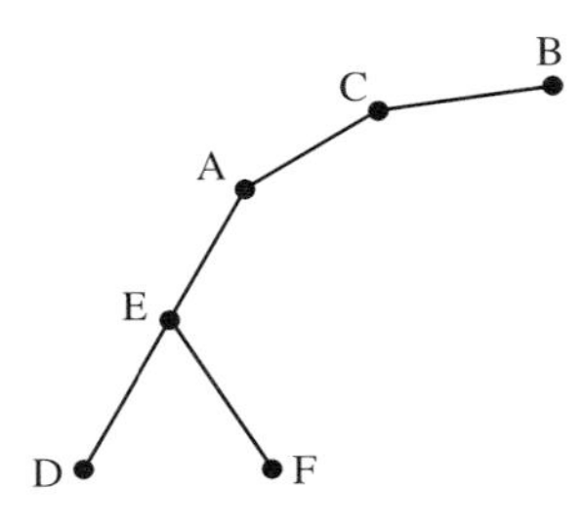

Figure 6.23 Schematic assembly and liaison diagram.

Some sub-sequences are not allowed because some liaisons must take place before others. In this case, referring to Figure 6.23, <A,E> occurs before <F,E>, and <A,C> before <B,C>. Other sub-sequences are not allowed because some liaisons must occur after others. In this case, <A,E> and <E,D> take place after <B,C>; <A,E> and <D,E> occur after <A,C>; <F,E> occurs after <A,E>.

To represent possible assembly states we use an ordered list specifying which of the liaisons have been created. Suppose liaisons are ordered: <A,E>, <E,D>, <A,C>, <B,C>, <F,E> and these are represented (Figure 6.24) as a row of five boxes. A filled box represents a completed liaison. The graph in Figure 6.24 shows all valid liaison sequences.

The search for assembly sequence plans is a matter of finding minimum paths. The weights associated with each transition are set by the nature of the assembly operation involved. Some partial assemblies are composed of two subassemblies. These are probably discarded in planning for simple assemblies, provided alternative sequences exist. In Figure 6.24 the state with liaisons (<E,D>, <A,C>) is an example. Removing the node reduces the graph and consequent search. Minimum path algorithms will search possible paths to determine a best assembly sequence.

Other methods of assembly sequence planning examine the possible ways of *disassembly*. This does not mean physically taking the assembly apart – just reversing the order of the assembly operations. For example, by reversing the assembly of a snap fastener, an extra operation such as holding a clip open can be eliminated. These methods can avoid search routes that do not lead to the final assembly. They look for coherent subassemblies which are decomposed into components or further subassemblies. This may be represented as a tree with the components as the leaf nodes. These are essentially the goal spaces considered earlier, and search in these goal spaces yields assembly plans.

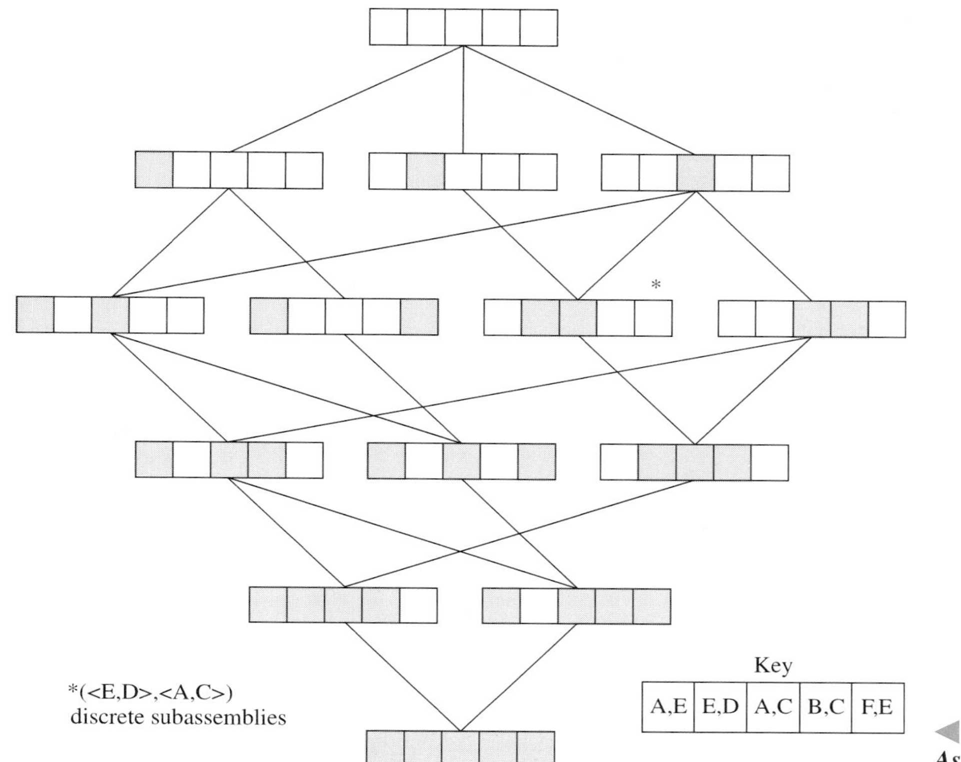

Figure 6.24
Assembly sequences.

6.7 Conclusion

Planning is necessary when a mechatronic system can make many possible state changes in the world but where only well-designed sequences of state changes will lead to specified goals. The ability to plan offers the mechatronic system usable flexibility.

Planning systems make choices among possibilities in order to achieve goals. They present the results of these choices as instructions to the execution system. Much planning involves searching for 'good' choices: that is, intelligent choices. Planning can thus allow a machine to exhibit intelligent behaviour.

CHAPTER 7
ACTUATORS

Anthony Lucas-Smith

7.1 Introduction

Actuators are devices which perform the final output stage of mechatronic behaviour. After the processes of perception and cognition have been successfully carried out, an actuator initiates an action, usually involving motion. This completes the cycle through which the external environment is sensed and understood, and the system pursues its specified goal, directing the actuator to make a change within the external environment. Just as sensors are *transducers* that change energy from an external source to another form, so actuators are transducers that follow an internal signal to carry out a transformation to produce external movement. In the process, an electrical, hydraulic, pneumatic (or other) power source produces kinetic energy, or movement. For example, the vision system of an autonomous vehicle inputs a signal which, after the process of cognition, results in execution – in which an electrical signal causes the motors of the vehicle to move in the appropriate direction.

The design of perception and cognition systems for mechatronics is strongly related to the needs of actuation, and the selection of sensors and actuators must be similarly interrelated.

In reading this chapter you should keep in mind:

- The role and performance requirements of actuators in achieving the goals of a mechatronic device.
- The links between perception, cognition and execution.
- The selection of appropriate actuator technologies to satisfy requirements, often within small tolerances and under severe constraints.

7.2 Actuator functions

7.2.1 The range of actuators

Almost all actuators produce some kind of movement. Movement is required for devices which travel and provide transportation, or have moving parts such as arms and grippers on robots. It may be necessary to provide forces to maintain stability by preventing movement, as for example in a support for building construction (though civil engineering has not yet widely adopted mechatronics).

An important class of actuators comprise *end effectors*, the manipulation mechanisms and tools attached to each 'limb' of a robot to provide its end function. These too may provide movement, as with grippers, or be associated with specialized tasks such as welding. The following list gives examples of familiar mechatronic applications and actuators.

(a) *Autonomous vehicles* These use electric motors for linear movement and steering mechanisms.

(b) *Remote guided vehicles* These are for use in dangerous or constricted areas (bomb and pipe examination). They are not autonomous but may have mechatronic elements.

(c) *Lifting and handling equipment* These often use hydraulic power to control heavy loads with precision.

(d) *Cameras and video recorders* These use miniature electric motors to move lenses small distances.

(e) *Automobiles* Many mechatronic operations such as steering and suspension control require actuators, e.g. electric motors and hydraulic pumps.

(f) *Animated models for entertainment and education* Humanoid and animal models use mainly pneumatic actuators to move limbs, eyes and other components to create realistic effects. Electric motors are likely to be used more in the future.

(g) *Manufacturing machinery* A wide range of electric, hydraulic and pneumatic actuators is used to move materials and parts for processing and assembling, often at high speed and with great precision. Many devices are referred to as robots, although there is no universal agreement on the definition of the term. A familiar example is the 'pick-and-place' machine which inserts electronic components into printed circuit boards.

(h) *Robotic joints* A wide range is available, including telescopic joints, geared and ungeared.

(i) *Robotic end effectors*, for example:

Grippers, which provide the capabilities of holding objects and rotating about any axis. They are typically used in holding tools for machining.

Specialized tools, which are used for a variety of tasks, and include welding torches, sanders, paint sprays, drills, laser and water jet cutters, and riveters.

Picking up and releasing effectors, which work by vacuum action.

7.3 Selection of actuators

An important part of mechatronic design is the selection of actuators to perform in a specified manner, and subject to certain constraints.

7.3.1 Performance requirements

Acceleration and deceleration

Movement from the stationary position must take place at an appropriate acceleration. A lift may be designed to travel at high speed in a very tall building. However, the acceleration and deceleration phases must not be too violent for human comfort.

Velocity

An appropriate velocity must be achievable. A robot with limbs which move at great speed might be too dangerous for use with operators nearby. However, a pick-and-place robot inserting components into a printed circuit board needs as high a velocity as can be achieved while maintaining accurate insertion, but must be able to stop quickly if a placement problem is sensed.

Responsiveness

Responsiveness of an actuator to the demands of a situation is partly a combination of its velocity and acceleration capabilities. It also depends on the effectiveness of feedback from the device's sensors and its perception/cognition/execution system. An autonomous vehicle which is needed to track a moving object to within three seconds of a predicted collision must have a response time sufficiently smaller, certainly less than a second, to take avoiding action.

Operational power

The power developed must be sufficient for the application. Transporters may need to move heavy loads. In contrast, a camera needs minimal power to focus a lens when moving it a fraction of a millimetre. Rotary motors have to develop sufficient torque during start-up as well as during high-speed operation.

Performance under varying loads

Power production may have to take into account variable loading. A lift is required to perform at consistent speeds and acceleration whether it has a light or heavy load.

Routeing achievability

Actuators have to produce movement along an appropriate route. This is partly a feature of the motor power but may also depend on the dimensions of the actuator and other parts of the whole device. It is technically easier for an end effector on a robot to move from one coordinate to another by moving specified distances along the *x*, *y* and *z* axes in turn. This would be unacceptably cumbersome for many applications. End effectors normally move in a straight line from one point to another, or along a curved route to avoid an object in the way. The precise routeing is the concern of the execution process and may be calculated at the time or specified in the form of stored data. It is often necessary to control a combination of actuators and determine their relative settings to achieve the required routeing.

Resolution, accuracy, repeatability

These are all aspects of the *precision* achievable by an actuator system.

Resolution is defined as the smallest interval of controlled motion that can be made. For example, a pick-and-place robot that is locating components in a printed circuit board with a separation of 2 mm minimum must have a resolution of 2 mm maximum (and preferably less).

This not to be confused with *accuracy* which determines how closely to the intended target an actuator is expected to locate. Accuracy depends on internal factors such as the performance of the system and the looseness of mechanical couplings, and on external factors such as skidding or slipping and the smoothness of the path followed.

Repeatability (or consistency) expresses the ability of a moving device to return repeatedly to the same point. Accuracy is endangered in a component if repeatability is inadequate. The rigidity of components may have an effect on repeatability, since bending under heavy loads can cause positional errors which are independent of the actuator's performance.

Compliance

This is an expression of the movement of a component in reaction to a force exerted on it. In colloquial terms it is the stiffness (low compliance) or sponginess/softness (high compliance) of the actuation. In many machining operations, such as drilling and milling (slot making), low compliance is essential otherwise the impact of side forces due to the positioning and composition of the material could push the machine cutting edge off course. In assembly processes where delicate objects are handled and in the movement of vehicles where safety is a consideration, relatively higher compliance is required. Human beings are skilled in varying compliance from movement to movement. Picking up and transporting a tool such as a chisel requires high compliance for safety reasons; using the chisel to shape hard wood requires a lower compliance, both of which muscle power automatically provides. Analogous changes in mechatronic actuation may have to be designed into the execution system. The designer should be aware of the compliance required and that the natural compliance of actuators often varies between the static and dynamic state. This variation is expressed, for example, in the specification of an electric motor for torque related to revolutions per minute.

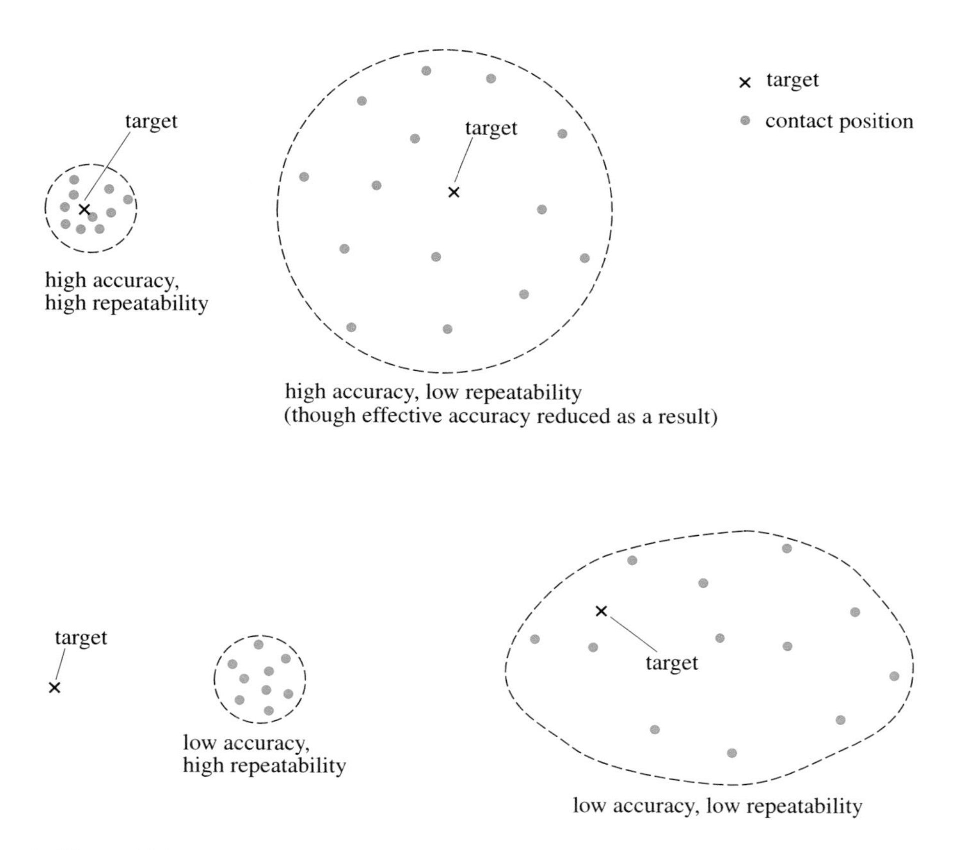

Figure 7.1
Illustrates the combined effect of accuracy and repeatability quality.

7.3.2 Selection constraints

In any application the designer will have to bear in mind a number of physical, financial, ergonomic, social and legal constraints. Some of these may be mandatory, such as safety regulations. Others may be the subject of trade-offs as, for example, in the choice of electric motors where small size and high accuracy may be strongly related to cost. The following is a list of typical constraints that affect the selection of an actuator.

weight and power-to-weight ratio
volume (bulk) and shape
cost
environmental requirements
air temperature (minimum to maximum range)
resistance to dust particles and other contaminants
noise emission
spark emission (in flammable surroundings)

vibration (emitted and received)

electrical supply (DC/AC, single phase/three phase)

electromagnetic compatibility (EMC) (this is an important issue following recent European Union rulings)

ergonomics

simplicity of installation, maintenance and repair

safety

reliability

power loss/efficiency

reversibility of motor.

7.4 Principal technologies

There are three primary sources of power for actuation: electric motors, pneumatics and hydraulics. Additional specialized technologies are discussed in Section 7.4.8.

7.4.1 Electric motors

There is a great range of motors available for performing many actuation tasks. The physical principle of all electric motors is that when an electric current is passed through a conductor (usually a coil of wire) placed within a magnetic field, a force is exerted on the wire causing it to move. Over the years many developments have resulted in performance improvements, while retaining the basic concepts of the traditional brush motor summarized in Figure 7.2.

The principal components are:

- North and south magnetic poles to provide a strong magnetic field (permanent or electromagnetic). Being made of bulky ferrous material they traditionally form the outer casing of the motor and collectively form the *stator*.
- An *armature*, which is a cylindrical ferrous core rotating within the stator. It carries a large number of windings made from one or more conductors.
- A *commutator*, which rotates with the armature and consists of copper contacts attached to the ends of the windings.
- *Brushes* in fixed positions and in contact with the rotating commutator contacts. They carry direct current to the coils, resulting in the required motion.

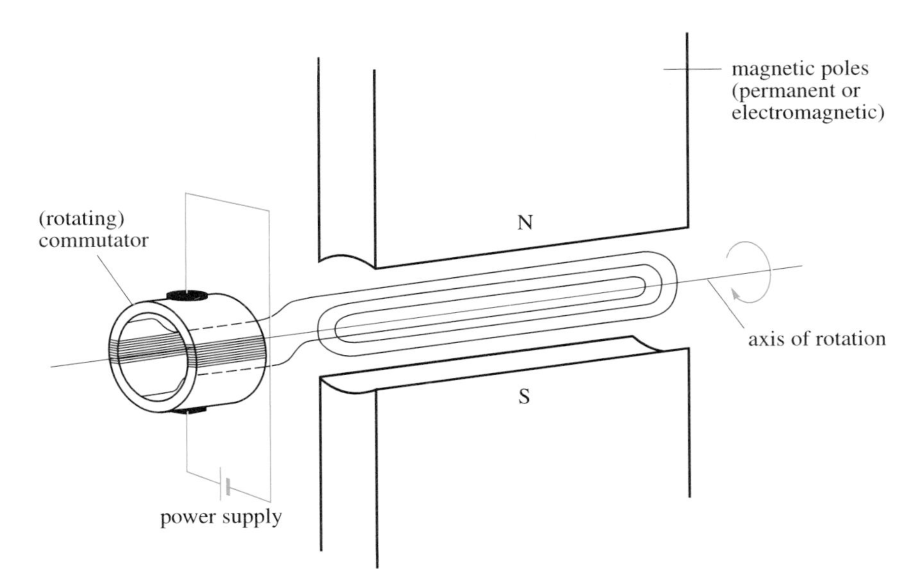

▲ ***Figure 7.2***
The basic structure of an electric motor.

I am going to consider four types of electric motor developments for possible mechatronic use, summarized in Figure 7.3 overleaf.

1. *AC synchronous motors* These are driven by a mains supply (50 Hz in the UK) at which they run at constant speed. The speed of the motor at constant mains frequency is determined by the design of the poles. In practice the mains supply varies around ±1–2% of 50 Hz in the UK but is 50 Hz when averaged over 24 hours (a legal requirement in the UK). Consequently, an AC motor does vary its speed slightly from time to time. A 2-pole design would typically operate at around 1500 rpm. Such motors have a limited role, being only suitable for continuous actuation at reasonably constant speed: for example in a refrigerator. Their use in a mechatronic system would be by means of rudimentary control, perhaps with external feedback to start and stop the motor.
2. *Variable speed motors* A simple variant of the synchronous motor includes circuitry for taking an AC input and outputting a variable frequency to drive the motor. This results in variable speeds and the capability of using feedback for closed-loop control. The addition of an encoder allows positional control by counting the number of rotations. However, it would be difficult to achieve precise control where fractions of revolutions are required. For this purpose a third type of motor is more suitable.
3. *Stepper (or stepping) motors* These possess the ability to move a specified number of revolutions or fraction of a revolution in order to achieve a fixed and consistent angular movement. By using a two-directional stepper motor to drive a continuous belt, we can achieve linear movement to any specified position, subject to the resolution of the motor. A familiar example is the print carriage on a typewriter.

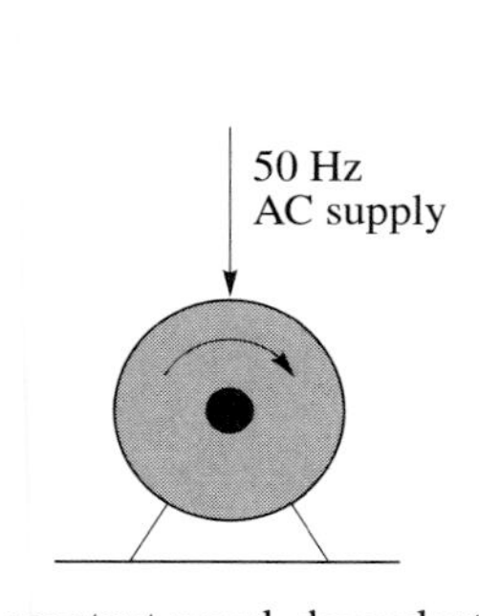

(a) AC synchronous motor

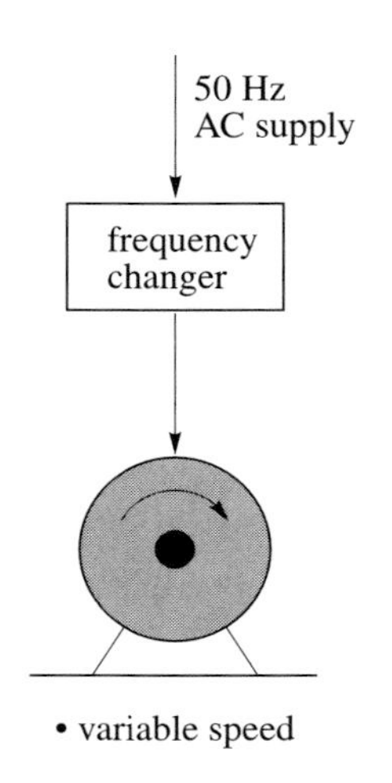

(b) Variable speed AC motor

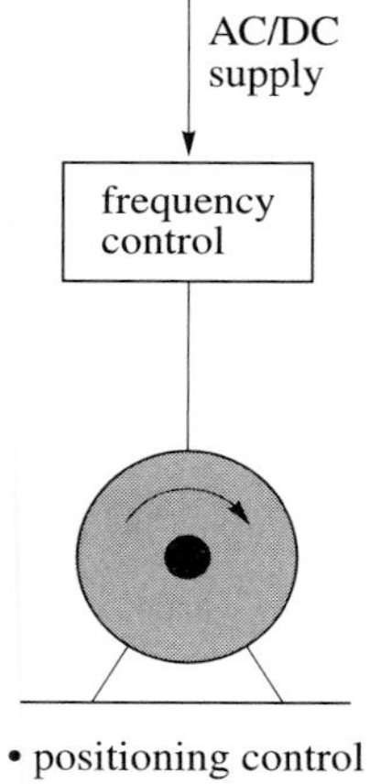

(c) AC/DC stepper motor

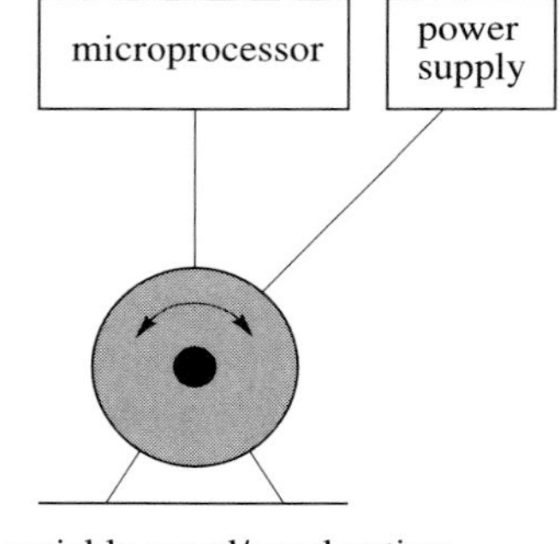

(d) Controlled drive motor

Figure 7.3
Development of electric motors.

The principle of stepper motor design is shown in Figure 7.4. It allows rotation of the central core by application of a direct current. In this version two pairs of poles can each change polarity by means of switching. The resulting four possible combinations of polarity give four possible rotor positions and step angles of 90°. Additionally, if switching allows unmagnetized poles, 45° positioning becomes possible. In practice, smaller resolution is needed, typically in the range 2–5°, which is achieved by the introduction of more poles.

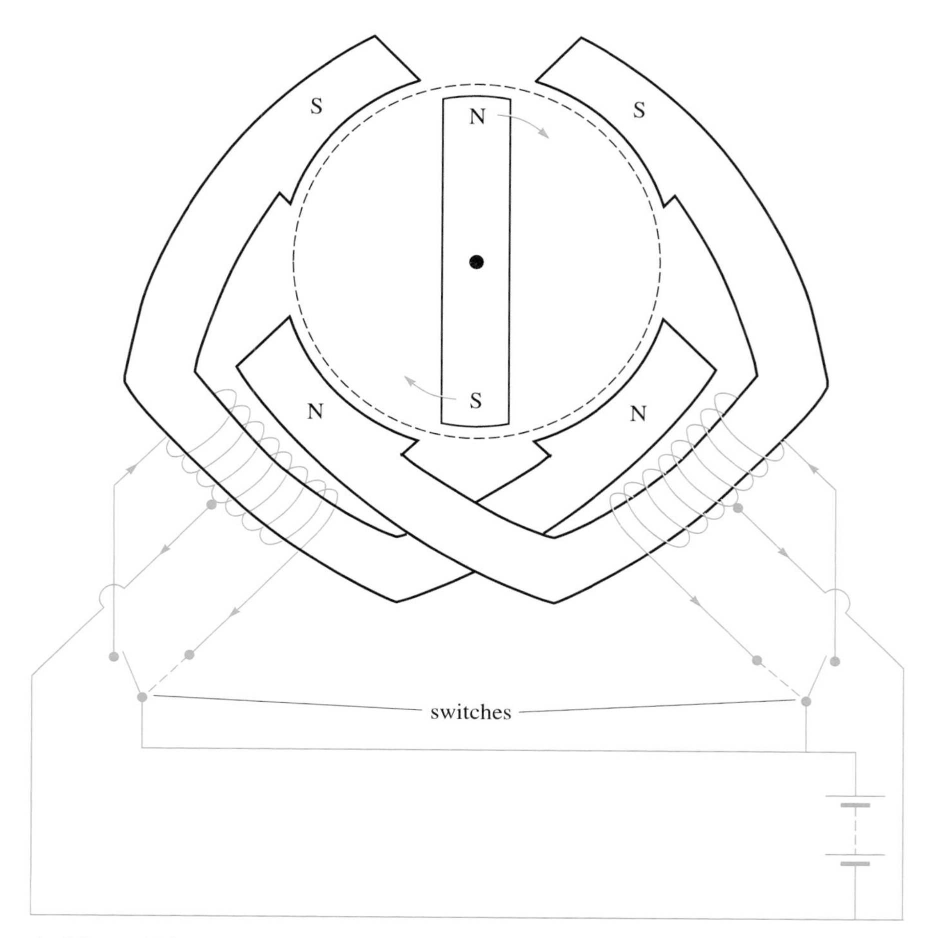

Figure 7.4
Simple form of a stepper motor with 90° turns.

Stepper motors have been developed extensively since the 1920s. They are often recognizable from their 'clicking' action: for example in clocks where the second hand clicks from position to position, rather than sweeping continuously. To some extent they have been superseded by controlled drive motors (described next), being less applicable to very high speed operation. However, in addition to relative cheapness, they do have the useful advantage that their degree of rotation is a function of their construction and is therefore consistent. No feedback is necessary for positional or speed control and any errors present are non-cumulative. Since they are activated by on–off switching, they are readily controlled by digital techniques. Stepper motors are brushless DC motors (with AC versions also available), so they avoid the maintenance and repair problems of brushes.

Figure 7.5 shows a more practical example of a stepper motor, arranged with four pairs of magnetic poles activated by a four-phase power supply. (For simplicity only two pairs of solenoids are drawn.) The slightly tighter spacing of the rotor teeth to those of the stator (typically in the ratio of 100

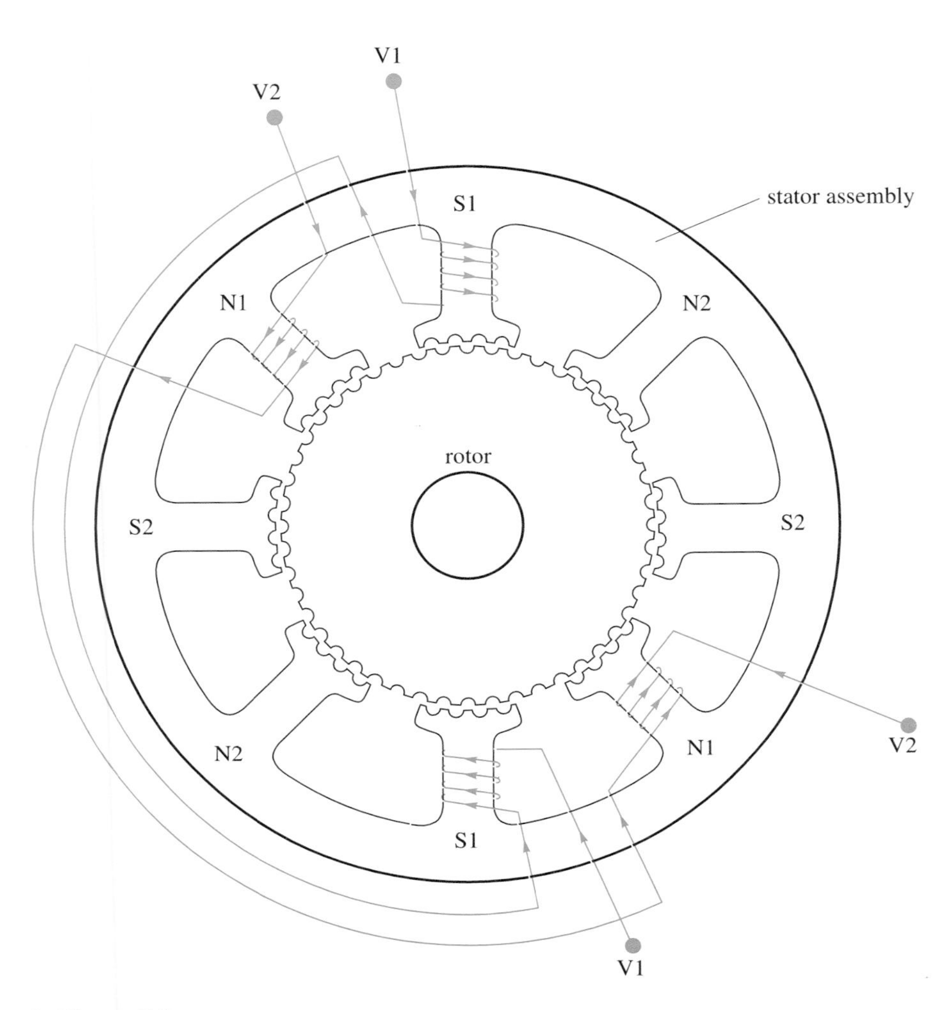

Figure 7.5
Stepper motor with 1.8 degrees per step.

teeth to 96) ensure that the two sets of teeth are close to each other but not quite aligned throughout. Movement is achieved when power is applied for short periods to successive pairs of electromagnetics in radial sequence $N_1 S_1 N_2 S_2$ (repeated). Where pairs of teeth are least offset, the electromagnetic pulse causes alignment and a small rotation is achieved, typically 1–2°.

4 *Controlled drive motors* These are a form of brushless DC motor, developed considerably since the early 1980s. They are a major enabling technology in the development of mechatronic products, allowing great precision and flexibility of actuation. They can:

provide large torques over a range of operating speeds;

reach high speeds almost instantaneously;

follow an intricate, predetermined sequence of movements, with speeds, accelerations and timings all precisely specified;

be programmed and reprogrammed using a microprocessor, resulting in great flexibility;

work in forward and reverse directions;

operate without producing electrical noise;

reduce the need for mechanical gearing and control mechanisms.

Such motors are more expensive than traditional types but make it possible to match the intelligent abilities of mechatronic devices with the need for more complex variable movements.

The technology that makes this possible depends on high-flux magnets resulting from the development of 'rare earth' materials such as samarium cobalt. Such materials allow the dimensions of magnets in motors to be considerably reduced while still increasing their contribution to torque. Instead of having a bulky, fixed magnetic stator, smaller motors become possible in which the stator rotates *within* the armature windings – a kind of inside-out design. Because the armature coils no longer revolve, they can be connected to the power source without the need for a commutator. No brushes are required as a result and the problems of sparking and contact degeneration disappear, eliminating electrical noise. Furthermore, reversing of current reverses the motor's direction.

The use of a microprocessor or PC connected to the motor gives more flexible control. The traditional equivalent has been a combination of circuitry and mechanical gearing. Gearing (and other mechanical components) can be dismantled and reassembled to change the manner of operation. However, this process of setting-up can be very time-consuming – hours or days with some machinery. An important advantage of the reduction of non-productive infrastructure machinery is that the remaining mechanical components can be less bulky. Much of the weight and solidity of traditional machinery is a design feature that is not so much necessary for carrying out its primary function but more for the machine to withstand the weight and momentum of its infrastructure, i.e. itself. A similar example is the design of a bridge where the limitations of the materials used result in most of its bulk supporting itself rather than the traffic it carries.

It is currently (1994) anticipated that there will be an increase in the use of controlled drive motors, particularly in manufacturing machinery. Many manufacturing processes have reached a plateau in their performance. Using traditional materials and components they cannot be redesigned to run much faster. Mechatronic intelligence combined with lighter, faster and more powerful actuators can, however, result in faster and more flexible production. The reduction in mechanical bulk will make the machinery itself cheaper and easier to manufacture. Furthermore, the programmability of actuation will avoid the lengthy set-up times that can add much to manufacturing costs. It is in such developments that dramatic improvements will be seen over the next few years.

Selection of controlled drive motors

The assistance of Keith Thornhill, Sales Engineer of Electro-Craft Ltd, suppliers of Electric Servo Motor Systems, is acknowledged in this and the following sections on the selection and application of controlled drive (brushless) motors.

The precise selection of motors for specific applications can be complex. Only the principles are outlined here.

- *Inertia matching* Both the motor and the actuator possess inertia, a function of each component's mass and its distance from the axis of rotation. Inertia is independent of the speed of rotation. Heavy rotating objects with large rotational diameters have high moments of inertia. These can be hand calculated for simple elements, and software can be used for more complex configurations. It is an important requirement that the inertia of the motor be similar to that of the actuated system. If not, the motor will perform erratically – for example, in not having fast enough response (overdamped), or showing unstable behaviour such as oscillation.
- *Specification of the performance profile* The actuation to be carried out must be specified in terms of acceleration, deceleration, speed (forwards, backwards or zero) and the periods for which they must be applied. In manufacturing machinery this might be accurately specified in advance in the form of a short cycle lasting less than a second and repeated continuously.

  An example is shown in Figure 7.6. In more autonomous machines the actuation might be specified as the output from an intelligent controller and vary from moment to moment. The performance profile shown in the figure implies that changes are instantaneous, which is strictly speaking impossible. However, the times involved are very small. For example, an acceleration of 0 to 3000 rpm can be achieved in 2.5 ms, an enormous improvement on traditional induction motors.
- *Torque profile requirements* Figure 7.7 shows a manufacturer's recommended torque–speed curve which applies to a specific motor. It will vary for other motors. For continuous running at various speeds the stall torque value must not be exceeded. This is the maximum torque value at which overheating wiil not occur. At greater values, up to the maximum indicated, overheating will not occur provided the motor is only run intermittently. As higher speeds are reached, the ability of the motor to deliver sufficient torque tails off and the designer must select a more powerful motor.

Control of motor

Figure 7.8 shows the components of a controlled-drive motor system. The amplifier unit provides power to drive the motor. A feedback loop from the motor is used to regulate the rotational speed, while an incremental encoder (see Volume 1, Section 2.4.5) is used to measure the speed. The terminal provides the user interface so that users can input control commands in the form of a simple parametric language via a keyboard. As a result the position controller outputs an analogue signal which defines the motion profile. Input could be from a standard

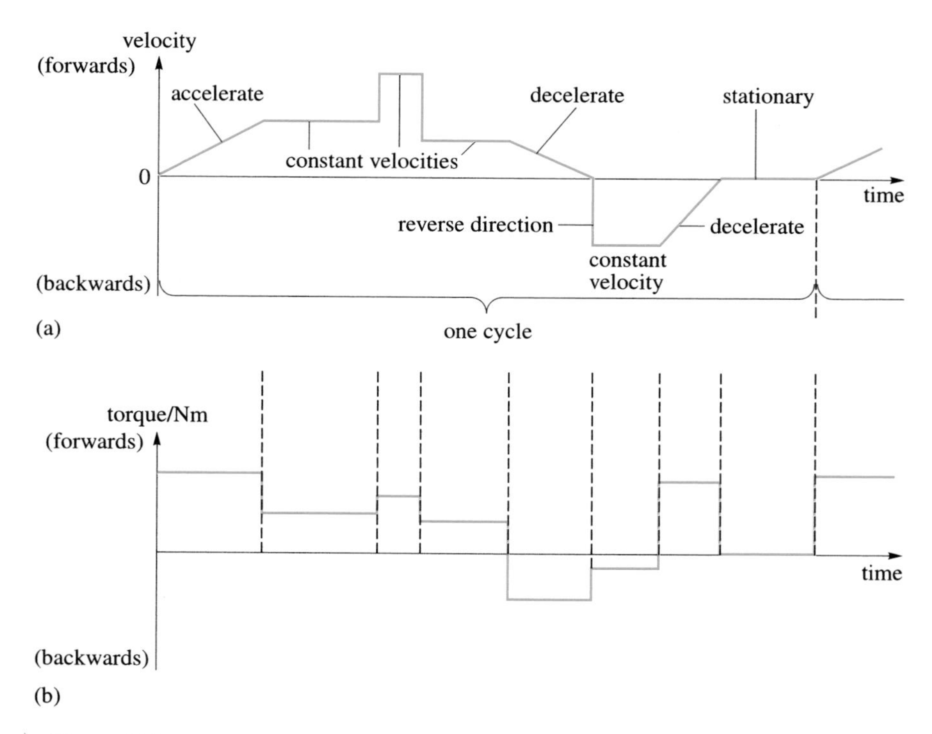

Figure 7.6
Performance profiles: (a) motion profile; (b) torque profile.

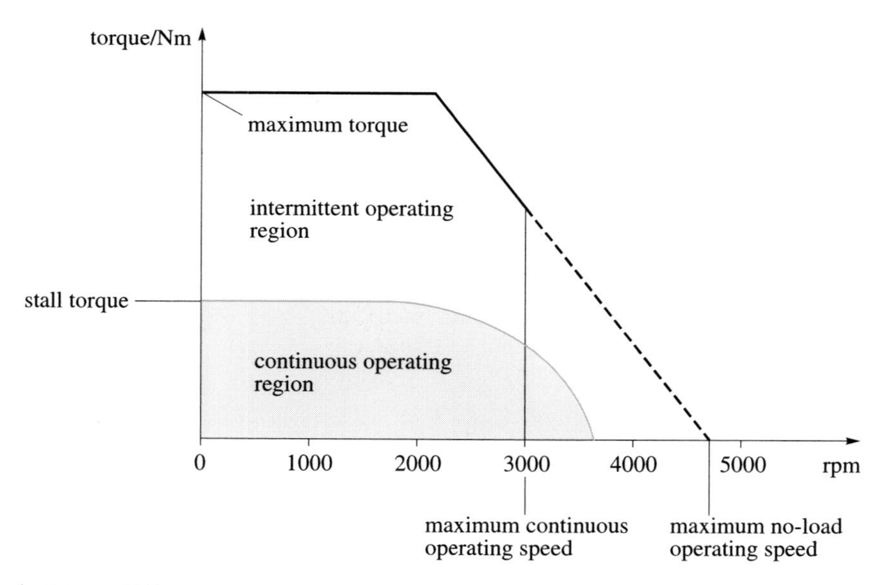

Figure 7.7
Operating curves (torque/speed relationship).

PC, though a much cheaper and less bulky method, particularly where a number of motors is used, is via a circuit board added to a standard industrial *programmable logic controller* (PLC).

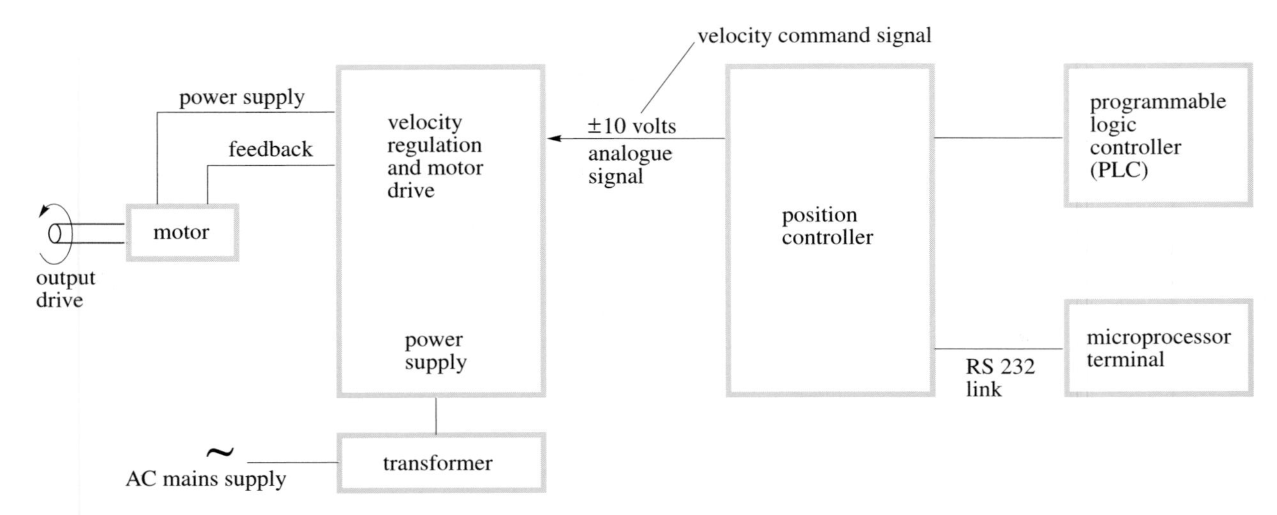

Figure 7.8
Operation of controlled drive brushless motors.

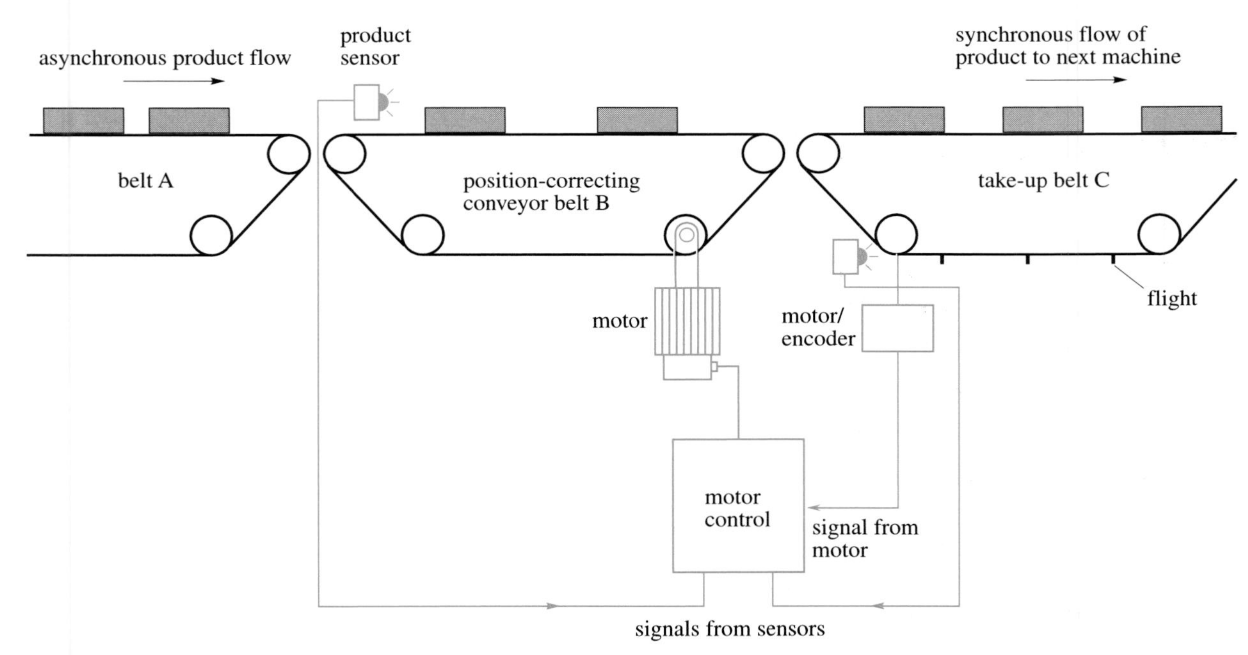

Figure 7.9
Random timing infeeds.

Controlled drive motors will find more and more applications where high speed, flexibility and precision are required. A typical use will be in autonomous vehicles. They are already widespread in manufacturing. One example is their use in random timing infeeds (Figure 7.9), as described below.

A conveyor belt **A** delivers products to a processing machine. The problem is that products arrive intermittently whereas the processing machine requires the products to be accurately positioned between markers (flights) on the take-up belt **C**. One solution is to use an intermediate position-correcting belt **B** driven by a servo-controlled motor, speeding up or slowing down the product in order to place it correctly for processing.

The speed of belts **A** and **C** are known to the controller from a drive motor encoder. The positions of the flights and of the arriving products are known from optical sensors. When a product is sensed, the positioning drive module calculates the time available to move it to the next available slot position. The corrected speed of belt **B** is calculated from this time value. Belt **B** is then accelerated or decelerated to accomplish the task. The system is flexible and can adjust to different products and feed speeds.

7.4.2 Actuators for the prosthetic hand

This section refers to Volume 1, Section 2.7.

The constraints on the actuators used to move the digits of the hand are quite strict, and aimed at producing movement as near as possible to realistic functioning. Movement has to be at a similar speed and acceleration and develop appropriate power otherwise the prosthetic hand might be dangerous or ineffective. Imagine a hand that had too powerful (or too feeble) a grip, that snapped open and shut at high speed (or reacted at very low speed). The designers considered that modern, light, efficient electric motors provided a better solution than pneumatic or hydraulic technology. However, in selecting a suitable motor they had to balance a number of parameters, namely:

- power output of the battery;
- volume and weight of the combined battery supply and motor;
- longevity of the battery;
- cost of the combination.

Trade-offs were necessary as these are all strongly interrelated. An appropriately powerful battery needing infrequent replacement or charging might be too bulky and expensive. Brushless commutation motors with samarium cobalt magnets would in themselves be lighter but would be more expensive and would require shaft encoders for positional control. Precision DC permanent magnet motors were finally selected.

Related to this case study, Hugh Steeper Ltd has developed very small, high-torque motors for use in commercially available children's prosthetic limbs (*Drive and Controls*, 1992). The 6-volt DC motors use rare earth samarium cobalt magnets.

7.4.3 Development of micromotors

With the development of very small mechatronic devices comes the need for smaller components. Traditional motors have provided much of the bulk of electromechanical devices and in this form could not be miniaturized sufficiently. Recent developments have resulted in micromotors with a cross-section of less than 5 mm (*Design Engineering*, 1992):

(a) Mitsubishi has developed a stepper motor 4.2 mm in diameter. It has potential applications in precision medical instruments, cameras and office equipment.

(b) Toshiba has developed a motor weighing 4 mg and measuring 0.8 mm in diameter, with operating speeds of 60 to 10 000 rpm.

7.4.4 Fluid power

In this section details have been supplied by Rexroth Pneumatics Ltd, suppliers of pneumatic and hydraulic systems. The help of Nigel Monk is gratefully acknowledged.

Fluids are materials that can be transported along pipes and, in the context of actuators, can be used to provide pressure and cause movement. Oil under pressure is used as a fluid to apply large forces at different locations by means of *hydraulic* technology. Similarly, compressed gases are used to apply smaller forces by *pneumatic* technology. Use of microprocessors and electromagnetically controlled valves means that movement can be applied with considerable precision and flexibility. In comparison with electric motors, fluid power can be directed from a single source to a number of positions by means of pipes and then relocated as required. End effectors can be attached and replaced according to needs. Electric motors are usually limited to single applications, although gearing and other mechanical linkages can provide some flexibility. In many situations they are likely to be more expensive than fluid actuation but may be essential to produce very high speed movement.

Hydraulics and pneumatics are similar in that they both produce linear motion using fluid pressure. Oil is negligibly compressible so that hydraulics can be used in situations where a total absence of movement is required or movement resisted under variable loads. In hazardous environments where electrical activity might produce sparking and combustion, hydraulics or pneumatics are often preferred to electric motors (though motors may be used remotely to compress gas or provide hydraulic pressure). If there is any possibility of an oil leak and combustion, pneumatics is the preferred option.

7.4.5 Pneumatic power

Pneumatics is widely used in manufacturing industry for powering end effectors on robots and jigs and other holding and manipulating devices. It provides a reliable and well-established technology which is simple to install and maintain. Based on the use of compressed gas to act as a source of stored energy, it is more efficient in energy storage than electrical batteries, weight for weight. For example:

> compressed butane in a cylinder, 4000–5000 $\mathrm{Wh\,kg^{-1}}$
>
> compared with:
>
> lead-acid battery, typically 500 $\mathrm{Wh\,kg^{-1}}$

In most situations a compressor is used to provide continuous pressure, which is then 'bled' away when changes in direction require a release of pressure. Gas in cylinders is alternatively used when a complete absence of motors is essential. Fast movement up to around $4\,\mathrm{m\,s^{-1}}$ is easily achieved, stopping at a specified position to within 0.1 mm.

Pneumatic systems can be assembled from a range of standard components. Linear movement is achieved by the piston arrangement shown in Figure 7.10(a), with rod-less components becoming more common. Partial-rotary movement is similarly achieved using a hinged piston (Figure 7.10b).

Examples of pneumatic actuators include:

(a) Manufacturing systems such as pick-and-place robots. Figure 7.11 shows the typical layout of a machine for putting components into a printed circuit board. Two complete assemblies, working independently, allow two circuit boards to be worked on simultaneously from a single controller and pneumatic power supply. Any point on the horizontal table can be reached by the y- and z-axis linear actuators. The x-axis actuator provides movement to pick up and then insert a component into the circuit board.

(b) Clamping system supplied by ITW Woodworth (*Design Engineering*, 1992). It consists of a chuck with an internal bladder that is inflated by pneumatic pressure.

(c) Pneumatic grips (Tillett, 1993).

(d) Animated models (humans, dinosaurs, extraterrestrials) are an increasingly familiar use as a technology known as 'animatronics'. The small dimensions and cheapness of the end effectors allow them to be used to simulate muscle functions and add to the realism of the effect and its value for educational and entertainment purposes. Microprocessor control makes it relatively easy to simulate the complex and variable movements required.

(e) Pneumatics is much used in food processing industries where relatively light materials must be handled at speed but 'compliantly'. Mushrooms can be picked; freshly pressed and moulded cheeses can be manipulated and date stamped; even gateaux can be cut and packaged.

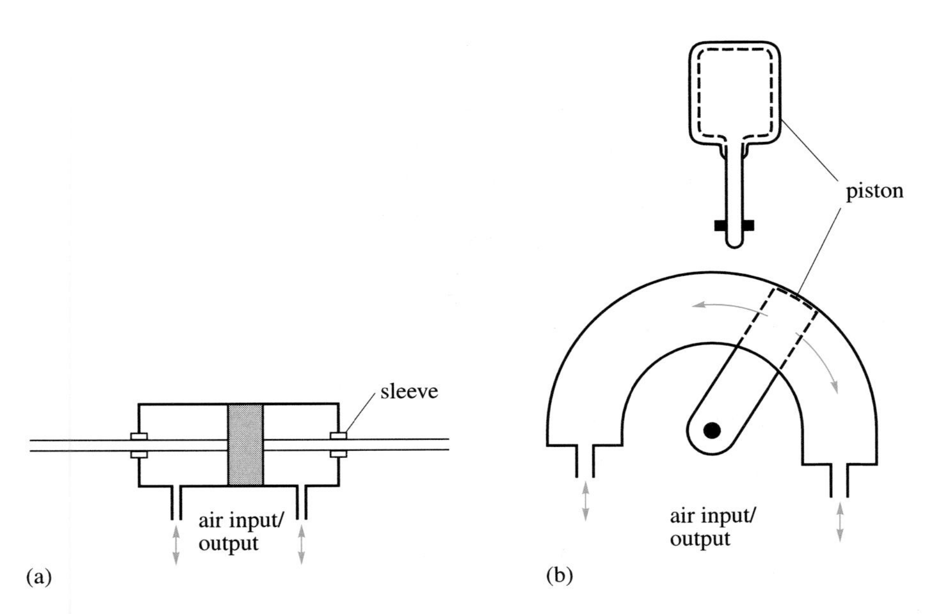

Figure 7.10
Pneumatic elements: (a) linear; (b) rotational.

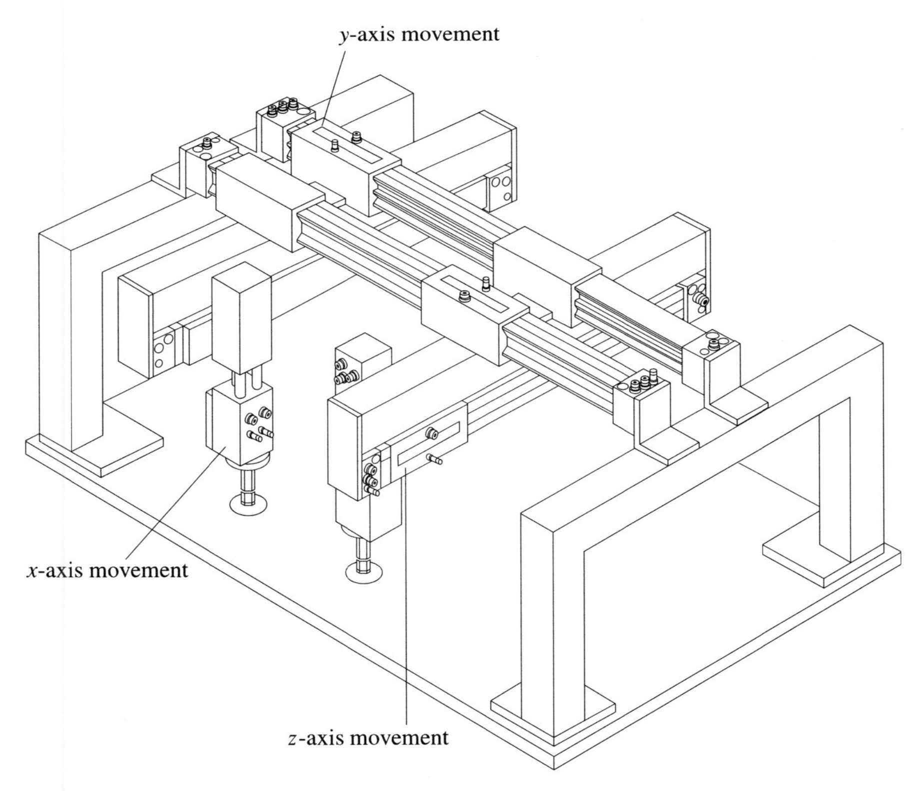

Figure 7.11
Pick-and-place pneumatic robot.

7.4.6 Hydraulic power

Hydraulics is associated by most people with high forces applied relatively slowly. The construction industry has many examples of human-operated machinery to push, dig, lift, support and manœuvre heavy weights with great precision. The basic principles are illustrated in Figure 7.12. Electromagnetically controlled valves are used to vary the flow of oil under pressure, thereby causing the extension or contraction of telescopic joints.

Examples of sophisticated hydraulics machinery abound, but are mostly operator controlled. As working in hazardous situations becomes less acceptable for social and legal reasons, machines will become autonomous and hydraulics more common in mechatronic applications. The following selection of current applications indicates areas where this likely.

(a) *Continuous hard rock mining machines* Heavy machinery (280 tonnes) required to cut 200 tonnes of ore per hour. Hydraulics is used to enable cutting, gripping, steering and hitching/anchoring.

(b) *Mining manipulators* Underground transporters to relieve miners of heavy work. Hydraulics is used to manœuvre bulky objects for assembly in cramped conditions.

(c) *Welding manipulator* Hoisting, rotating and positioning of heavy work-pieces. Hydraulics allows linear and curved welding paths to be followed with precision.

(d) *Simulators for motion systems* Familiar uses include aircraft, vehicle and sailing simulation for safety testing and operator training. Less familiar applications are wave generators for hydrodynamic model testing, rocking compensators for the dredging and offshore industries, and test benches for investigating the dynamic behaviour of large constructions.

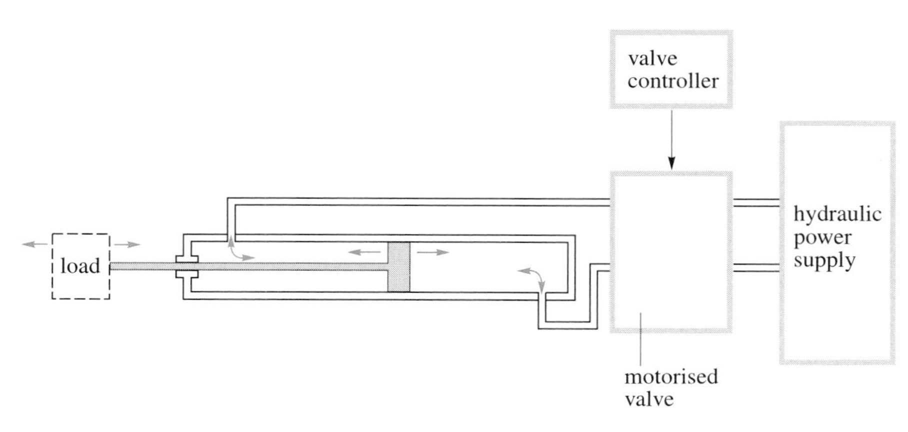

▲ ***Figure 7.12***
Schematic view of a hydraulic actuator.

(e) *Timber handling* Because of the sensitive environmental and political issues involved in forestry, machinery has been developed within severe constraints such as low soil compaction, use of biodegradable fluids, protection of unfelled timber, and low exhaust emission. Hydraulic machinery must satisfy these constraints and yet be economically competitive in the felling, trimming and cutting of a wide range of timbers.

(f) *Telescopic lifters* These include aerial ladders for firefighting and rescue purposes. Computer control ensures that sensitive motion positioning is achievable in hazardous conditions, such as high winds, while working within load tolerances.

(g) *Bridges and dams for flow control* From the Thames Barrier to Columbia's Guavio power station and beyond, gates weighing many tonnes and under great pressure must be opened to regulate water flow.

7.4.7 Integration of actuators with sensors

Since the late 1960s, sensors and actuators have been connected (for example, within a manufacturing system) by programmable logic controllers (PLCs) so that they operate and interact in a synchronized manner. The move towards international communications standards, developed throughout the 1980s as *open systems interconnection* (OSI) resulted from the need to be able to connect a range of devices in a flexible and manufacturer-independent manner. Among a number of implementations is the Fieldbus network protocol known as *Profibus* (PROcess FIeldBUS). Since its introduction in 1990 it has become widely used and at the time of writing (1994) is still developing. A number of sensors and actuators can be connected along a single 'bus' in the form of a twisted pair of wires, or to multiple buses interconnected by controllers such as PLCs (Figure 7.13).

7.4.8 Emerging technologies

Piezoelectric actuation

Piezoelectric crystals respond to an applied voltage by expanding. Thornley *et al.* (1993) report on multilayer devices measuring 18 mm long by 5 mm square, expanding by around 0.015 mm, exerting a force of 800 N, and responding within 0.001 ms. The small movement produced is insufficient for most applications but can be amplified by such methods as hydraulics, levers and impulse transfer. Applications under consideration include ink jet printer heads and fast selection and movement of threads in textile manufacture. In both applications there is a need for high-speed precise movement.

It is also likely that piezoelectric devices will find uses within mechatronic devices other than as the primary source of movement. Thornley *et al.* (1993) describe a high-speed clutching mechanism which uses piezoelectric actuation.

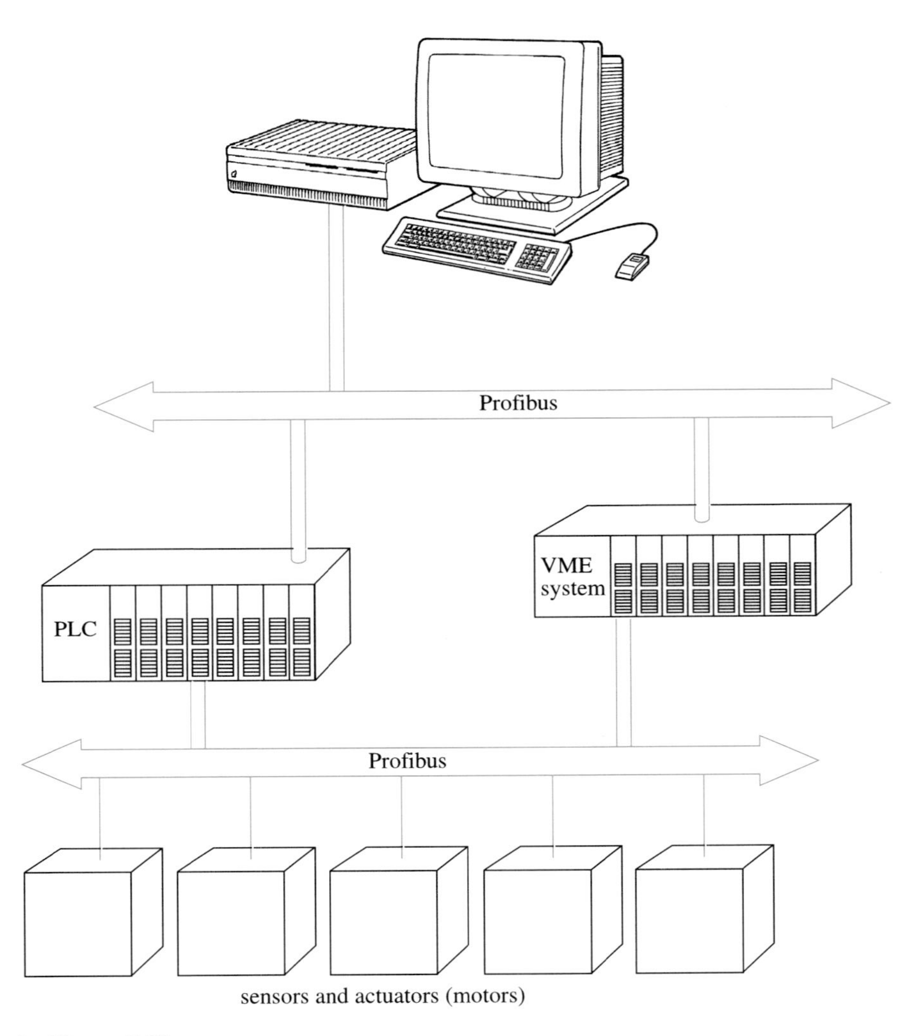

Figure 7.13
Profibus connections. VME, Versa Module Europe.

Magnetostrictive actuators

In many ways these resemble piezoelectric actuators, as both involve dimensional changes and are likely to have similar applications. In magnetostriction a current-controlled magnetic field is applied to rare earth iron-based alloys. Large forces can be developed and positioning obtained to a high degree of accuracy.

Muscle-like materials

Experiments with natural and artificial muscle elements indicate that more human-like actuation is achievable compared with traditional methods. The direct conversion of chemical energy into mechanical energy has been carried out using strips of artificial polymer 'muscle' acted upon by acetone (Caldwell,

1993). Using a test rig, Caldwell showed that the process was able to drive a robot gripper effectively, though with a slower response rate than human muscle.

Electrochemical actuators

These are based on electrically excited chemical reactions which transform chemical into kinetic energy, usually within a gas. Applications are limited by the slowness of the reaction and the low power developed. However, the technology uses low voltages and is free of mechanical wear, so it might be of use in medical applications.

Electroviscous fluids

As with piezoelectric actuators, the technology of electroviscous (or electrorheological) liquids is unlikely to provide primary movements in mechatronics. These fluids are oils with suspended micrometre-sized particles and have high dielectric constants. Their drag characteristics (i.e. viscosity) change when an electric potential is applied, making them suitable for use in clutch mechanisms. Monkman (1993) describes a possible application. The significance of the effect is that it may enable the development of flexibly compliant movement.

7.5 Conclusion

Actuation can be the weak link in a mechatronic system through lack of precision. However well the processes of sensing, cognition and control are carried out, successful operation depends on the precision and effective operation of actuators. Important choices have to be made in selecting from the electrical, pneumatic, hydraulic and other technologies available. High forces and torques previously satisfied only by hydraulics can be obtained from the latest motors. The speed and precision required in many picking and placing operations can be satisfied by electric motors or pneumatic actuators.

References

Caldwell, D. G. (1993) 'Natural and artificial muscle elements as robot actuators', *Mechatronics*, Vol. 3 (June), No. 3, pp. 269–283.

Design Engineering (1992) 'A big step for micro motors', *Design Engineering*, November 1992, p. 12.

Drive and Controls (1992) 'Child's artificial hand uses miniature mosfets', *Drive and Controls*, June 1992, p. 40.

Monkman, G. J. (1993) 'Dielectriophoretic enhancement of electrorheological robotic actuators', *Mechatronics*, Vol. 3 (June), No. 3, pp. 305–313.

Prior, S. D. *et al.* (1993) 'Actuators for rehabilitation robots', *Mechatronics*, Vol. 3 (June), No. 3, pp. 285–294.

Thornley, J. K. (1993) 'A very high-speed piezoelectrically actuated clutching device', *Mechatronics*, Vol. 3 (June), No. 3, pp. 295–304.

Tillett, N. D. (1993) 'Flexible pneumatic actuators for horticultural robots – a feasibility study', *Mechatronics*, Vol. 3 (June), No. 3, pp. 315–328.

CHAPTER 8
CONTROL

Chris Bissell

8.1 Introduction

In mechatronics applications, it is often necessary to control the dynamic behaviour of a system so as to comply with a set of specifications. Some of the most common examples are the control of the speed of rotation of a motor driving a load, or the position of some mechanical part such as a camera lens, a robot arm gripper, or a machine tool. Often the controlled variable – speed or position in the preceding examples – has to remain within specification even when the system is subject to external disturbances or perturbations. A robot arm may have to follow precisely the same trajectory even when the load it carries, and hence its dynamic behaviour, changes considerably. A lift has to come to rest exactly at floor level irrespective of whether it is empty or has its full complement of passengers. And hand-held video or cinema cameras for professional use have to point and focus correctly even when being moved around in quite a violent manner.

This chapter will look at some of the strategies used by mechatronics designers to ensure that such performance specifications are met. The topic of control is an enormous one, and an important engineering discipline in its own right. Automatic control systems can be found in the home (in compact disc players, central heating systems, washing machines, and elsewhere); in vehicles (aircraft flight control, train safety systems, and automobile steering and braking, for example); in industry (chemical plant, generating stations, gas and water supply systems) – indeed, in almost every area of modern industrialized society. There is certainly not the space here to deal with the subject in depth, so if you are interested in learning more about control engineering you are referred to the books listed at the end of the chapter.

What you *should* get out of this chapter is a feel for why particular control strategies are adopted in particular circumstances. You will *not* learn to design control systems yourself, but you should be able to discuss the control of a simple mechatronic system with a specialist engineer. Every so often in what follows I shall highlight an important general principle of control. Like most general principles, though, there will be occasions when they do not hold. So please treat my general remarks with caution!

8.2 Control strategies

As a first example, let's look at some ways in which the speed of an electric motor might be controlled. All the techniques considered in this chapter are based upon having available a mathematical model of the device or system being controlled – often called the ***process*** or the ***plant*** by control engineers. A very simple model of a motor is represented by the block diagram of Figure 8.1. Here it is assumed that the angular speed of the motor Ω is directly proportional to the applied electrical voltage, v, such that we can write

$$\frac{\Omega}{v} = G$$

where G, the motor *gain*, has the units rad s^{-1} V^{-1}. So, for example, if an applied voltage of 5 V results in an angular speed of 10 rad s^{-1}, $G = 10/5 = 2$ rad s^{-1} V^{-1}.

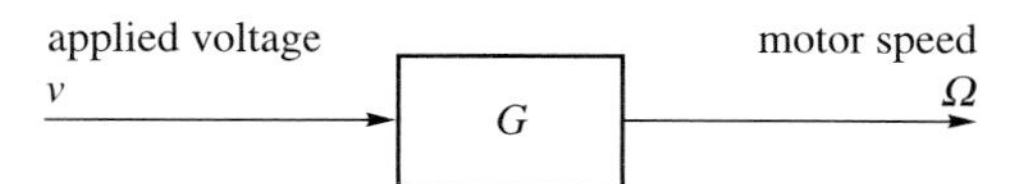

Figure 8.1 Model of a motor as a constant gain, G.

Clearly this is a greatly oversimplified model. For a start, it completely neglects all the various mechanical and electrical effects (friction, inertia, induction, and so on) which result in the motor taking a finite time to change its speed when the voltage input is suddenly changed. Nevertheless, even this very simple model enables us to come to some important preliminary conclusions about control systems.

Suppose, first of all, that we try to control the speed of the motor simply by using a preset value of voltage for each particular desired speed. This could be done by a control knob, or by a computer program sending appropriate instructions to a power supply. Problems soon arise, however, in the presence of disturbances of the type mentioned in the last section. If the load on the motor increases, its speed will decrease for the same applied voltage. If the friction in the motor bearings changes as the motor warms up, so will the speed. So this simple type of control – known as *open-loop* control for reasons which will become clear below – cannot compensate for either external disturbances to the system (load change) or changes to the process model (variation in G due to a change in friction, for example). The basic problem is that there is no perception involved – an open-loop system operates 'in the dark'.

A partial way round the problem is shown in Figure 8.2. The effect of a load change on the motor speed is just as though the input voltage to the motor had changed – represented in the diagram by the subtraction from the supplied voltage of an amount determined by the magnitude of the disturbance. So if we can measure the change in load, and calculate the correct adjustment to the supplied voltage in order to compensate, then the motor should run at the correct speed even as the load changes. This strategy is known as ***feedforward control***: the disturbance is measured and used to adjust the voltage before it has any effect on the motor speed. Note, however, that feedforward can only cope with disturbances that can be easily measured – and it can't help if G itself changes in some unknown way.

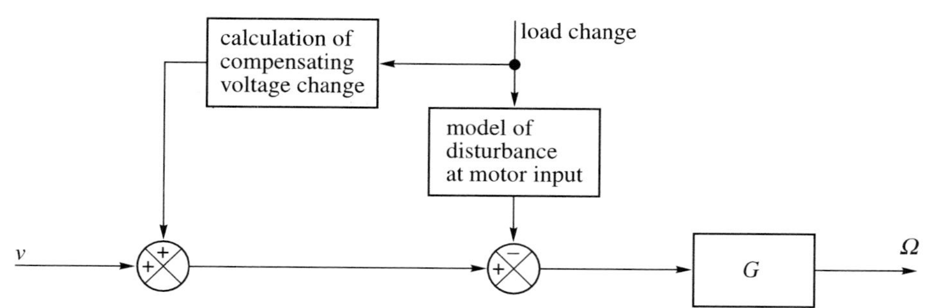

Figure 8.2 Feedforward to compensate for a disturbance to the motor load.

The third – and perhaps the most important – control strategy is shown in Figure 8.3. Here we assume that the motor speed itself is measured and compared with the desired speed. These values are then used by the control computer to generate a voltage input to the motor. If the speed is greater than it should be, the voltage to the motor is reduced; if the speed is too low, the voltage is increased. This strategy is known as ***feedback*** or ***closed-loop control***, and it can compensate for both external disturbances or changes to the process being controlled. Whatever the cause of the deviation of the actual speed from the desired speed, the controller generates a new voltage tending to restore the speed towards its required value.

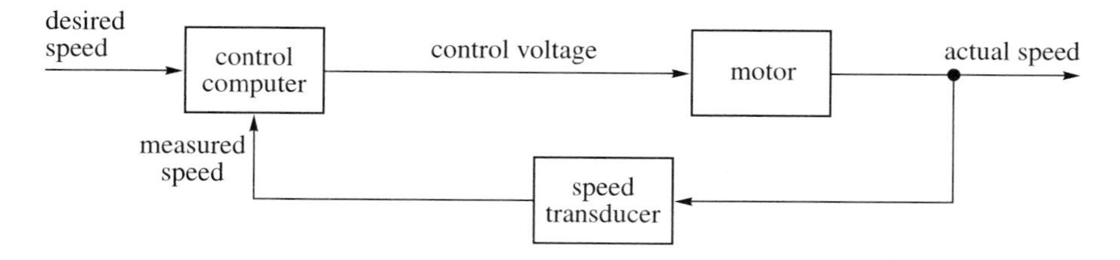

Figure 8.3 Feedback control of motor speed.

Figure 8.4 shows a simple example of such a feedback loop in a more standard form. Note that although a transducer is needed to measure the motor speed, it is not modelled explicitly in the figure – this is called the ***unity feedback convention***.* We simply assume that the actual speed Ω_o can be measured quickly and accurately and that it can be compared – using software, perhaps – with the desired speed Ω_i. The difference between the desired and actual speeds, known as the error signal e, forms the input to the controller. In this example, the latter is a *proportional controller*, which generates a motor voltage $v = Ke$ directly propor-

* In this way it is often possible at the design stage to ignore scaling factors due to transducers, data converters, and so on, and to take them into account later. The unity feedback convention will be adopted in almost all the examples of this chapter. You should be aware, though, that some of the general results presented have to be modified slightly when the desired and actual values of the controlled variable are *not* compared directly as they are in Figure 8.4.

tional to the error. The multiplication carried out by the proportional controller is now commonly implemented by microprocessor or computer.

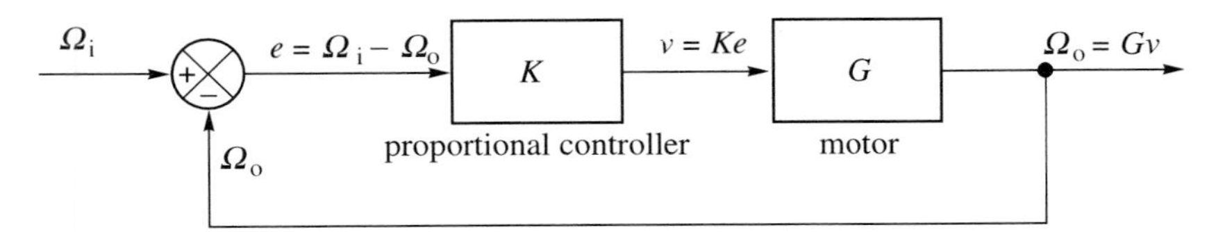

Figure 8.4
A speed control feedback loop in unity feedback form.

Another important convention implied in Figure 8.4 is that all the variables are defined as deviations from some operating or reference point. Because of this convention, a positive or negative value of v (as a result of a positive or negative value of error e) is interpreted as a *change* of motor voltage above or below its reference level. This feature of the standard modelling process is an important point to bear in mind when implementing control systems.

Working from the block diagram, we can now derive a general equation relating desired and actual speed for the closed-loop system. By considering the combined effects of the various blocks we write

$$\Omega_o = KGe = KG(\Omega_i - \Omega_o)$$

Hence

$$\Omega_o(1 + KG) = KG\Omega_i$$

and the relationship between desired and actual speed is given by

$$\frac{\Omega_o}{\Omega_i} = \frac{KG}{1 + KG}$$

This is an extremely important expression. If the controller gain, K, is sufficiently high, then $KG \gg 1$ and Ω_o/Ω_i approaches 1. In other words the error between actual speed and desired speed becomes, in theory, arbitrarily small as K is increased. Applying a precisely similar argument to Figure 8.5 we can also model the effect of a disturbance, represented by a reduction in the voltage supplied to the motor of d volts. After some algebraic manipulation we have

$$\Omega_o = \Omega_i \frac{KG}{1 + KG} - d\frac{G}{1 + KG}$$

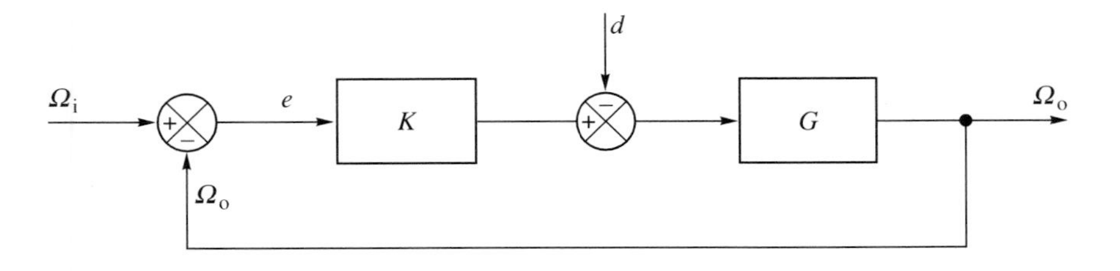

Figure 8.5
The feedback loop model including a disturbance.

The first term on the right-hand side models the effect of the loop in the absence of a disturbance. The second term models the additional effect of the disturbance on the actual speed. If K is sufficiently high, the term $G/(1+KG)$ – and hence the effect of the disturbance – can be made arbitrarily small, again in theory.

This analysis has made a lot of assumptions and left a lot of questions unanswered. We have neglected any consideration of the dynamic behaviour of the system, modelling the motor (and, indeed, the complete loop) as a pure proportional gain. These oversimplified models will be made more realistic in the next section. Nevertheless, provided that the motor settles down to a steady speed after a change in demanded speed, the analysis holds in this *steady state*. The expression

$$\frac{\Omega_o}{\Omega_i} = \frac{KG}{1 + KG}$$

can then be interpreted as the closed-loop steady-state gain, and the first of our general principles of feedback control can be stated as:

> Increasing controller gain tends to reduce the steady-state errors in response to (1) demanded changes in controlled variable and (2) steady disturbances entering the loop.

8.3 Some standard models of system dynamics

I now want to present a number of standard models widely used in mechatronics (and elsewhere). Apart from the first example, I shall not even outline the derivations of the models, but simply ask you to accept them or else refer to standard texts. The models are all simple examples of *linear differential equations with constant coefficients*.* I have chosen to give them in *standard forms*, without reference to the physical parameters of motors, loads and so on. This means that the same models can be applied in a range of control applications – both in mechatronics and elsewhere. Think, perhaps, of a 'library' of standard

* If x is the system input, and y the output, an nth-order system model of this type has the general form:

$$a_n \frac{d^n y}{dt^n} + a_{n-1} \frac{d^{n-1} y}{dt^{n-1}} + \ldots + a_1 \frac{dy}{dt} + a_0 = b_m \frac{d^m x}{dt^m} + b_{m-1} \frac{d^{m-1} x}{dt^{m-1}} + \ldots + b_1 \frac{dx}{dt} + b_0$$

where a_n, b_m are constants and $n > m$. Note that the presence of terms like

$$\sin x, \quad \left(\frac{dy}{dt}\right)^2 \quad \text{or} \quad \left(\frac{dx}{dt}\right) \times \left(\frac{dy}{dt}\right)$$

would lead to a *non-linear* differential equation. Such non-linear differential equations are often difficult to handle mathematically, and will not be considered in this chapter.

models, to be selected and used as required. The appropriate model depends on the particular application, as well as on the physical device(s) being modelled: there is no single model of a given system which is suitable for all circumstances.

8.3.1 An integrator model: the position/torque relationship of an ultrasonic motor

Ultrasonic motors are used to position the lens in the autofocus systems of some small cameras. The motor consists of two concentric rings connected to the outer edge of the lens system. When excited ultrasonically, one ring moves relative to the second, fixed, ring, hence moving the lens to the required place. Because the moving ring is so light, its inertia can be neglected in comparison with the friction between the rings, and the motor torque T generated ultrasonically is almost entirely used to overcome this friction. Such a system is often called a *friction-dominated system.*

A model of friction found to hold well in the ultrasonic motor is known as the *viscous* friction model (because it has the same form as the friction of a body moving through a viscous fluid).* In this model the frictional force is assumed to be directly proportional to the speed of the body, and acts in the opposite direction to the motion. For rotational motion, then, the frictional torque will be proportional to the angular velocity.

Because we are dealing with a position control system, we are interested in obtaining a model relating the angular position θ of the motor ring to the torque generated by the motor. Neglecting the inertia of the ring, then, we equate the motor torque to the frictional torque and we can write down an appropriate differential equation model immediately:

$$T = c\,\frac{\mathrm{d}\theta}{\mathrm{d}t}$$

where $\frac{\mathrm{d}\theta}{\mathrm{d}t}$ is the angular velocity and c is a constant known as the friction coefficient.

By integrating both sides of the equation with respect to time we can write this equation in the alternative form

$$\theta = \int \frac{T}{c}\,\mathrm{d}t$$

* The viscous friction model does not include static friction or 'stiction' – the extra push usually needed to start a body moving from rest. Including stiction explicitly gives a non-linear model which is not easy to use for control system design. So control engineers tend to adopt the linear viscous friction model, and consider non-linear effects separately if and when necessary.

This is more convenient for a block diagram representation, as shown in Figure 8.6, since the convention is to write inside the block a representation of the operation carried out on the input to convert it into the output. The model is therefore termed an ***integrator***: the 'input' torque can be viewed as being integrated (and multiplied by the constant $1/c$) to determine the 'output' position.

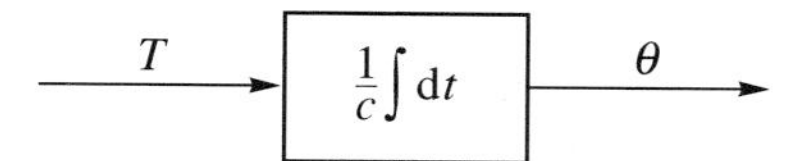

Figure 8.6
The integrator model of an ultrasonic motor.

Integrator step response

A common method used by engineers to characterize system behaviour is the system ***step response***: in other words, the way the system output changes when the input is suddenly changed from one steady value to another. (The output is also assumed to have been unchanging with time before applying the input step change.) Most commonly the *unit step response* is considered – that is, the input is assumed to change from 0 to 1 at time $t = 0$. For the *linear* system models we are considering here, the unit step response can then be scaled appropriately to deal with step inputs of different magnitudes. So in the case of our model of the ultrasonic motor, the step response describes the way the angular position changes when the input torque is suddenly changed from 0 to 1 (measured in some appropriate units).

The step change in torque, and the resulting step response in position, are shown in Figure 8.7. Initially $T = 0$ and $\theta = 0$. When T is changed to a constant, non-zero value, the system model implies a constant angular velocity of T/c. The unit step response is therefore a straight line, or *ramp*, of slope $1/c$.

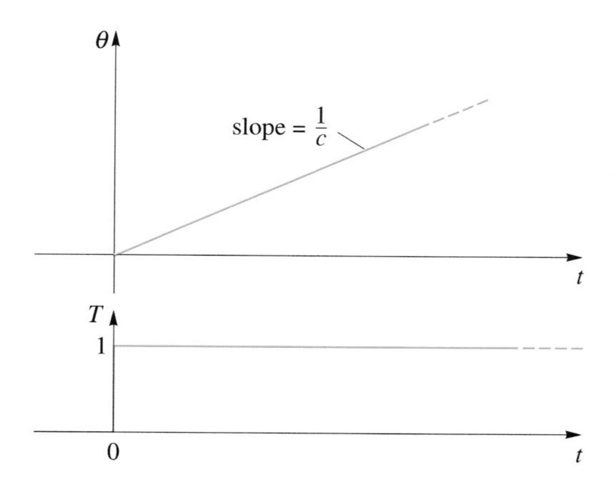

Figure 8.7
Unit step response of an integrator model.

The characteristic feature of an integrator is that the system output continues to increase so long as there is a (positive) input present. Integrators form an important functional component of controllers, as well as being a useful model in some circumstances of systems being controlled.

8.3.2 A first-order model: the speed/voltage relationship of a DC motor

The general approach adopted for the ultrasonic motor can be applied to other physical systems. From now on, though, I shall simply present standard results without their derivation. Let's begin by looking again at the electric motor considered earlier, and assume – as will often be the case – that the inertia of the motor plus load cannot be neglected. One particular type of motor, known as an armature-controlled DC servomotor, is carefully designed so that when driving a load its dynamic behaviour corresponds closely to the following standard differential equation relating speed Ω and input voltage v:*

$$\tau \frac{\mathrm{d}\Omega}{\mathrm{d}t} + \Omega = kv$$

where k and τ are constants. This is a first-order linear differential equation, and the model is widely referred to as the *standard first-order model.*

The constant k is known as the steady-state gain: if the motor is running at a constant speed Ω_{SS} with an applied steady voltage v_{SS}, then $\mathrm{d}\Omega/\mathrm{d}t = 0$ and $k = \Omega_{SS}/v_{SS}$. The constant τ has the dimensions of time. It is known as the time constant, and it is best thought of as a measure of how fast the system can respond to a change in the input. The gain and the time constant for a particular physical system can be found either by analysing the individual components (using mechanical, electrical and hydraulic principles as appropriate), or by subjecting the system to various practical tests. Very often a combination of the two approaches is best.

The unit step response is again a convenient way of characterizing dynamic behaviour. The general form of the first-order response can be found by solving the differential equation for $v = 0$, $t < 0$; $v = 1$, $t \geqslant 0$, but the details of this solution are not important here. More important for the designer is to be able to use the standard first-order step response curve shown in its normalized form in Figure 8.8. Readers familiar with electronics will recognize the standard first-order model as identical to that of an ideal low-pass RC or RL circuit, but it can also be used for many other physical systems.

One of the most important points to note about Figure 8.8 is that both axes are dimensionless. The horizontal axis is labelled in multiples of the time constant, while the vertical axis is labelled as a proportion of the steady-state value. This enables the graph to be used easily for any values of k and τ.

* The voltage v is in practice the input to a power amplifier which in turn drives the motor. I shall not go into the details of such power amplifiers, but just use a model relating voltage and speed.

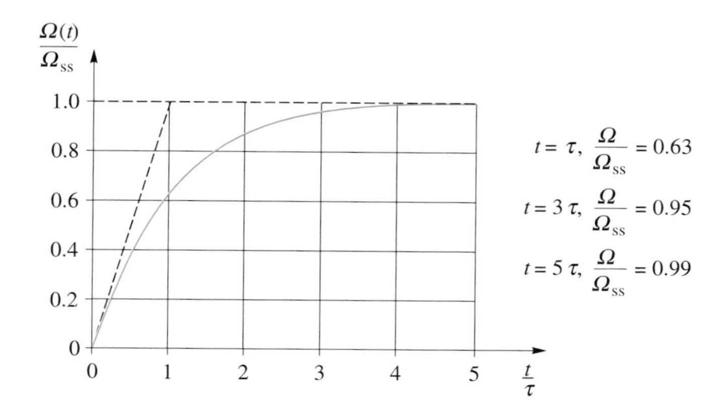

Figure 8.8
Normalized unit step response of a first-order model.

A DC servomotor has a time constant of 0.1 s and a steady-state gain of 5 rad s^{-1} V^{-1}. Use the normalized curve to sketch its response to a change in input voltage from 0 to 2 V at time $t = 0$. Estimate the speed of the motor at time $t = 0.2$ seconds.

The steady-state speed is equal to $5 \times 2 = 10$ rad s^{-1}. Since the time constant is 0.1 s, the denormalized step response becomes that of Figure 8.9. A time of 0.2 s after the input step corresponds to two time constants, so the speed is about 85% of its steady-state value or around 8.5 rad s^{-1}.

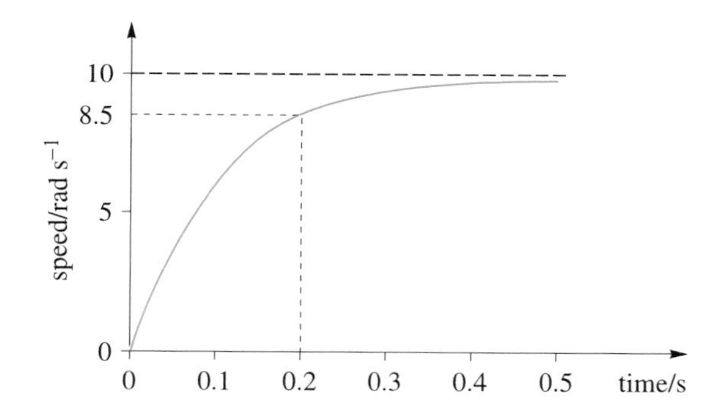

Figure 8.9
First-order unit step response with k = 5 and τ = 0.1.

8.3.3 The standard second-order model

An important point to note about the first-order step response curves of Figures 8.8 and 8.9 is that the system response approaches its final value steadily, without overshooting it or oscillating about it before settling. Many mechatronic systems have the potential for oscillation and vibration, so the final entry in our 'library' of standard models is one which can be used in such cases. Readers familiar with electronics may recognize it as a model of an RLC circuit, while those with an engineering mechanics background will probably think of it as the 'mass-spring-damper'. The model is a useful way of modelling the dynamics of a variety of flexible structures which can oscillate when moved or otherwise disturbed from equilibrium. Again, the derivation will not be given for any particular instance: the important thing, as before, is to learn to work with the standard form, which in this case is:

$$\ddot{\theta}_o + 2\zeta\omega_n\dot{\theta}_o + \omega_n^2\theta_o = k\omega_n^2\theta_i$$

As before, k is the steady-state gain, this time relating output θ_o to input θ_i when neither input nor output are changing with time (so $\ddot{\theta}_o = \dot{\theta}_o = 0$ in the equation). The parameters ζ and ω_n are known respectively as the *damping factor* (or *damping ratio**), and the *undamped natural angular frequency.* The standard parameters k, ζ and ω_n are used to characterize second-order system behaviour just as k and τ did in the first-order case.

The general shape of the second-order step response varies with the damping factor, so instead of just one standard, normalized step response curve we need a whole family, as illustrated in Figure 8.10. Note that as before both axes are

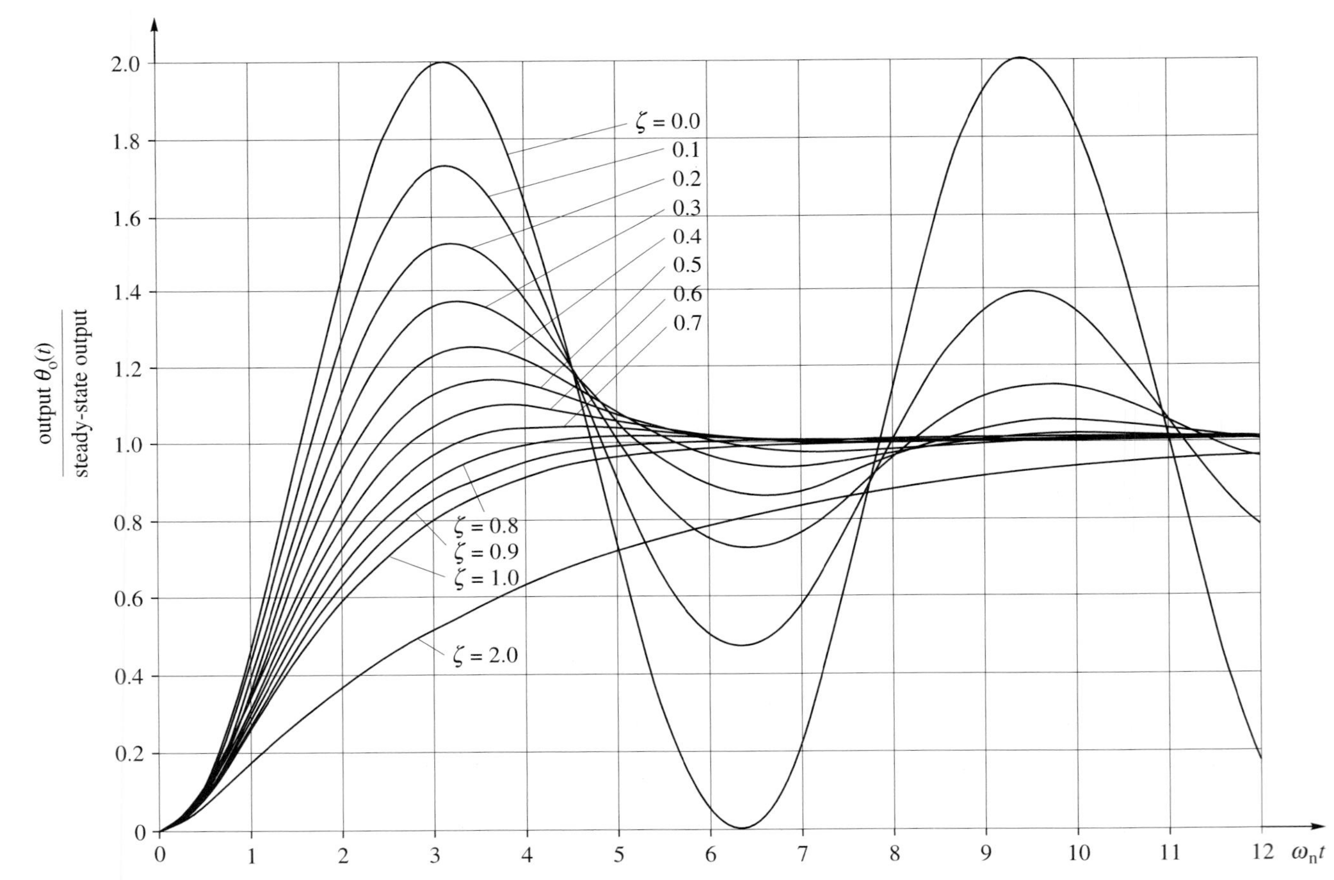

Figure 8.10
Normalized, second-order, unit step response curves.

* Control engineers tend not to distinguish between the two terms, although some other branches of engineering adopt a different definition for damping ratio.

dimensionless. Here the horizontal axis is labelled in multiples of ω_n, while the vertical axis, as before, is given in terms of the proportion of steady-state output.

Systems with $0<\zeta<1$ are known as *underdamped* and are characterized by an oscillatory step response. The lower the damping factor, the more pronounced the oscillation, and the closer the frequency of oscillation is to ω_n in rad s^{-1}. Systems with $\zeta>1$ are known as *overdamped*, and their step response rises to its final value without oscillation or overshoot. A system with $\zeta=1$ is termed *critically damped* and, for a given value of ω_n, has the fastest rise in step response without overshoot. A system with $\zeta=0$ is said to be *undamped*, oscillates continuously at the undamped natural frequency (hence its name), and can often be thought of as representing the borderline of instability.

A position control system can be modelled by a standard second-order differential equation relating the output (actual position) to the input (desired position). The system's undamped natural frequency is 8 Hz and its damping factor is 0.6. Using Figure 8.10, estimate the percentage peak overshoot of the step response. How long does it take the step response to reach 90% of its final value after the application of the step input?

The overshoot can be estimated directly from the curve for $\zeta=0.6$ as around 10%. From the same curve, it can be seen that the 90% rise time is given by $\omega_n t \approx 2.3$. An undamped natural frequency of 8 Hz is equivalent to $8 \times 2\pi \approx 50$ rad s^{-1}. Hence $50t \approx 2.3$ and $t \approx 0.046$ s or 46 ms.

Many mechatronic systems are designed to behave like standard first- or second-order models, so it is important for designers to be able to manipulate such models with confidence. Step responses are often used in performance specifications and to compare competing designs, even for systems whose output may never actually be required to change suddenly from one value to another.

8.3.4 Transfer functions and the *s*-plane

Linear differential equation models – even the standard first- and second-order ones just described – are fairly unwieldy to write down and manipulate. Because of this, control engineers usually transform them into alternative, completely equivalent forms which are particularly easy to use. The mathematical basis of the technique is known as the ***Laplace transform***, and it is discussed in some detail in the texts listed at the end of the chapter. Here I shall explain how to derive the transformed models, but not why it is mathematically valid to do so. Neither shall I go into details of exactly when and how it is valid to use such models – so you should not try to adapt the procedure to circumstances other than those described here unless you already have a good knowledge of Laplace transform techniques.

The key to the technique is to convert the linear differential equation, as a function of time, into an algebraic equation as a function of a new variable, the Laplace variable, $\boldsymbol{s}$. To do this, you simply carry out the following transformations on the differential equation:

- A function of time – $v(t)$ for example – becomes a new function of $\boldsymbol{s}$, written $\boldsymbol{V}(\boldsymbol{s})$. Similarly, if k is a constant, $kv(t)$ becomes $k\boldsymbol{V}(\boldsymbol{s})$.
- The first derivative of a function of time $\mathrm{d}v/\mathrm{d}t$ becomes $\boldsymbol{s}\boldsymbol{V}(\boldsymbol{s})$ (and $k\,\mathrm{d}v/\mathrm{d}t$ becomes $k\boldsymbol{s}\boldsymbol{V}(\boldsymbol{s})$).
- The second derivative $\mathrm{d}^2v/\mathrm{d}t^2$ becomes $\boldsymbol{s}^2\boldsymbol{V}(\boldsymbol{s})$, and so on.
- The integral of a function of time $\int v(t)\,\mathrm{d}t$ becomes $\dfrac{\boldsymbol{V}(\boldsymbol{s})}{\boldsymbol{s}}$

Note the convention of denoting a time variable by a lower case letter, and the transformed variable by the corresponding bold upper case letter (except for some Greek letters).

So, for example, the standard first-order model is transformed as follows:

$$\tau\frac{\mathrm{d}\Omega}{\mathrm{d}t} + \Omega = kv \leftrightarrow \tau\boldsymbol{s}\Omega(\boldsymbol{s}) + \Omega(\boldsymbol{s}) = k\boldsymbol{V}(\boldsymbol{s})$$

The convenience of this new model is that, unlike the differential equation on which it is based, it can be manipulated algebraically in a very useful way. For example, by dividing both sides of the transformed equation by the term $(1+\tau\boldsymbol{s})$ we can write

$$\Omega(\boldsymbol{s}) = \boldsymbol{V}(\boldsymbol{s})\,\frac{k}{1+\tau\boldsymbol{s}}$$

In other words, to get the transform of the system output $\Omega(\boldsymbol{s})$, just multiply the transform of the input $\boldsymbol{V}(\boldsymbol{s})$ by the term $k/(1+1\tau\boldsymbol{s})$. Or, to put it another way, the term $k/(1+\tau\boldsymbol{s})$ behaves much like the gains used to relate system input and output in Section 8.2. It is known as the ***transfer function*** of the system, and is often written $\boldsymbol{G}(\boldsymbol{s})$.*

Write down the transfer function model of a standard second-order system with steady-state gain 10, damping factor 0.5 and undamped natural angular frequency 2 rad s^{-1}.

---

* I'm glossing over a lot of mathematical background here. For example, the whole procedure only applies to linear models. Furthermore, transfer functions can only be interpreted in this way if both the system input and output, *and* their derivatives, are zero before the input is applied. (This is sometimes termed the quiescence condition before time $t = 0$.) However, the same condition applies to the standard step responses, so it is not really a significant restriction as far as control engineering is concerned. It *is* important, however, if transform techniques are used to solve differential equations in circumstances when the quiescence condition does not hold. For more details consult any standard text on Laplace transforms.

The standard differential equation model, when transformed, becomes

$$s^2\Theta_o(s) + 2\zeta\omega_n s\Theta_o(s) + \omega_n^2\Theta_o(s) = k\omega_n^2\Theta_i(s)$$

Hence the transfer function is given by the expression:

$$G(s) = \frac{\Theta_o(s)}{\Theta_i(s)} = \frac{k\omega_n^2}{s^2 + 2\zeta\omega_n s + \omega_n^2}$$

Substituting the given parameter values leads to a transfer function for this particular case of

$$G(s) = \frac{40}{s^2 + 2s + 4}$$

The s-plane and stability

You may be thinking that these transfer functions do not seem to be much simpler than the differential equations they were derived from! The beauty of them, though, is that they can be used in a pictorial way which includes the important features of system dynamic behaviour in a concise form. The key to this pictorial representation is to consider the values of the Laplace variable which make the transfer function infinitely great. These values are called *poles*.

Consider a standard first-order system

$$G(s) = \frac{k}{1 + \tau s}$$

The value of s which makes $G(s)$ infinitely great is the value for which $1 + \tau s = 0$. We say, therefore, that a first-order system has a single pole at $s = -1/\tau$. In general, s can take on complex values, so it is normal to plot the poles of a system on a two-dimensional complex plane known as the ***s-plane***. Figure 8.11 shows this for the first-order system; as is usual, the position of the pole is marked with a cross.

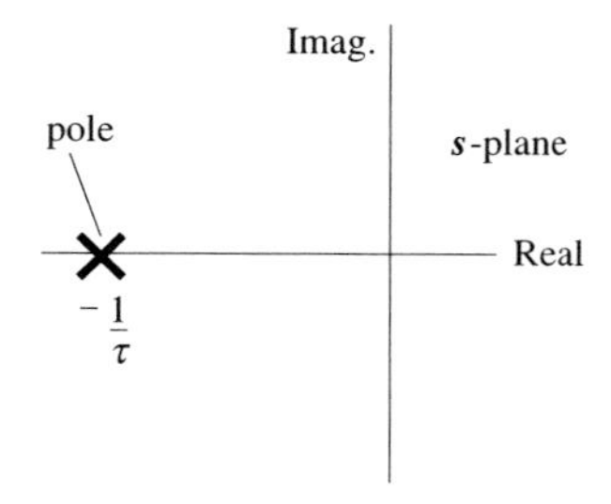

Figure 8.11
The s-plane model of a first-order system.

Plot the s-plane poles for the second-order system of the previous example, for which

$$G(s) = \frac{40}{s^2 + 2s + 4}$$

We want the values of s for which $s^2+2s+4=0$. Applying the usual formula for solving a quadratic we have

$$s = \frac{-2 \pm \sqrt{(4-16)}}{2} = -1 \pm j1.73$$

There is therefore a complex conjugate pair of poles, as plotted in Figure 8.12.

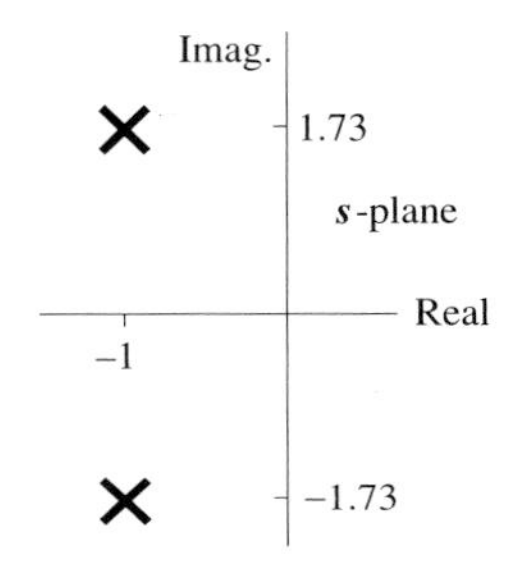

Figure 8.12
A pair of complex conjugate poles in the s-plane.

A second-order system will have two poles because the denominator of the transfer function is a quadratic equation in s. The various possibilities for the standard system are summarized in Figure 8.13. Note:

- For an overdamped system, with $\zeta>1$ (Figure 8.13a), there are two real poles, located symmetrically about the point $s=-\zeta\omega_n$.
- For a critically damped system, with $\zeta=1$ (Figure 8.13b), there is a double pole located at $s=-\omega_n$. (The quadratic equation in s has a double root.)
- For an underdamped system, with $0<\zeta<1$ (Figure 8.13c), there are two complex conjugate poles at a radial distance ω_n from the origin. The angle marked ψ is known as the damping angle. By simple trigonometry it follows that $\sin\psi=\zeta$.

Apart from the steady-state gain, then, the s-plane plot summarizes all the information contained in either the system differential equation or its transfer function. Control engineers learn to interpret their systems in terms of such plots, as will be described in the next section.

Before leaving the s-plane, I want to stress one final important point. From the theory of differential equations and Laplace transforms it follows that there is a close relationship between the poles of a system and the general form of its

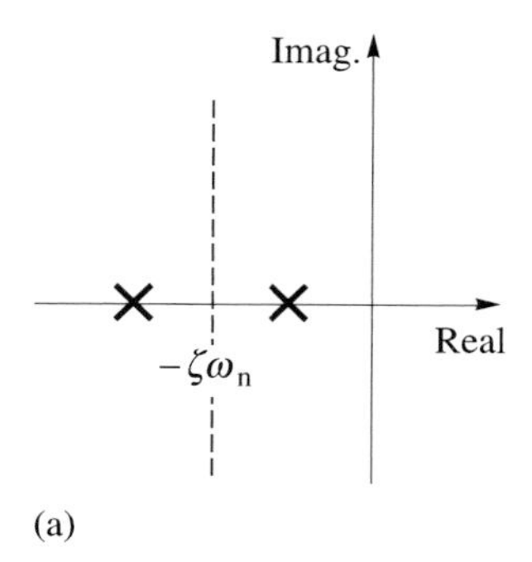

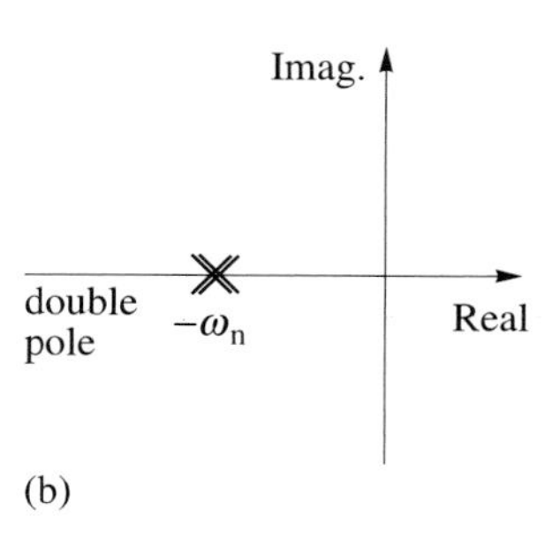

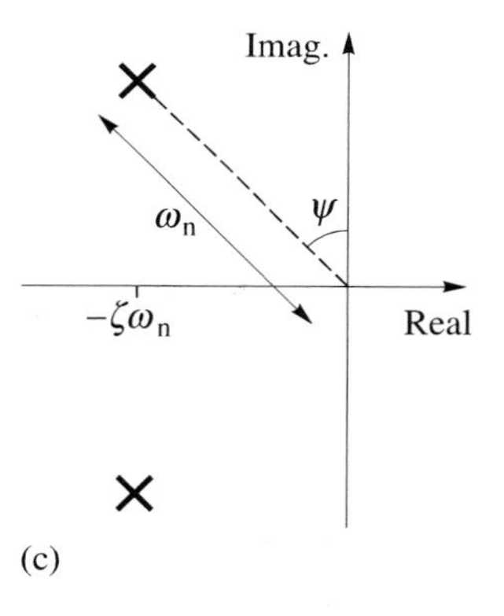

Figure 8.13
Diagram of s-plane models of three categories of second-order system:
(a) $\zeta > 1$ (overdamped);
(b) $\zeta = 1$ (critically damped);
(c) $0 < \zeta < 1$ (underdamped).

behaviour when it is disturbed from rest. There isn't space in this chapter to prove it, but the relationship can be stated as follows:

> A pole at $s = a$ is associated with a time-varying component Ce^{at} in the response of the system to a disturbance from rest. (C is a constant depending on the particular system and the way it is disturbed.) A complex conjugate pair of poles at $s = a \pm jb$ is similarly associated with a time varying term $Ce^{at}\sin(bt + \phi)$ in the response.

Figure 8.14 illustrates typical examples of these time-varying waveforms. If a system has all its poles to the left of the imaginary axis in the s-plane, then all the components associated with its poles die away eventually with time. The further to the left the poles are, the faster these *transients* die away, and the faster the system will respond to changes in input. Such systems are said to be *stable*. If, on the other hand, a system has a pole or poles to the right of the imaginary axis, then following a disturbance certain components in the response will not die away with time, but will grow indefinitely until some physical limit is reached. For example, at high speeds a racing cyclist can develop a front wheel wobble which grows to such a point that she is thrown off. Or an inexperienced tightrope walker might begin to swing from side to side, the amplitude of the swing increasing until he falls. Such systems have become *unstable*. An important part of control system design (not to mention cycle riding, and tightrope walking!) is to ensure that the system does *not* become unstable. In the light of the previous discussion, a control engineer might say that we have to ensure that all the system poles remain in the left half of the s-plane. In general, then:

> A linear system is stable if all the system poles lie in the left half of the s-plane.

We can relate the waveforms of Figure 8.14 to what we already know about first- and second-order step response. A first-order system has a real pole at $-1/\tau$. The step response therefore includes a decaying exponential term, but not an oscillatory one: it approaches its final value without oscillation. (The mathematical form of the step response is $k(1-e^{-t/\tau})$.) The further to the left the pole lies, the shorter

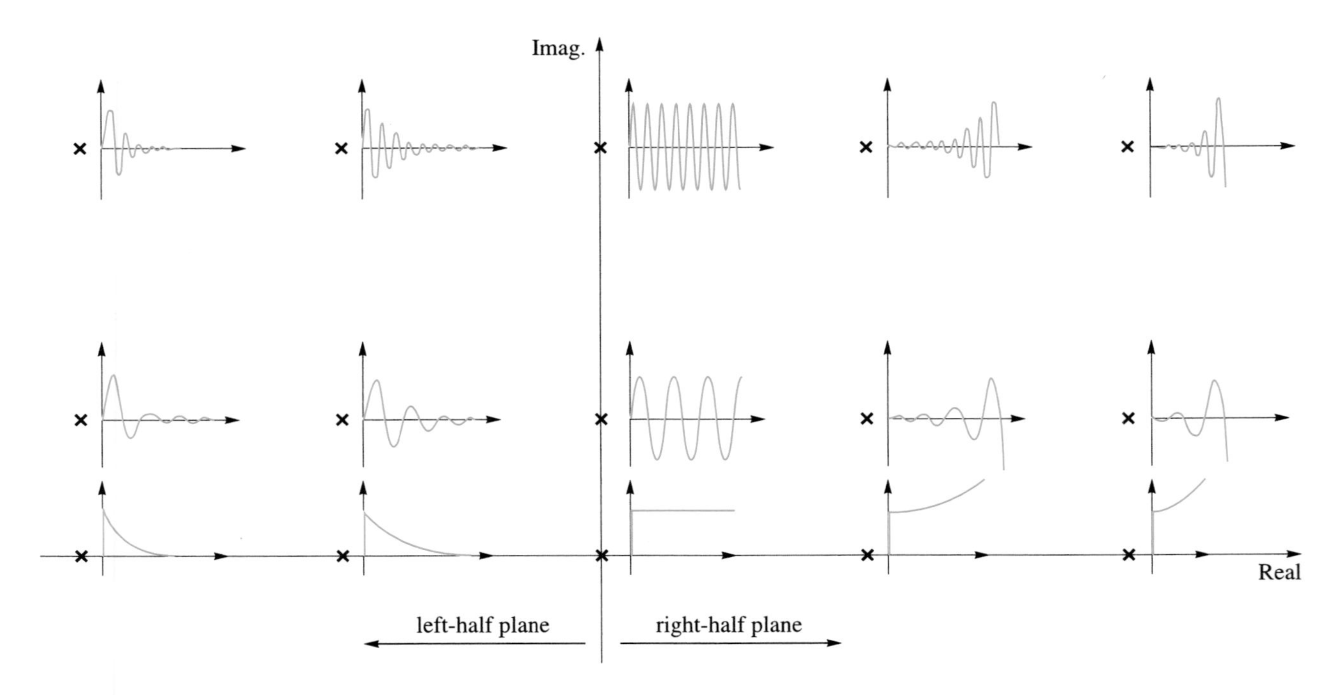

Figure 8.14
Time-varying waveforms corresponding to poles located in various regions of the s-plane (upper half of the plane only).

the time constant, and the faster the step response reaches its steady-state value. In the second-order case, there are various possibilities. Overdamped systems have two, negative, real poles, and again no oscillatory component in the step response; underdamped systems have a complex conjugate pair of poles (with negative real parts), and hence an oscillatory component.* Again, the further to the left the poles lie, the faster the system response.

8.4 Back to control systems

The transfer function models described in the previous section are particularly convenient to use with system block diagrams. Look back for a moment at the closed-loop system of Figure 8.4, where we modelled the speed/voltage relationship of the motor as a pure gain. Using our transform models we can convert each variable in Figure 8.4 (Ω_i, e, v, Ω_o) into the corresponding transform, and the motor gain G into its transfer function $\mathbf{G}(s)$ – as shown in Figure 8.15. Then we can carry out algebraic manipulations identical to those in Section 8.2, but instead of deriving a closed-loop steady-state gain, we obtain the *closed-loop transfer function* ¢

$$\frac{\boldsymbol{\Omega}_o(s)}{\boldsymbol{\Omega}_i(s)} = \frac{K\mathbf{G}(s)}{1+K\mathbf{G}(s)}$$

Using this general expression we can now investigate the dynamic behaviour of a whole range of feedback loops.

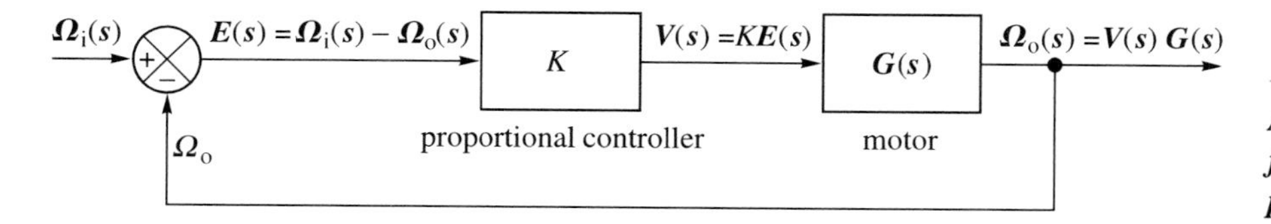

Figure 8.15
A Laplace transform feedback model with a proportional controller.

8.4.1 A speed control system

In this case we can write $\mathbf{G}(s) = G/(1+s\tau)$, where I am using G (rather than k) to represent the steady-state gain of the motor. Hence the closed-loop transfer function becomes

$$\frac{\boldsymbol{\Omega}_o(s)}{\boldsymbol{\Omega}_i(s)} = \frac{KG/(1 + s\tau)}{1 + [KG/(1 + s\tau)]}$$

* The frequency and decay rate of the decaying sinusoid can be linked to the s-plane pole positions, as can be inferred from Figure 8.13, but I shall not go into details here.

Multiplying numerator and denominator by $(1+s\tau)$ we have

$$\frac{\Omega_o(s)}{\Omega_i(s)} = \frac{KG}{1 + s\tau + KG}$$

This is still a first-order system, but with a different steady-state gain and time constant. To convert into standard first-order form we have to divide numerator and denominator by $(1+KG)$

$$\frac{\Omega_o(s)}{\Omega_i(s)} = \frac{\dfrac{KG}{1+KG}}{1 + s\dfrac{\tau}{1+KG}}$$

That is, the closed-loop system has a steady-state gain of $KG/(1+KG)$ and a new time constant $\tau/(1+KG)$. The closed-loop steady-state gain agrees with the expression derived in Section 8.2; the implication, as before, is that increasing K reduces the steady-state error (in principle) to an arbitrarily small value. Here we concentrate on the dynamics of the closed loop. The expression for the closed-loop time constant, $\tau/(1+KG)$, shows that by increasing the controller gain the closed-loop time constant is reduced. The output will settle to its final value sooner – in other words, the system closed-loop response is speeded up. The effect is clearly illustrated in the s-plane, as shown in Figure 8.16. The pole of the motor alone is shown at $s=-1/\tau$. For small values of K the closed-loop pole will be very close to this position, but as K is progressively increased (and the time constant decreases) it will be located progressively further to the left. In such a case the design activity might be described by a control engineer as: 'choosing the gain K so as to position the pole at a suitable point in the s-plane.'

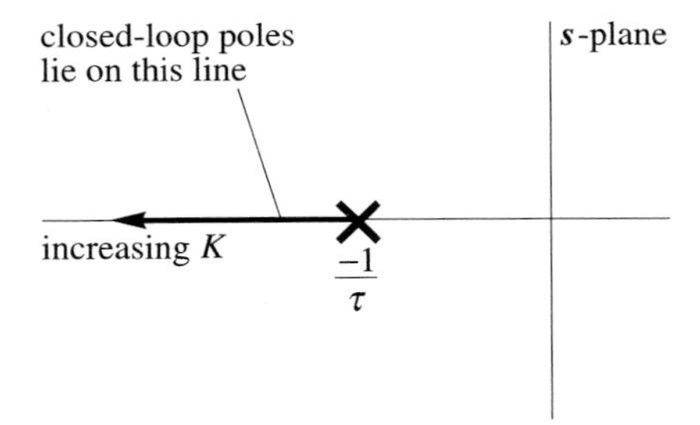

Figure 8.16 ***Locus of closed-loop pole for the motor speed control system.***

8.4.2 A friction-dominated position control system

Figure 8.17 shows the ultrasonic motor used in a closed-loop position control system with a proportional controller. We can use the same approach as before to determine the general form of the closed-loop response. The integrator model relating motor input to position output has been converted into a transfer function

$$G(s) = \frac{1}{cs}$$

Here the gain K represents the combination of the proportional controller and the ultrasonic generator of motor torque.

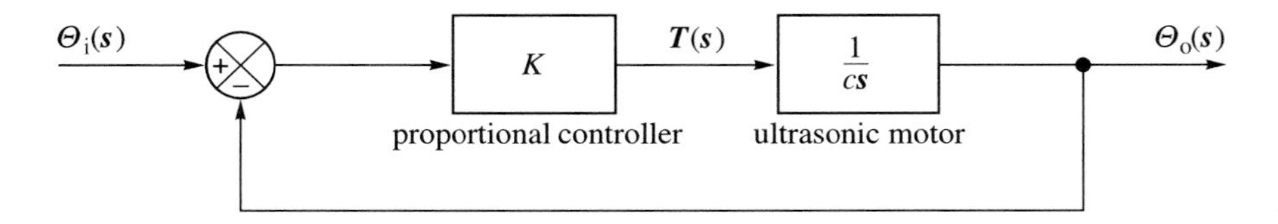

Figure 8.17
The ultrasonic motor position control system.

We can write the closed-loop transfer function as

$$\frac{\Theta_o(s)}{\Theta_i(s)} = \frac{KG(s)}{1 + KG(s)} = \frac{K/cs}{1 + (K/cs)} = \frac{K}{cs + K} = \frac{1}{1 + s\frac{c}{K}}$$

The final form shows that the closed-loop response is a standard first-order response with unity steady-state gain and a time constant equal to c/K. Increasing the controller gain does not affect the steady-state error (which is always zero because the closed-loop steady-state gain is always 1), but as before it does speed up the response. The corresponding s-plane plot is shown in Figure 8.18. Again, as the gain is increased, the closed-loop pole moves steadily to the left, corresponding to a faster, first-order response.

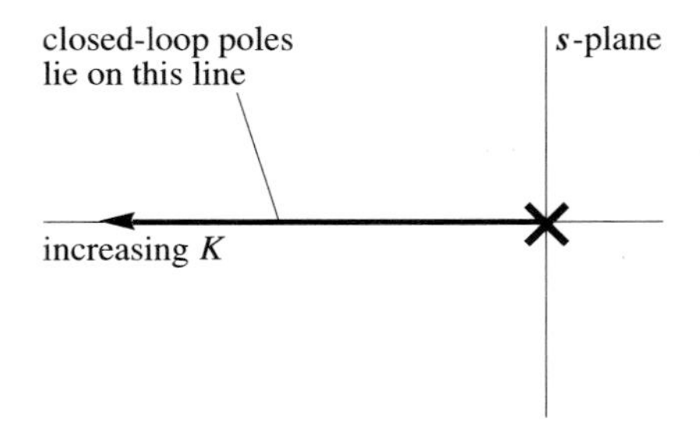

Figure 8.18
Locus of closed-loop pole for the ultrasonic motor position control system.

It's time, though, to sound a note of caution. The models we have been using are simplified, and the mathematical analysis is only valid so long as the models themselves are. But the models cannot remain valid indefinitely if we try to build a system which responds faster and faster. Effects that we chose to neglect in setting up the models may become significant, invalidating our conclusions – and ultimately the physical limitations of the motor parts mean that there is a limit to the speed of response. Part of the skill of the mechatronics engineer is to be aware of the circumstances in which such models do and do not hold.

8.4.3 Integral action and steady-state error

Perhaps the most striking difference between the two closed-loop systems considered so far is the nature of the steady-state error. For the speed control system with a proportional controller, changing the demanded speed to a value different from the operating point results in a finite steady-state error in the actual speed obtained. It is easy to see why this must be so. To drive the motor at a new speed requires a change in voltage, and this change in voltage is equal to Ke. And for Ke to have a finite value, so must e. In the case of the camera autofocus system, however, things are different. The motor continues to drive the focusing ring towards its required position as long as there is an error. When the error is zero (providing there is no load disturbance) the motor stops at the required position without needing to generate additional torque. This characteristic can be viewed as a consequence of the integrating behaviour of the ultrasonic motor used for position control. In contrast, the velocity/speed model of the electric motor does not have this integrating characteristic, so with a proportional controller there is a finite steady-state error.

8.4.4 A second-order position control system

My first example of a position control system, using the ultrasonic motor, was particularly straightforward because inertia could be neglected, allowing a basic model to be derived for the motor. Many mechatronic products, though, require the precise positioning of loads possessing a great deal of inertia. The DC servomotors of the type described above in the speed control example can be used for such purposes. In such cases, though, we need a slightly different model of the motor, this time relating input voltage to output *position* rather than *speed*. The new model is shown in Figure 8.19. Part (a) of this figure shows how the angular position can be viewed as the integral of the angular velocity: a separate pure integrator block, with transfer function $1/s$, is included to model this. Part (b) shows how this integrator block can be incorporated into a single model directly relating output angular position to input voltage:

$$G(s) = \frac{\Theta_o(s)}{V(s)} = \frac{k}{s(1 + s\tau)}$$

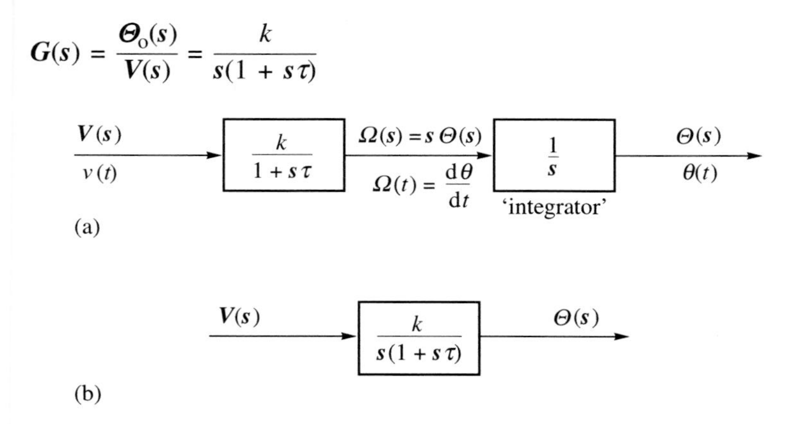

Figure 8.19 **A position/voltage model of a DC servomotor.**

A position control system for one joint of a robot arm, using such a motor, is shown in Figure 8.20. To keep the algebra simple I am assuming the motor has a steady-state gain of 1, but this does not make the model any less general. (We can view the proportional gain K as incorporating the motor gain – and, in practice, there will be other scaling factors depending on the type of transducers and data conversion involved.) Applying the usual formula for the closed-loop transfer function we have

$$\frac{\Theta_o(s)}{\Theta_i(s)} = \frac{KG(s)}{1 + KG(s)} = \frac{K}{s(1 + s\tau) + K} = \frac{K}{\tau s^2 + s + K}$$

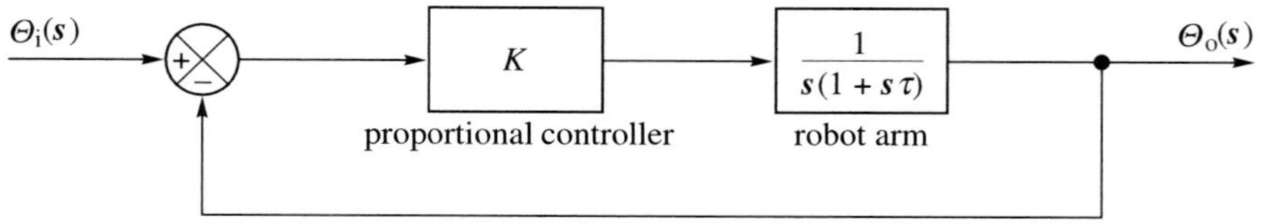

$\Theta_i(s)$ — K (proportional controller) — $\frac{1}{s(1 + s\tau)}$ (robot arm) — $\Theta_o(s)$

Figure 8.20
A robot arm position control system.

The closed-loop therefore behaves like a standard second-order system with poles at

$$s = \frac{-1}{2\tau} \pm \frac{\sqrt{(1 - 4K\tau)}}{2\tau}$$

Just as we did in the first-order case we can show in the s-plane the locus of possible closed-loop pole positions as K is increased from a small value. This ***root locus*** is shown in Figure 8.21: for small values of K the roots are located near those of the motor itself, at the origin and $s = -1/\tau$. When $4K\tau < 1$ there are two real roots (an over-damped system), while for $4K\tau = 1$, there is a double root located at $s = -1/2\tau$ (a critically damped system). If the gain is increased further, so that $4K\tau > 1$, there is a complex conjugate pair of roots with a real part equal to $-1/2\tau$ and an imaginary part which increases steadily as K is increased. In other words, increasing the gain results in an increasingly underdamped system, with a progressively more oscillatory response to control inputs and disturbances – very different closed-loop behaviour from that of the previous two feedback systems.

Figure 8.21 is a classic example of how the s-plane can convey a great deal of information about system behaviour in a concise way: I suggest you take some

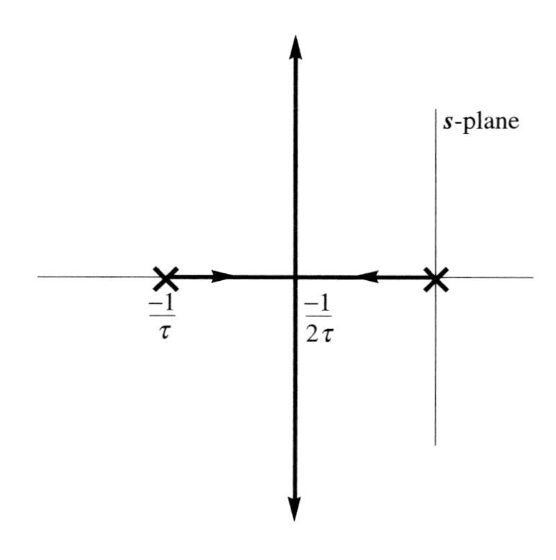

Figure 8.21
Locus of closed-loop poles for the robot arm position control system.

time to compare it carefully with Figures 8.13 and 8.14 and the standard second-order step response curves of Figure 8.10.

A system like that of Figure 8.20 uses a motor with a time constant of 1 s. What value of K will give a closed-loop step response with a 25% peak overshoot? How long after the application of the step does the peak overshoot occur?

What is the closed-loop steady-state gain? Can you think of an explanation for this value?

The closed-loop transfer function with $\tau = 1$ is

$$\frac{K}{s^2 + s + K}$$

Comparing this with the standard form

$$\frac{k\omega_n^2}{s^2 + 2\zeta\omega_n s + \omega_n^2}$$

we can identify $k = 1$, $K = \omega_n^2$ and $1 = 2\zeta\omega_n$. From Figure 8.10 we see that for 25% peak overshoot we require $\zeta = 0.4$. Hence $\sqrt{K} = \omega_n = 1/2\zeta = 1.25$ and $K \approx 1.6$.

Again using Figure 8.10, the peak overshoot occurs at normalized time $\omega_n t \approx 3.5$ – that is to say, at time $t \approx 3.5/1.25 = 2.8$ s.

The closed-loop steady-state gain is 1, whatever the value of K, implying zero steady-state error. As was the case with the ultrasonic motor, in the absence of load disturbances the DC servo continues to drive the joint towards its required position for as long as there is an error.

Many closed-loop systems used in mechatronics exhibit behaviour similar to the position and speed control systems just considered. To generalize from the discussions so far in this section we can say:

1. Increasing proportional gain in a feedback control system tends to speed up system response, reduce steady-state error, and decrease the effects of steady disturbances. But increasing the gain too much can result in an oscillatory closed-loop response – and even instability.

2. If the system possesses integrating action (has an s multiplier in the transfer function denominator) then the closed-loop steady-state error in response to a step change in demanded value is zero.

8.4.5 Tracking performance of a position control system

In our discussion so far we have looked at features of closed-loop performance like speed of response (as measured by the time constant of a first-order system, or the natural frequency of a second-order system); degree of oscillation (as measured by the damping factor); and steady-state error when the demanded value of the controlled variable is changed from one steady value to another. But this is by no means the whole story. Many mechatronic position control systems are also required to ensure that the position error is less than a specified value – or even zero – *while the position itself is changing at a given rate*. For example, a robot arm or machine tool may need to follow a particular trajectory to within a fine tolerance to ensure the quality of the parts being produced. Weapons systems obviously have to be built to track their targets closely. Control systems designed to achieve this sort of behaviour are often known as ***trackers***, in contrast to ***regulators*** which are primarily designed to keep a controlled variable at a constant value in spite of disturbances – many temperature control systems, for example.

Figure 8.22 shows the *steady-state* signals in the position control system of Figure 8.20 as it tracks a demanded position changing at a constant rate. The first thing to note is that both θ_i and θ_o must change at the same rate, otherwise the error would simply get indefinitely larger (with a positive or negative sign depending on which of θ_i and θ_o were greater). The second is that this tracking behaviour is only possible with a proportional controller because of the integrating action of the motor. A constant error e_{ss} (and hence a constant voltage input to the motor Ke_{ss}) allows the actual position θ_o to increase continuously in the steady state, tracking the demanded position θ_i, but lagging it by e_{ss} radians.

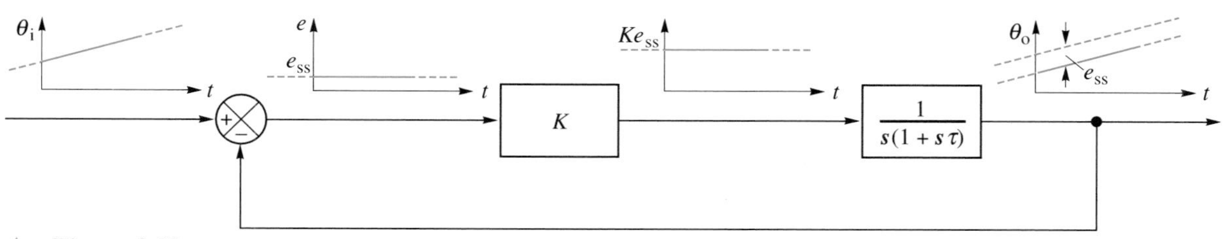

Figure 8.22
Steady-state tracking performance of the robot arm position control system.

Working from the block diagram we can write the relationship between θ_o and e as:

$$\frac{\Theta_o(s)}{E(s)} = \frac{K}{s(1 + s\tau)}$$

Cross multiplying, we have

$$s\Theta_o(s) + \tau s^2\Theta_o(s) = KE(s)$$

Converting this back to a differential equation by reversing the procedure described in Section 8.3 gives

$$\dot{\theta}_o + \tau\ddot{\theta}_o = Ke$$

In the steady-state tracking condition, therefore, $\ddot{\theta}_o = 0$ and $\dot{\theta}_o = \dot{\theta}_i$ is the demanded angular velocity. Tracking performance is usually specified as the value of e_{ss} for demanded angular velocity of 1 rad s^{-1}, in which case $\dot{\theta}_o = 1$ and $e_{ss} = 1/K$. In fact, like the steady-state error in response to a step change in input, we can say in general that:

Steady-state error under tracking conditions decreases as controller gain K increases, providing the system remains stable.

8.5 Other controller options

The only controller considered so far has been the proportional controller. Very often, however, proportional control alone cannot satisfy all aspects of the system specification.

A position control system like that of Figure 8.20, with a motor time constant of 1 s, is to be permitted a step response overshoot of 10% at most. It also has to have a value for ω_n of 2 rad s^{-1} or greater to satisfy the speed of response requirements. Can a proportional controller satisfy the specification?

The closed-loop transfer function is, from Section 8.4.4,

$$\frac{K}{s^2 + s + K}$$

Comparing this with the standard form we see that $\omega_n = \sqrt{K}$ and that K needs therefore to be at least 4 to fulfil the speed of response requirement. But also from the standard form we have $2\zeta\omega_n = 1$, so for $\omega_n \geqslant 2$ rad s^{-1}, ζ can be at most 0.25. But a peak overshoot of less than 10% requires a damping factor greater than 0.6, so a proportional controller cannot satisfy the specification.

In this section I shall describe some of the other options available to the mechatronics designer to improve control system performance.

8.5.1 Improving system damping: the use of derivative action

In the previous example, the value of K which would satisfy the speed of response requirement gives the excessively oscillatory step response with the general form shown in Figure 8.23. This figure provides the clue to a remedy. With proportional control, exactly the same input to the motor is generated at point A as at point B, since the error is the same. Yet at point A the joint is moving away from the required value, while at point B it is moving towards it. So it would make sense to increase the remedial control action at A and reduce it at B. This would hopefully reduce the size of the overshoot and improve the damping.

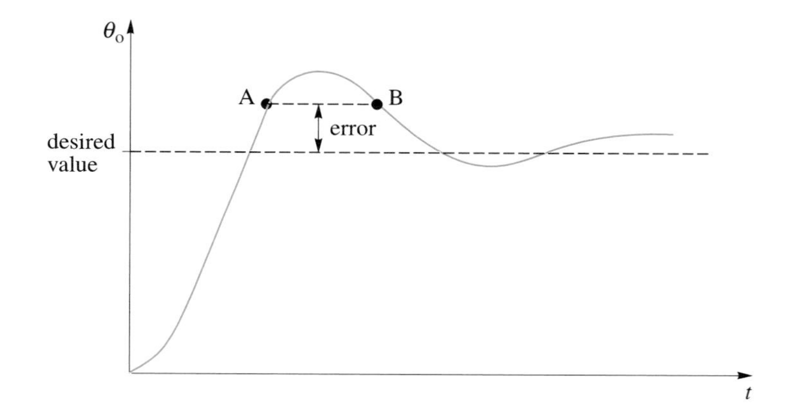

Figure 8.23 Lightly damped, second-order step response.

One way of doing this is illustrated in Figure 8.24. Instead of feeding back just the measured value of position, the complete feedback signal consists of the actual position plus a value proportional to the rate of change (derivative) of the position. This extra component increases the control action at A and reduces it at B according to the sign of the derivative. The technique is known as ***velocity feedback*** or *rate feedback*. The velocity can either be measured directly, using a suitable transducer such as a tachogenerator, or it can be calculated by the control computer from the time-varying behaviour of the robot arm position.

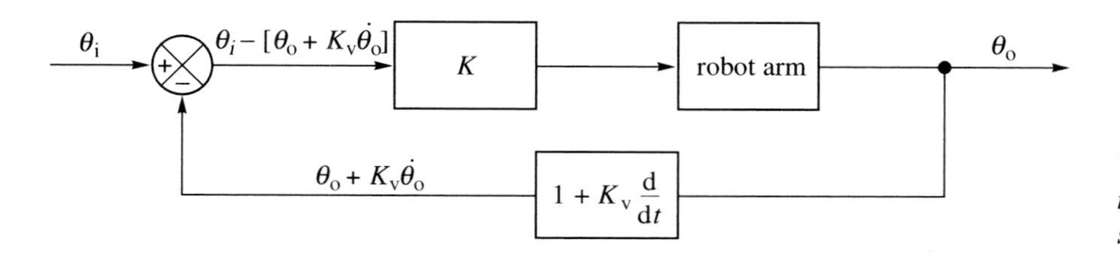

Figure 8.24 Velocity feedback in the robot arm position control system.

The total feedback signal with velocity feedback can be written as

$$\theta_o + K_v \frac{d\theta_o}{dt}$$

where K_v is a constant. This would correspond to a feedback path transfer function $(1+K_v s)$. To explain the effect on system behaviour I shall first redraw Figure 8.24 as Figure 8.25. (It doesn't matter whether the position and velocity elements in the feedback signal are subtracted together or separately from θ_i.)

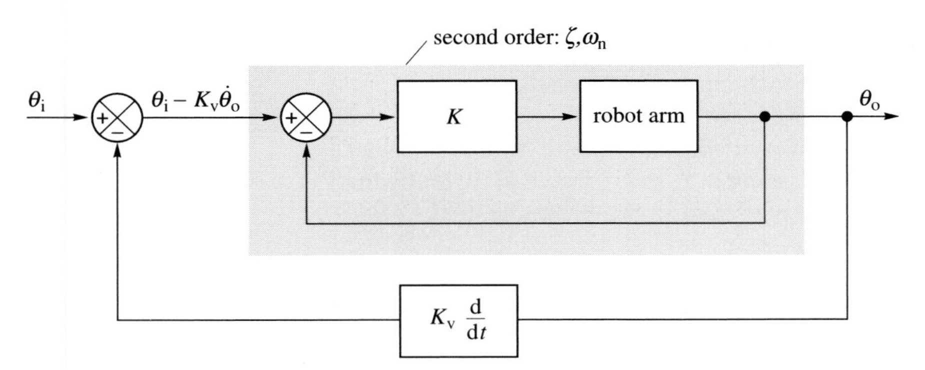

Figure 8.25 Modelling the effect of the velocity feedback by redrawing Figure 8.24.

The position control loop under proportional control alone – the part within the shaded box outline of Figure 8.25 – was modelled as a standard second-order system with unity steady-state gain. With the addition of velocity feedback, the input to this boxed part becomes $\theta_i - K_v\dot{\theta}_o$ instead of θ_i. Hence the new system differential equation is a standard second-order equation with θ_i replaced by $\theta_i - K_v\dot{\theta}_o$:

$$\ddot{\theta}_o + 2\zeta\omega_n\dot{\theta}_o + \omega_n^2\,\theta_o = \omega_n^2(\theta_i - K_v\dot{\theta}_o)$$

or

$$\ddot{\theta}_o + (2\zeta\omega_n + K_v\,\omega_n^2)\dot{\theta}_o + \omega_n^2\theta_o = \omega_n^2\theta_i$$

That is, the system with velocity feedback is still a standard second-order system with the same ω_n, but now it has a new damping factor ζ_{new}, where

$$2\zeta_{new}\omega_n = 2\zeta\omega_n + K_v\,\omega_n^2$$

or

$$\zeta_{new} = \zeta + \frac{K_v\omega_n}{2}$$

What value of K_v would give the system of the previous example a damping factor of 0.6 and thus a step response overshoot of 10%? What is then the closed-loop transfer function?

Substituting into the above expression, we have

$0.6 = 0.25 + K_v$

so $K_v = 0.35$.

Substituting into the standard form using $\zeta_{new} = 0.6$, together with $\omega_n = 2$ and $k = 1$, we find the closed-loop transfer function is:

$$\frac{4}{s^2 + 2.4s + 4}$$

The specification of the last example did not include a tracking requirement. Many systems do have such a requirement, however, and unfortunately velocity feedback makes tracking performance worse! We can apply the same argument as at the end of Section 8.4 to see why. Referring to Figure 8.25, it can be seen that, because of the velocity feedback, the controller output is $K(e - K_v\dot{\theta}_o)$ instead of Ke. Hence the differential equation relating error and actual position becomes

$$\dot{\theta}_o + \tau\ddot{\theta}_o = K(e - K_v\dot{\theta}_o)$$

In the steady state, therefore, for a constant tracking rate $\dot{\theta}_o$ of 1 rad s^{-1} we have

$$1 = K(e_{ss} - K_v)$$

and

$$e_{ss} = \frac{1}{K} + K_v$$

rather than $1/K$ as was the case without velocity feedback. So the effect of velocity feedback is to *increase* the steady-state tracking error.

Proportional + derivative (P+D) control

An alternative to velocity feedback is to include derivative action in the controller itself, so that the controller output for a given error is

$$K(e + T_d \frac{de}{dt})$$

This corresponds to a controller transfer function

$$K(1 + T_d s)$$

This type of controller is known as a *proportional + derivative* or ***P+D controller***, and T_d is a constant known as the *derivative time*. The position control system with a P+D controller is shown in Figure 8.26. The tracking error with a P+D controller is clearly the same as with a proportional controller: in the steady state, the tracking error is constant and so the extra derivative term in the controller has no effect.

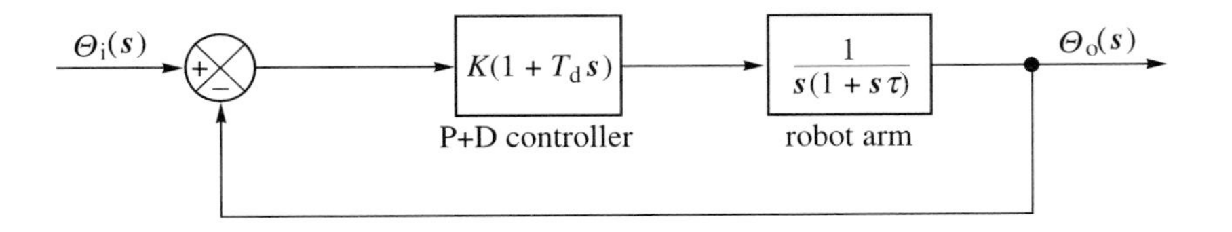

***Figure 8.26* Proportional + derivative control of robot arm position.**

But what about the step response? We can use the familiar expression

$$\frac{KG}{1 + KG}$$

to derive the closed-loop transfer function, providing we substitute in the appropriate transfer function $\boldsymbol{G(s)}$ for G and the controller transfer function $K(1+T_d s)$ for K. Hence we have

$$\frac{\Theta_o(s)}{\Theta_i(s)} = \frac{K(1 + T_d s)G(s)}{1 + K(1 + T_d s)G(s)}$$

Using a value of $T_d = 0.35$, so that the 'amount' of derivative action is the same as the 'amount' of velocity feedback in the previous example, and assuming $K = 4$ and $\tau = 1$ as before, we obtain:

$$\frac{\Theta_o(s)}{\Theta_i(s)} = \frac{4(1 + 0.35s)}{s^2 + 2.4s + 4}$$

The denominator of this transfer function is identical to that of the system with velocity feedback – that is, it is the same as the denominator of a standard second-order form with $\zeta = 0.6$ and $\omega_n = 2$. It is tempting to think that the step response will therefore be identical. This is not the case, however; we cannot use the standard curves of Figure 8.10 for this second-order system with its non-standard transfer function numerator. The step response can easily be simulated by computer, however, using one of the many software packages available. Figure 8.27 shows the result.

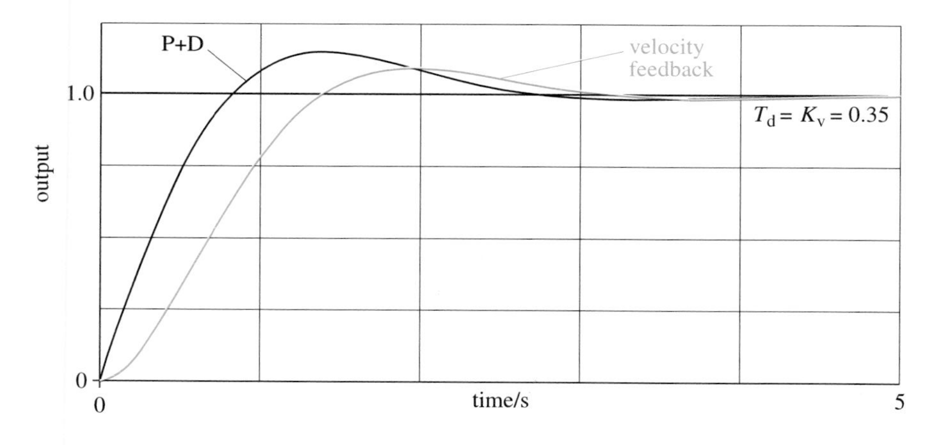

***Figure 8.27* Comparison of robot arm step responses with a P+D controller and velocity feedback.**

The effect of the extra term $(1 + 0.35s)$ in the numerator is that the system with the P+D controller has a faster rise time, but a rather greater overshoot, than the velocity feedback design. As might be expected, an increase in T_d increases the damping further; trial and error with the computer package shows that a value of $T_d = 0.45$ results in a step response with an overshoot limited to 10%, as shown in Figure 8.28.

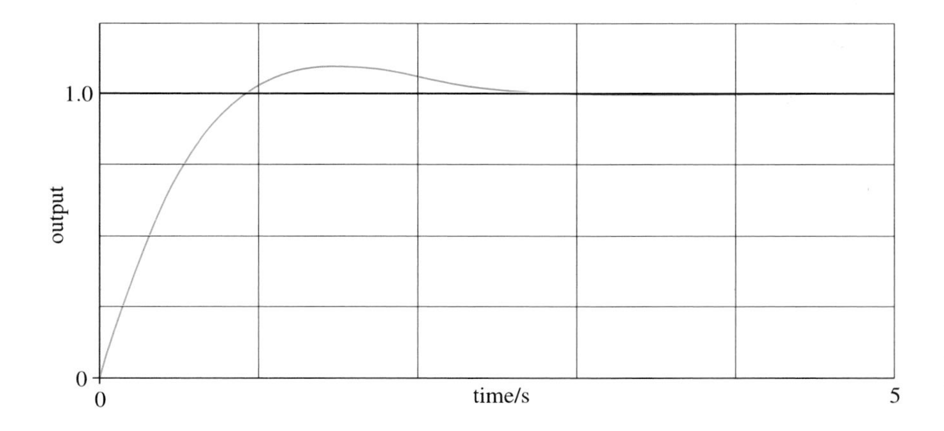

Figure 8.28
P+D control fulfilling the step-response specification by increasing T_d to 0.45.

We must interpret such simulations, though, in the light of our knowledge of system practicalities. The greater the derivative action (the greater the values of T_d or K_v), the greater the voltage applied by the controller to the motor in response to the change in demanded position. Ultimately, the motor (or other actuator) will not be able to respond in time to force the process to behave in the required way, and the models and their predictions break down. So we can't just increase derivative action indefinitely to improve system response.

To summarize, then:

- Both P+D control and velocity feedback increase the damping of a lightly-damped system.
- P+D control has no effect on tracking performance, whereas velocity feedback worsens it.
- The step response of a system using P+D control will be faster, and possibly exhibit greater overshoot, than an otherwise comparable system using velocity feedback.

The choice between velocity feedback and P+D control (or, indeed, some other solution) will be influenced in practice by many considerations other than those mentioned so far. The degree to which a measured variable is contaminated by noise is one important factor, particularly in deciding whether to sense the rate variable directly (using a tachogenerator, for example) or to differentiate the position variable using computer software. A noisy signal with unwanted high-frequency components changes rapidly by definition, and is likely therefore to generate even larger unwanted components in its derivative. The greater the derivative action (the greater the value of T_d in a P+D controller), the worse this noise amplification becomes. Numerical approximations to differentiation used in control software may also need to be designed carefully to avoid unwanted side effects. As is always the case in engineering, the final design solution will involve a judicious mix of theory, practical constraints and experience.

8.5.2 Improving steady-state performance: controllers with integrating action

You should remember from Section 8.4.3 that one way of interpreting the zero steady-state error of the position control systems considered so far is as a consequence of the integrating effect of the motors. Incorporating ***integral action*** into the controller can have a similar effect. Look at Figure 8.29 for a moment, which illustrates an attempt to control a first-order system so that there is zero steady-state error in response to a change in demanded output.

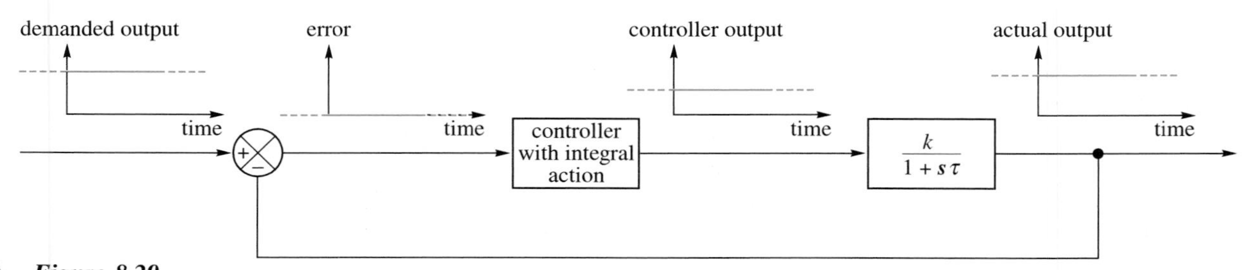

Figure 8.29
Integral action in the controller eliminates steady-state error.

Since the plant does not possess integrating action this can only be achieved by providing a steady controller output even for zero steady-state error. Figure 8.30 shows how integrating action in the controller can achieve this. The time integral of the error can remain at a finite value even after the error has become zero and remains there.*

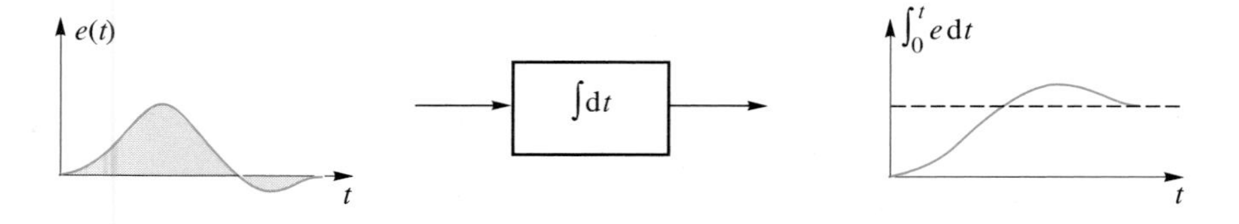

Figure 8.30
Integrating an error signal.

Although there are some control systems which use pure integrators as controllers, integral action is normally provided in conjunction with other controller terms. In industrial process control, *proportional + integral (P+I) action* is common, where the controller output for an error e is given by the expression

$$K\left(e + \frac{1}{T_{\mathrm{i}}}\int e\,\mathrm{d}t\right)$$

* In appropriate circumstances, of course, the time integral of the error can also remain at zero. In general, though, it takes on a constant value even when $e_{ss} = 0$.

T_i is a constant known as the integral time. In mechatronics, integral action – when used – is often combined with proportional *and* derivative action, leading to the *three-term* or ***PID controller*** whose output is

$$K\left(e + \frac{1}{T_i}\int e\,dt + T_d\frac{de}{dt}\right)$$

The designer thus has available three controller parameters, which gives great flexibility in controller design. (It should be noted, though, that too much integral action can make a closed-loop system slow to respond or even unstable. In this respect integral action has the opposite effect to derivative action.)

Integral action in the controller has other desirable steady-state properties. If both the plant and the controller possess integrating action, as in the position control example of Figure 8.31, then the steady-state *tracking* error can be reduced to zero. And finally, the error in response to a steady *load disturbance* can also be eliminated by a controller with integral action: once again the controller can maintain a steady compensating output to the plant even when e_{ss} is zero.

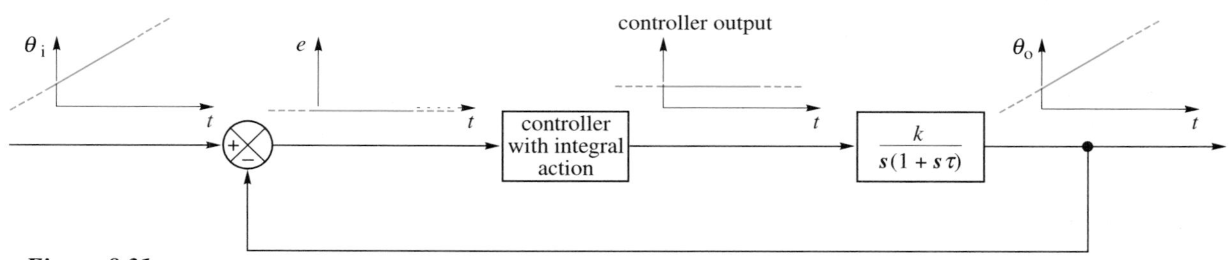

Figure 8.31
Integral action in both controller and plant eliminates steady-state position error when tracking.

Modelling the dynamic behaviour of such control systems is beyond the scope of this chapter, but note that all the steady-state deductions from diagrams like Figures 8.29 and 8.31 only hold providing the closed-loop system is stable. In more complex systems than the simple examples presented here, a poor choice of controller parameters can mean that closed-loop poles move into the right half of the s-plane, resulting in instability. Control engineers have various tools at their disposal for checking the stability of such systems: details can be found in the texts listed at the end of this chapter.

Table 8.1 sums up the (theoretical) effects of integrators in control loops of the kind considered so far – assuming the loop is stable.

TABLE 8.1 EFFECTS OF INTEGRATORS IN CONTROL LOOPS

	e_{ss} **in response to step change in loop input**	e_{ss} **in response to step change in disturbance**	e_{ss} **under steady tracking conditions**
Integrator in plant only	zero	finite	finite
Integrator in controller only	zero	zero	finite
Integrator in both controller and plant	zero	zero	zero

8.5.3 Feedforward control

Integral action and/or high gain in the controller is not the only way of reducing the effects of load disturbances to an acceptable level. As was mentioned in Section 8.2, an alternative technique is to measure the disturbance itself, and use this measurement to make an appropriate change to the control action. This is known as *feedforward*, and has the advantage that significant changes can be made to the control action, in order to compensate for time-varying disturbances, without waiting for the disturbances to have an effect on the variable being controlled.

Feedforward is often combined with feedback control. This is a common approach in mechatronic systems for position control and in active suspension systems. Examples of the latter are: isolation of the passenger compartment of a helicopter from vibration; or provision of a stable loading deck on a boat in a choppy sea.

Figure 8.32 illustrates the general idea. In addition to the closed-loop control strategy – which may include velocity feedback or derivative action to provide a suitably damped response – disturbances to the system are sensed and fed into the control computer. Provided that a suitable model exists of the effect of the disturbance on system behaviour, the computer can then calculate a compensating additional component in the control action. In the case of a robot arm or a lift, for example, a sensor might be used to detect the load torque – or a sophisticated, computer-based algorithm might be used to detect the changing moment of inertia as a heavy load is moved around in space. In the case of vibration isolation, the time-varying disturbance is measured and processed to generate additional control action.

Such techniques often involve considerable signal processing, and have become much more widespread with the availability of cheap, high-performance microprocessors which can be built into mechatronic products. In the next and final section of this chapter we look at some of the distinctive features of control using such digital methods.

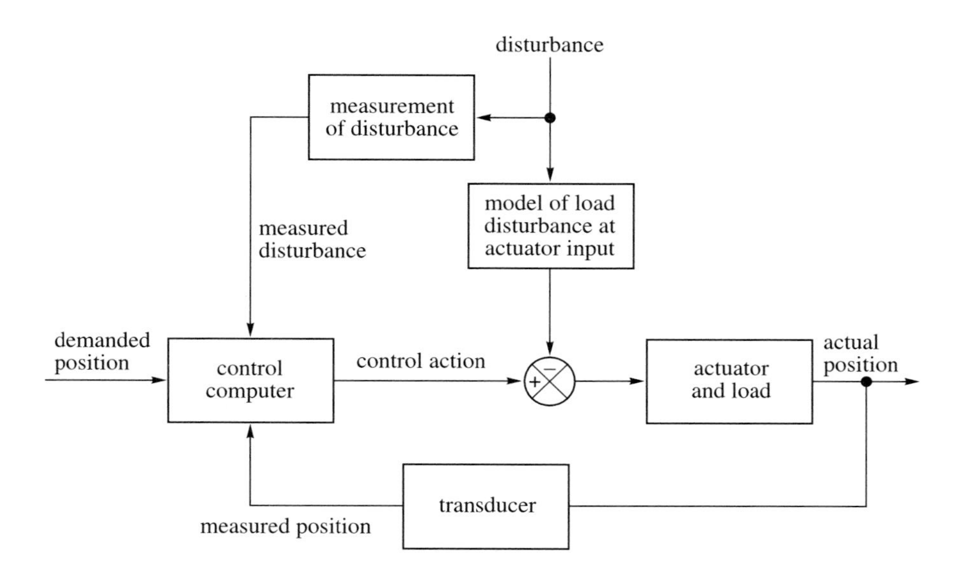

Figure 8.32
Feedback and feedforward control combined.

8.6 Digital control

So far in this chapter I have been assuming that all the signals in a control system – a signal representing robot arm position, for example – can vary continuously with time. Similarly I have assumed that the result of processing these signals by the control computer is exactly the same as the result of carrying out the ideal mathematical operations of multiplication, differentiation, integration, or whatever. But in a digital control system, the variables of interest are sampled and converted to finite-length binary numbers (or the transducers themselves produce a digitally coded output) and the computer processes these values using numerical algorithms. In this section I shall look at some of the implications of this for control system design and behaviour.

8.6.1 Sampling and data conversion

When a continuous signal is sampled, its value is measured at particular instants in time only – usually equally spaced. Clearly, if the sample values are to accurately reflect the time-variation of the signal, they have to be sufficiently closely spaced in time. What has to be avoided at all costs is giving the signal time to fluctuate significantly between sample values, as illustrated in Figure 8.33. In part (a) of the figure there are no such fluctuations, and the samples accurately reflect signal behaviour. In part (b), however, the fluctuations are 'missed' by the sampler. To avoid this phenomenon, more than two samples have to be taken per cycle of the highest frequency component present in the signal. Or, as ***Shannon's sampling theorem*** is more usually stated:

> Assuming that the frequency components of a continuous signal extend from zero to B Hz, then the signal must be sampled at a rate of greater than $2B$ samples per second if all the information in the signal is to be preserved in the samples.

In a control system it is common to sample at a rate many times that of the highest significant frequency component.

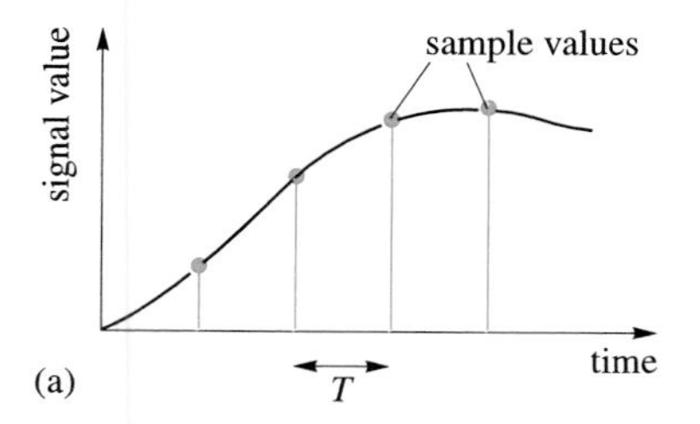

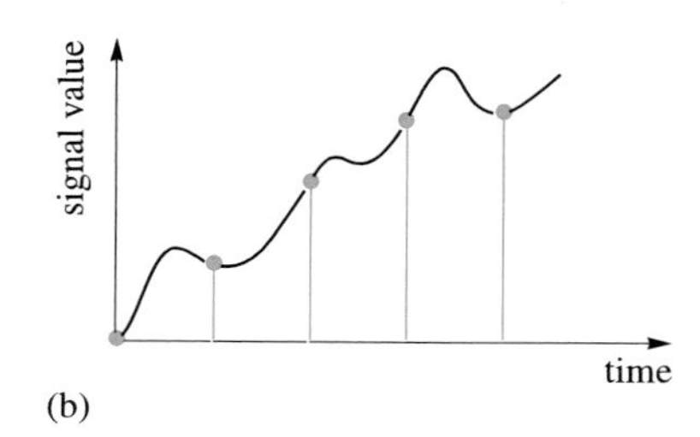

Figure 8.33
A continuous signal must be sampled more than twice per cycle of the highest frequency component to preserve all the information.

Sampled values have to be coded digitally for processing by the control computer. This means that they are rounded to the nearest of a set of quantization levels: 2^n levels for an n-bit analogue-to-digital converter (ADC). The output values generated by the computer for transmission to the actuator are similarly restricted to a finite set of levels by the digital-to-analogue converter (DAC). The control signal generated by the computer is normally held constant between sampling instants, taking on the 'stepped' form illustrated in Figure 8.34 rather than the continuously varying signal tacitly assumed in the discussion of previous sections. These effects of data conversion can often be ignored if the sampling rate is sufficiently high and the converters have a high enough resolution. If it is impossible or undesirable to sample at such a high rate, special digital design techniques may have to be adopted.

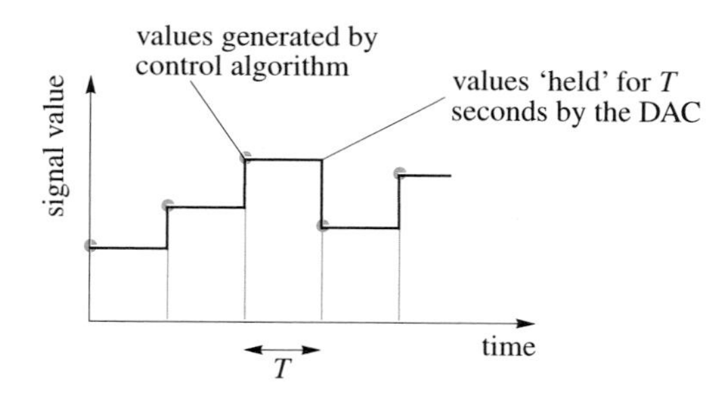

Figure 8.34
General form of control action in a digital control system.

8.6.2 Digital processing

The control computer will have available samples of the error signal (or of the demanded and actual values of the controlled variable, which amounts to the same thing). Using these sample values, approximations to the operations of differentiation and integration can be computed. Figure 8.35 shows a digital approximation to differentiation, in which the derivative of the error is approximated by the slope of the line joining two adjacent samples. The operation can be represented by the following ***difference equation***:

$$d_n = \frac{e_n - e_{n-1}}{T}$$

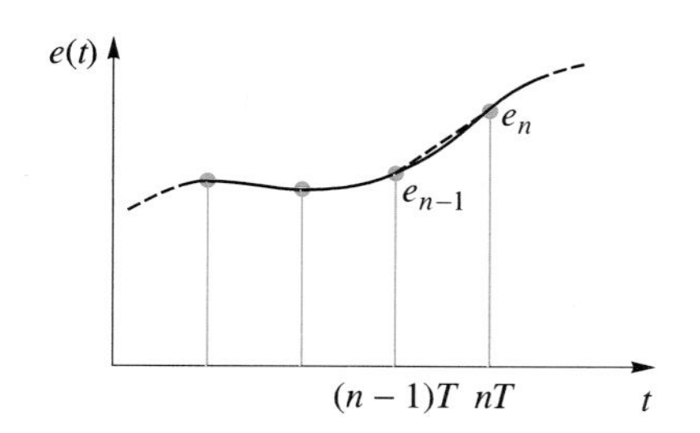

Figure 8.35
A digital approximation to differentiation.

Here d_n represents the nth approximation to the derivative, generated at time nT (where T is the sampling interval). Similarly e_n and e_{n-1} represent the current and previous samples of the error used to calculate the slope.

A similar approach may be adopted to integration, as illustrated in Figure 8.36. The continuous integral $y(t)$ with respect to time of an error signal $e(t)$

$$y(t) = \int_0^t e(t)\,\mathrm{d}t$$

corresponds to the area under the curve of part (a) of the figure. One way of approximating this area, given samples e_n of $e(t)$, is to sum the areas of the rectangular strips shown in Figure 8.36(b). The approximation y_n to the integral at time nT is then given by the expression

$$y_n = \sum_{i=0}^{n} e_i T$$

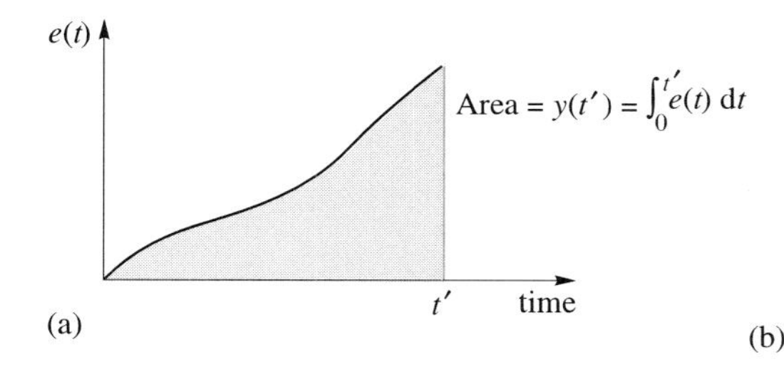

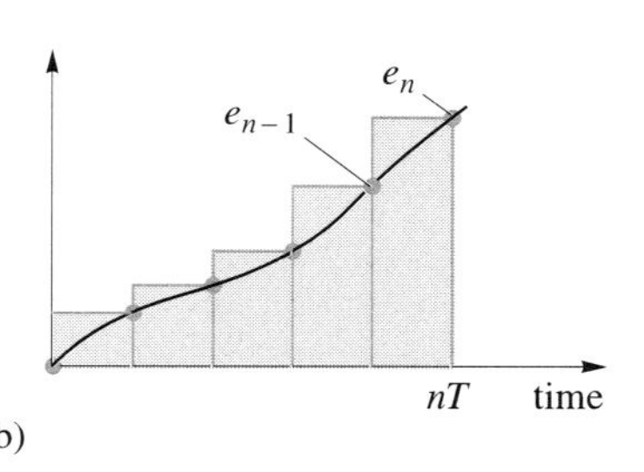

Figure 8.36
A digital approximation to integration.

Clearly, the smaller the sampling interval T, the closer this approximation is to the value of the integral.

A difference equation representing this form of numerical integration may be obtained by noting that

$$y_n = y_{n-1} + e_n T$$

That is, at each sampling instant the new value of the integral is calculated by adding the area of the new strip $e_n T$ to the previous value of the integral y_{n-1}.

There are other forms of numerical differentiation and integration, with slightly different difference equations.

Using approximations of this type, P+I, P+D and PID controllers (as well as many other strategies not discussed in this chapter) may be implemented in computer software.

8.6.3 Choice of sampling rate

The sampling theorem sets a lower limit to the sampling rate in a control system – albeit a rather ill-defined one, since there is often no clear upper limit to the frequencies present in a control loop. Control engineers tend to use rules of thumb, often expressed in terms of the closed-loop '3 dB bandwidth'.* Depending on the design strategy being used, the sampling frequency would generally be selected between 10 and 50 times this 3 dB bandwidth.

A practical *upper* limit to the sampling rate is set by a number of factors. First there are the physical limitations set by the speed of operation of the control computer or microprocessor which, if the system is to operate in real time, has to carry out all the calculations required by the control algorithm within one sampling interval. If several control loops are involved, there may be a trade-off between using a separate computer for each loop at a high sampling rate, or a single computer for the whole system but restricted to a lower rate. Second, sampling too fast can lead to problems with numerical algorithms, as a result of the finite word lengths of data converters and processors. For example, if numerical differentiation is involved, a high sampling rate might imply dividing one very small number by another, leading to inaccuracies. It also turns out that the higher the sampling rate, the more precise the controller parameters need to be in order to implement the control algorithm properly. At very high sampling rates, it may just not be possible to realize these parameters accurately enough with a processor of a realistic word length. For high-performance systems a range of techniques specific to digital control have been developed to cope with such problems.

* If the input to a linear system is a pure sinusoid then the output is also a pure sinusoid, but differing, in general, in both amplitude and phase. Suppose the input to the loop is a sinusoid of amplitude A, and the output a (phase-shifted) sinusoid of amplitude B. The (frequency-dependent) ratio B/A is known as the closed-loop amplitude ratio. The 3 dB bandwidth is then the frequency above which the amplitude ratio remains at least 3 dB below its low-frequency value. This bandwidth can be calculated from the system transfer function or measured by frequency-response testing.

8.6.4 A simple example

Just to give a flavour of how the performance of a digital control system can differ from that of an analogous continuous system, consider Figure 8.37, showing an ultrasonic motor used in a digital position control system. To keep the algebra simple, the motor is modelled as a pure integrator, $1/s$, with all other factors combined into a single gain element K. The samples of the demanded position are denoted x_n, and those of the actual position y_n. We can simulate the behaviour of this system using a difference equation. (If you find the details of the difference equation and its solution difficult to follow, concentrate on the graphs and conclusions at the end of the section.)

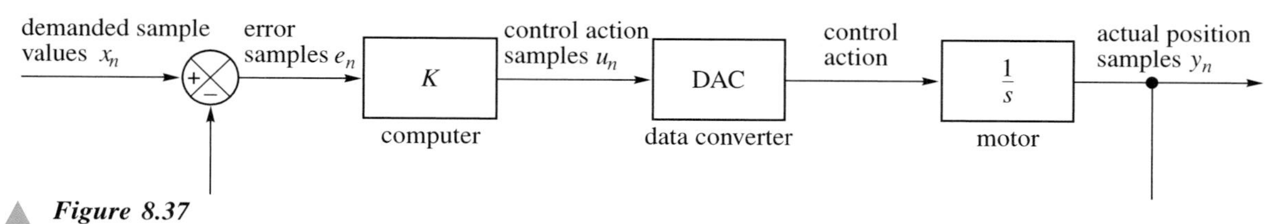

Figure 8.37
A digital position control system using an ultrasonic motor.

The output of the DAC to the motor consists of a stepped waveform as discussed above. These steps are 'integrated' by the motor into a succession of short ramps, shown in Figure 8.38, where the successive levels of the DAC output are denoted u_n.

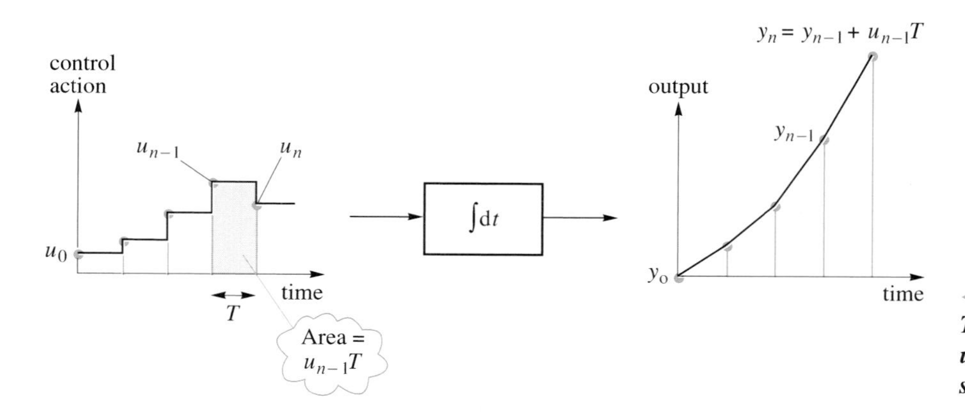

Figure 8.38
The response of the ultrasonic motor to the stepped input.

Using our knowledge of the integrator step response we can write

$$y_n = y_{n-1} + u_{n-1}T$$

But

$$u_n = Ke_n = K(x_n - y_n)$$

so

$$y_n = y_{n-1} + KT(x_{n-1} - y_{n-1})$$

or

$$y_n = (1 - KT)y_{n-1} + KTx_{n-1}$$

We can use this difference equation to simulate the closed-loop behaviour – manually or by computer. Suppose, for example, that $K = 5$ and the sampling rate is 10 Hz ($T = 0.1$). Then $KT = 0.5$ and

$$y_n = 0.5y_{n-1} + 0.5x_{n-1}$$

We can calculate the closed-loop step response by setting up Table 8.2. Initially the system is quiescent, with output and input samples (and previous output and input samples) all equal to zero. The input is then suddenly changed to a value 1 and held there, so that after a certain instant the input samples are those of the unit step: 1, 1, 1, 1, 1, 1, 1 ... The output sample values can then be calculated iteratively, using the difference equation $y_n = 0.5y_{n-1} + 0.5x_{n-1}$. Notice how entries are carried over diagonally from the column representing x_n to that representing x_{n-1}, and from the column representing y_n to that representing y_{n-1}.

TABLE 8.2

Previous input samples x_{n-1}	Input samples x_n	Previous output samples y_{n-1}	Output samples $y_n = 0.5y_{n-1} + 0.5x_{n-1}$
⋮	⋮	⋮	⋮
0	0	0	0
0	0	0	0 (input step applied here)
0	1	0	0
1	1	0	0.5
1	1	0.5	0.75
1	1	0.75	0.875
⋮	⋮	⋮	⋮

increasing time ↓

This is all rather time consuming, so I've used a computer spreadsheet to automate the procedure, as shown in Figure 8.39. The step response resembles the first-order step response of the continuous system in Section 8.3.2. Note, however, that the position between samples follows a straight line (because of the nature of the ultrasonic motor response to the DAC output), in contrast to the smooth variation of the standard first-order model (with a continuously-varying controller output). Figure 8.40 shows what happens if the gain is increased to 8: as might be expected from our earlier discussion of feedback loops, the response speeds up. But look what happens as the gain is increased further – to a value of 15 in Figure 8.41 and to 21 in Figure 8.42. The step response first becomes oscillatory and then completely unstable!

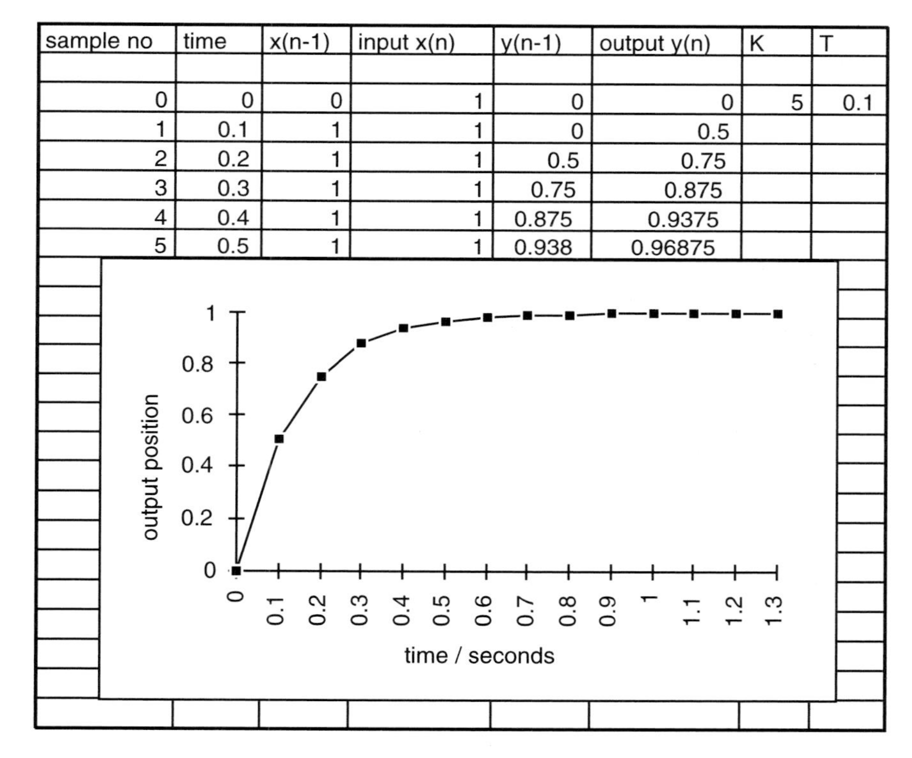

sample no	time	x(n-1)	input x(n)	y(n-1)	output y(n)	K	T
0	0	0	1	0	0	5	0.1
1	0.1	1	1	0	0.5		
2	0.2	1	1	0.5	0.75		
3	0.3	1	1	0.75	0.875		
4	0.4	1	1	0.875	0.9375		
5	0.5	1	1	0.938	0.96875		

Figure 8.39
Simulation of digital loop with K = 5;
($y_n = 0.5y_{n-1} + 0.5x_{n-1}$).

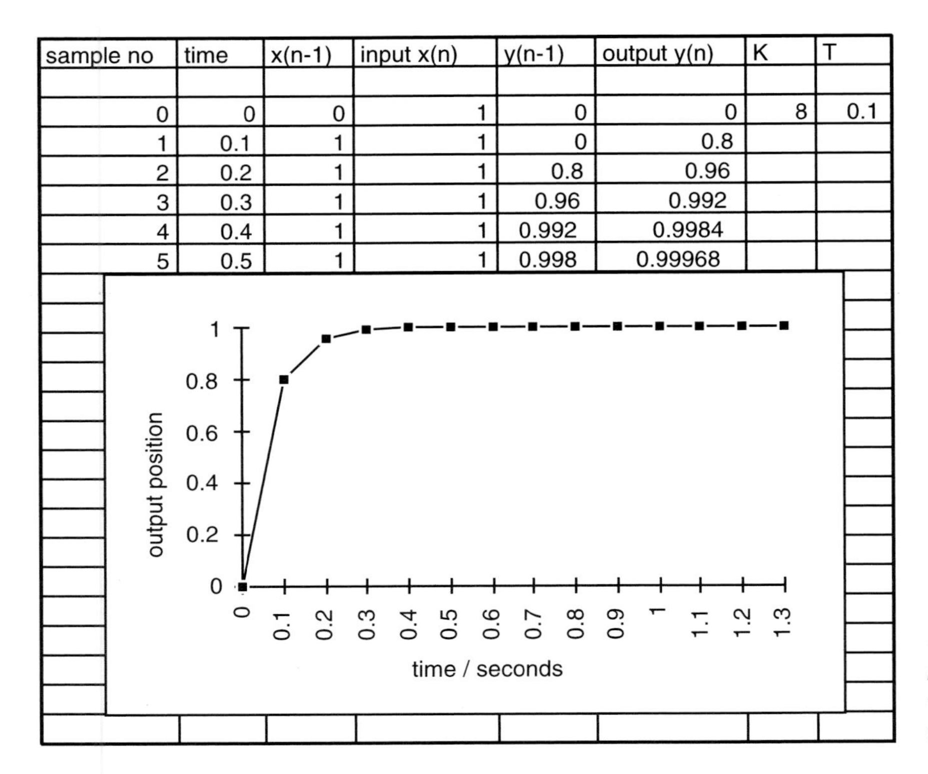

sample no	time	x(n-1)	input x(n)	y(n-1)	output y(n)	K	T
0	0	0	1	0	0	8	0.1
1	0.1	1	1	0	0.8		
2	0.2	1	1	0.8	0.96		
3	0.3	1	1	0.96	0.992		
4	0.4	1	1	0.992	0.9984		
5	0.5	1	1	0.998	0.99968		

Figure 8.40
Simulation of digital loop with $K = 8$; ($y_n = 0.2y_{n-1} + 0.8x_{n-1}$).

This is very different behaviour from that of the continuous model (although in line with my general principle that increasing the gain too much can result in instability). And what is more, since *K* and *T* appear in this particular difference equation as the term *KT*, increasing the sample interval (reducing the sampling frequency) has a similar effect – apart from the scaling of the time axis due to the change in *T*. If the sampling rate is reduced below a certain value for a given gain, oscillatory behaviour is observed, and if it becomes *too* low, instability results. This isn't too surprising if we consider what happens in the loop as the sampling rate is decreased. During any given sampling interval, no information is fed back to the controller about changes in the output. So if this sample interval is too long in comparison with the loop dynamics, then the controller is unable to take

sample no	time	x(n-1)	input x(n)	y(n-1)	output y(n)	K	T
0	0	0	1	0	0	15	0.1
1	0.1	1	1	0	1.5		
2	0.2	1	1	1.5	0.75		
3	0.3	1	1	0.75	1.125		
4	0.4	1	1	1.125	0.9375		
5	0.5	1	1	0.938	1.03125		

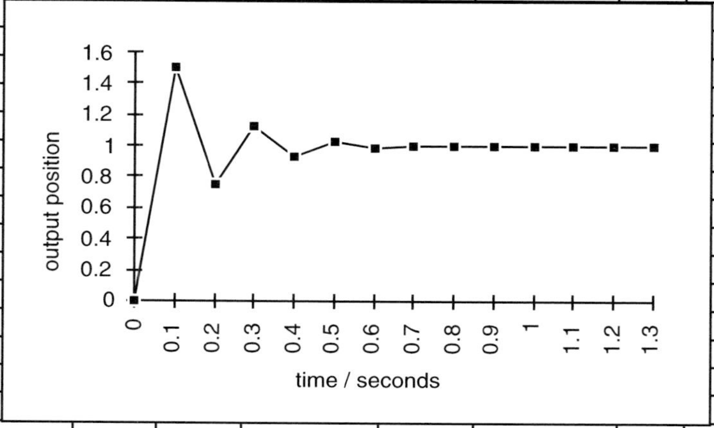

Figure 8.41 *Simulation of digital loop with K = 15; ($y_n = -0.5y_{n-1} + 1.5x_{n-1}$).*

sample no	time	x(n-1)	input x(n)	y(n-1)	output y(n)	K	T
0	0	0	1	0	0	21	0.1
1	0.1	1	1	0	2.1		
2	0.2	1	1	2.1	-0.21		
3	0.3	1	1	-0.21	2.331		
4	0.4	1	1	2.331	-0.4641		
5	0.5	1	1	-0.46	2.61051		

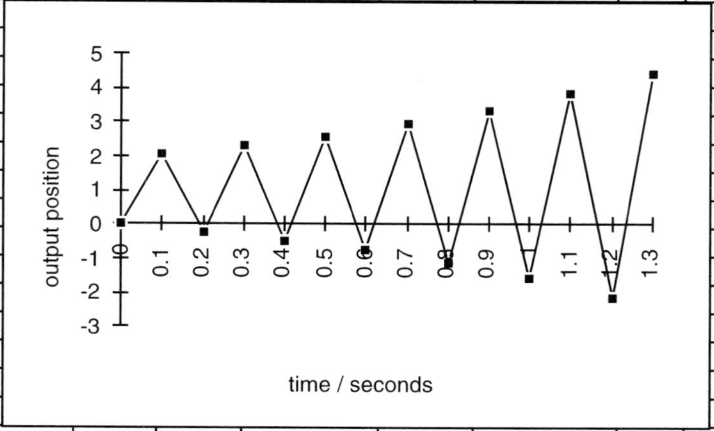

Figure 8.42 *Simulation of digital loop with K = 21; ($y_n = -1.1y_{n-1} + 2.1x_{n-1}$).*

appropriate remedial action in time to stop the output oscillating indefinitely. In fact, it is generally observed that:

If the sampling rate is sufficiently high (in the context of loop dynamics), a digital implementation of a continuous control system design often behaves very like the original.
If the sampling rate is too low, a digital system can exhibit oscillatory behaviour or instability – even when the original continuous model does not predict this.

Higher order digital control systems can be modelled in a way very similar to our approach to the ultrasonic motor system. Higher order difference equations are needed, and the behaviour between samples is determined by the dynamics of the process being controlled.

Sketch the step response of the digital position control system of Figure 8.37 for $KT = 1$. Can you give a physical explanation for this behaviour?

With $KT = 1$ the difference equation reduces to $y_n = x_{n-1}$. The step response is illustrated in Figure 8.43. The output reaches exactly its required position in one sampling interval and stays there, so the output sample values (0, 1, 1, 1, 1, ...) are equal to the input ones delayed by one sampling interval.*

Many mechatronic control systems use high sampling rates, together with tried and trusted control strategies which predate digital technology. Such systems can often be modelled without taking into account explicitly the digital nature of the loop. Other systems use a sampling rate which is (1) low enough for sampling and data conversion to have a significant effect of performance, but (2) high enough for the sampling not to result in significant loss of information about the measured variables, and hence for good control to be possible. Such digital systems require a new set of modelling and design tools beyond the scope of this chapter.

* This is an example of what is known as a 'deadbeat' response, and is occasionally used in digital control systems. In general though, even if the *sample* values stay at the required value after a short time delay, the controlled variable itself may fluctuate between samples, giving rise to 'hidden oscillations'.

sample no	time	x(n-1)	input x(n)	y(n-1)	output y(n)	K	T
0	0	0	1	0	0	10	0.1
1	0.1	1	1	0	1		
2	0.2	1	1	1	1		
3	0.3	1	1	1	1		
4	0.4	1	1	1	1		
5	0.5	1	1	1	1		

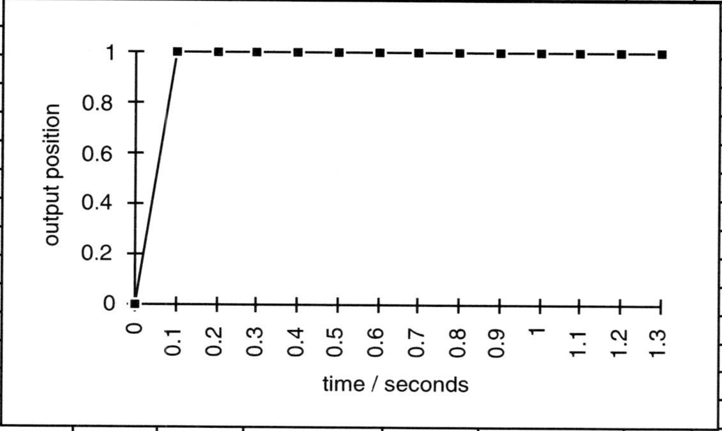

***Figure 8.43* Simulation of digital loop with $K = 10$; ($y_n = x_{n-1}$).**

8.7 Conclusion

This chapter has presented many of the characteristics of feedback control, drawing a number of important general principles from the behaviour of simple first- and second-order models. It is worth repeating these here – always bearing in mind that exceptions to the 'rules' can be found in certain circumstances.

1. Increasing proportional gain in a feedback control system tends to speed up system response, reduce steady-state errors, and decrease the effects of disturbances. But increasing the gain too much can result in an oscillatory closed-loop response – and even instability.
2. If the controller possesses integrating action (has an s multiplier in the transfer function denominator) then the closed-loop steady-state error in response to a step change in demanded value, or a steady load disturbance, is zero. If both the controller and the process possess integrating action, then the position error under steady tracking conditions is also zero. Too much integral action in a controller, however, can render closed-loop performance sluggish or even drive the loop into instability.

3 Both P+D control and velocity feedback can be used to increase the damping of a lightly damped system. P+D control has no effect on tracking performance, whereas velocity feedback worsens it. The step response of a system using P+D control will be faster, and possibly exhibit greater overshoot, than an otherwise comparable system using velocity feedback. Too much derivative action, however, can lead to problems with noise, or make too great demands on actuators.

4 If the sampling rate of a digital control system is sufficiently high (in the context of loop dynamics), it often behaves like the continuous one on which it is based. If the sampling rate is too low, the digital implementation can exhibit oscillatory behaviour or instability – even when the original continuous model does not predict this.

Control system designers often have to reconcile the apparently conflicting requirements of low (or zero) steady-state errors, fast speed of response, little or no oscillation, and good rejection of disturbances. Controllers combining proportional, derivative and integral action are often used in mechatronic products to achieve such design aims. Such control algorithms are easy to implement in computer or microprocessor software. In mechatronic systems a feedback loop is often combined with feedforward to further improve disturbance rejection.

Further reading

If you want to learn more about control engineering I recommend either of the two following books:

Bissell, C. C. (1994, 2nd edn), *Control Engineering*, Chapman & Hall.
Golten, J. & Verwer, A. (1991), *Control System Design and Simulation*, McGraw Hill.

(The latter comes with useful simulation software which will run on almost any personal computer.)

System modelling is considered in more detail in:

Doebelin, E. O. (1985) *Control System Principles and Design*, Wiley.
Meade M.L. and Dillon, C.R. (1991) *Signals and Systems*, Chapman and Hall.

CHAPTER 9
ARCHITECTURES

George Kiss

9.1 Introduction

Mechatronic design is a complex process. The desire to get to grips with this complexity is what motivates an interest in architecture issues.

In this chapter I first discuss what I mean by architecture, based on an analysis of common everyday usage and its specialization in some engineering disciplines like electronics and software engineering.

Issues about structures and their description are discussed in detail in Section 9.3. Abstraction and modularity are identified as two major tools.

Structural descriptions are always given from some point of view. Section 9.4 is about structure from the point of view of time and control (causality) and their close connection with each other.

Finally, we look at two case studies of architecture design for intelligent machines that have been influential in the history of the subject.

9.2 Architecture in mechatronic design

9.2.1 Architecture and architects

The words 'architecture' and 'architect' were originally introduced in relation to buildings, but more recently have been both generalized and specialized in various areas of science and technology. Let's start by looking at how the *Shorter Oxford Dictionary* summarizes the general uses of these words.

Architecture

1 The art or science of constructing edifices for human use (occasionally regarded merely as a fine art).
2 The action or process of building (now archaic).
3 Structure.

4 A special method or style of structure and ornamentation.
5 Construction generally (figurative use).

Architect

1 A master-builder, specifically one whose profession it is to prepare plans of edifices and exercise a general superintendence over their erection.
2 One who designs and frames any complex structure.
3 One who so plans and constructs, as to achieve a desired result; a builder-up.

The general concepts that these dictionary entries encompass are: designing, constructing, building and framing as activities; complex structure and edifice as the objects of these activities; and method and style as attributes of the activities and their objects.

These concepts could all be regarded as in some way relevant to mechatronics, especially if we replace 'edifices' by 'machines'. More on this in a moment, after a brief discussion of how architecture is used as a technical term in the areas close to mechatronics.

Architecture as a technical term is now fairly commonly used in electronics, software engineering, artificial intelligence and cognitive science. As is often the case, the original sense of the term becomes somewhat modified, while retaining some commonality. Let us look at just two examples briefly, electronics and software engineering.

In electronic engineering, architecture means the description of a complex system in a form that abstracts away fine detail. Sometimes the description is said to be global or high level. It is common to use block diagrams for such descriptions. As an illustration of this, Figure 9.1 reproduces a diagram from Texas Instruments' (1991) *User's Guide* to their TMS320C3x devices (x stands for one of the extra identifying numbers). This is a family of digital signal processors which are of great use to the mechatronics designer in processing information from sensors and executing control algorithms. The only way in which a device with such ***complexity*** (containing many millions of transistors and other components) can be designed, built and explained to a prospective user is to take it in stages, by looking at a simplified illustration first. The architecture diagram copes with complexity through *abstraction*: it leaves out fine detail and describes the structure of the device 'at a high level of abstraction from detail' (see also Section 9.3.1). The level chosen is determined by such factors as our short-term memory and attentional capacity, the convenient size of a sheet of paper, and the resolution of our printing.

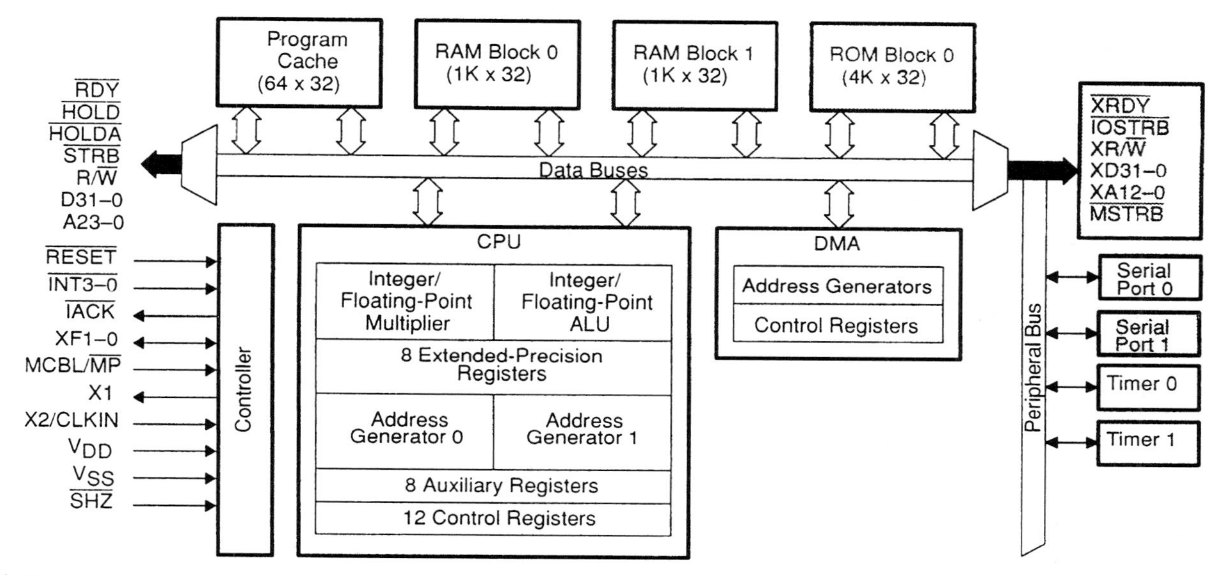

Figure 9.1
The architecture of the TMS320C3x device.

The information that is hidden in the high-level diagram can be revealed in lower level diagrams, as shown in Figure 9.2 (overleaf). This process of adding more detail can be continued almost indefinitely by 'zooming in' to a chosen area of each higher level diagram. For very large scale integration (VLSI) engineering use, the expansion usually stops at the level that depicts the structures etched onto a semiconductor surface.

This example illustrates that the description of a complex device or system can be given at any one of a whole range of degrees of detail. In electronics, by the architecture of a system we mean the highest level description in this range.

In software engineering, the term ***architecture*** is similar. It is a high-level view of the machine, or of a whole class of machines, for which programs are being written. The architecture usually has a strong influence on the manner and style of programming (at least in some stages of developing a program). Examples of architectures in this context are the von Neumann architecture, multiprocessor architectures and distributed system architectures. (Each of these classes has subclasses, but this introductory section is not the place to go into these.) Let us sketch the differences between these architectures.

- *Von Neumann architecture* This is the oldest and is named after the designer of one of the first digital computers. Its abstract computational model is a *random access sequential machine*. The architecture is simple. There is a processor and a memory. The processor can read and write any part of the memory at any time (hence the name 'random access'). The processor executes operations on the information read from memory and writes the results back into memory. The operations to be executed (often called the 'instructions') are specified by the contents of some memory locations, as are the operands (often called the 'data'). The processor is called a sequential

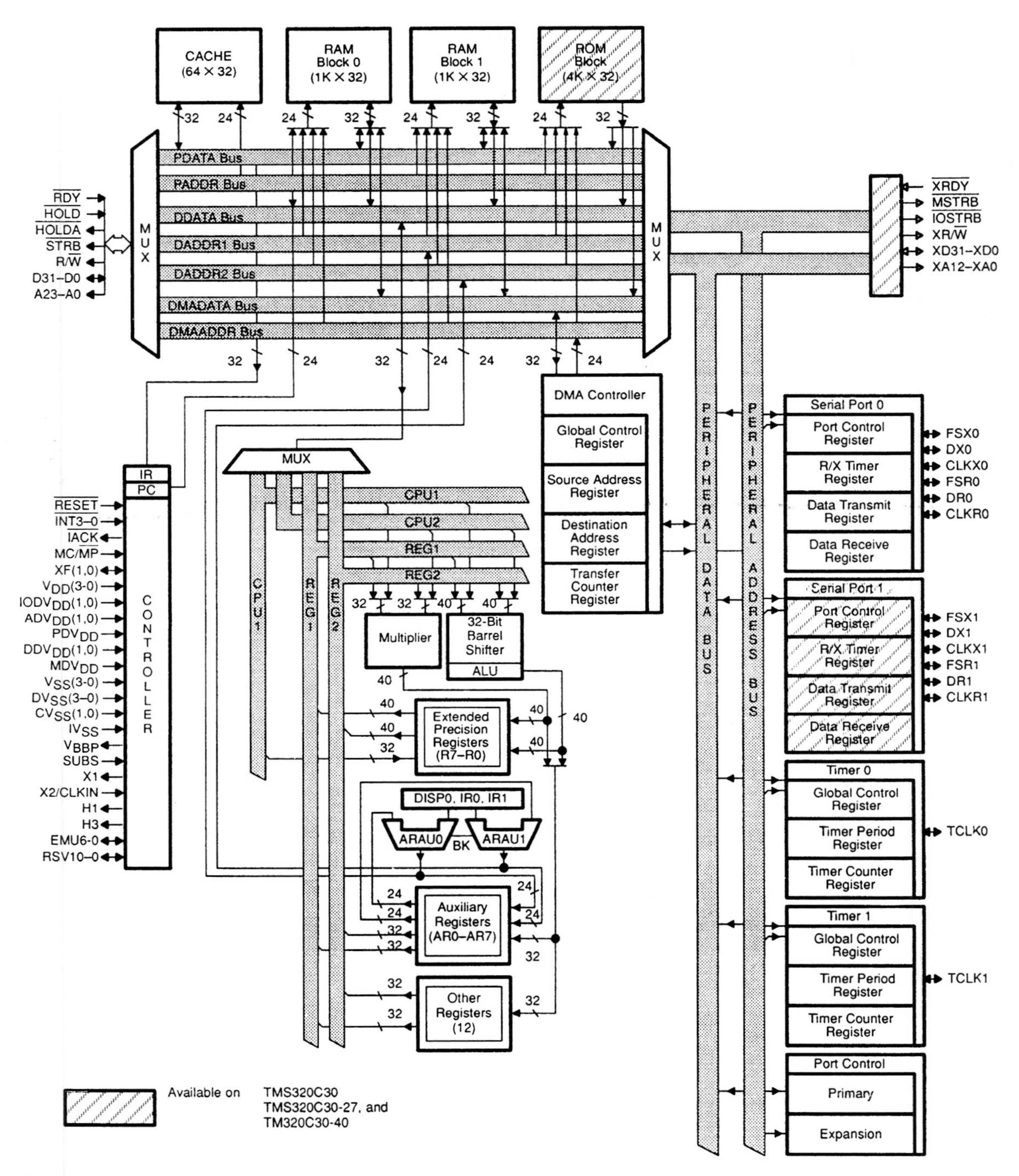

▲ ***Figure 9.2***
More detailed block diagram of the TMS320C3x device.

machine because the operations are done one at a time in a temporal sequence (no 'concurrency').

- *Multiprocessor architecture* This differs from the von Neumann architecture in that there is more than one processor and they share a common memory.
- *Distributed system architecture* In this there are several processors, each with its own memory, and the processors communicate by sending messages over communication channels.

I shall have more to say about these architectures later. Here they merely serve as illustrations of what computer scientists mean by architecture. Clearly, the usage is similar to that in electronics (in giving a high-level description of structure), but the description is given from the point of view of how the structure impacts the programs that will be written.

In summary, when I mention the term 'architecture' in this chapter, I shall mean a description of the structure of a complex system, formulated at a high level of abstraction from detail. (The concept of structure is left intuitive at this point, but will be discussed in more detail later.)

9.2.2 Architecture in design

Many different approaches to design exist. Some of them have been codified into 'methodologies'. There is top-down and bottom-up design, structured design, object-oriented design, evolutionary design, concurrent design, and so on. Some of these are discussed in Chapter 10.

Architecture design is not a methodology but a *stage* of the design process. In a top-down design approach the architecture is designed first and then, through successive refinements, the design is elaborated to the detailed implementation level. This process forms a stage of moving from initial requirements through specification to implementation. Architecture design is most often associated with the early stages of stating the functional specifications.

The architecture design usually divides the product into large modules that bring together a set of *conceptually related concerns*, with relatively *narrow interfaces* between them. A narrow interface minimizes the interaction between the modules it connects. The advantage of narrow interfaces is that if a change has to be made in one module during the design process, little change is needed in other modules. This is useful because changes can be made locally. In a design team working on different aspects of the design in parallel, the changes need not be communicated to other team members.

The modules in the architecture may not and need not relate directly to the modules that will be used in the implementation. The modules in the architecture are conceptual groupings, chosen to facilitate the design process. The modules chosen in the implementation are dictated by physical constraints and the advantages to be obtained by grouping certain components together.

Architecture design is desirable in mechatronics because it enables the designers to isolate themselves from the details of the eventual implementation technologies. The modules in the architecture can be, and should be, abstract enough so that they can be described in terms of function without getting involved with the technologies. The appropriate time for consideration of the various technologies is the implementation stage.

Let me explain this further. The architecture can be described abstractly in terms of a set of *state* variables (see also Section 9.3.1). For an autonomous vehicle (AV), examples of such variables might be the speed and momentum of the vehicle, its position in the workspace, its orientation, the instantaneous loading of its power source, the information it has about obstacles, its task goal, etc. We might choose to implement such variables electrically, mechanically, optically, chemically, 'computationally', or in other technologies. (I have placed computationally in quotation marks, because it is open to debate whether computer technology is a separate technology, rather than just electronics or optics or whatever technology implements the computation.) Such choices belong to the implementation stage.

In summary, architecture design enables the mechatronics engineer to abstract away from the large variety of technologies available and enables the initial design effort to be concentrated on producing a correct functional specification. Some methodologies (especially in software engineering) even expect the designer to prove the correctness of the functional specification, using symbolic logic.

Finally, let me point out that architecture design is not some rarefied activity that only a few designers need to engage in. It is true that new major classes of architectures are not invented every day. That needs a von Neumann to come along. However, in my view each and every object designed has an architecture in the sense that high-level structural descriptions of it can be given. After all, architectures are just descriptions, or *views*, of systems. It is good practice to formulate such a description during design because it facilitates conceptual clarity, intelligibility, communication to others, and possibly provability.

As an example of this, let us look at some structural descriptions of LUCIE (the Lancaster University Computerized Intelligent Excavator) offered by its designers in diagrammatic form (Bradley and Seward, 1992). The authors use five diagrams to describe the architecture of their system, shown in Figure 9.3. Each of these reflects a different view of LUCIE. The first, part (a), is a general mechatronic architecture diagram. They call part (b) the control system hierarchy, part (c) the electronic system view, part (d) the hydraulic system view, and part (e) the software view. Notice carefully that the meaning attached to the connecting arrows and lines in the five diagrams varies widely. Some of them represent a control relationship, others are communication links, and yet others represent energy flows. Similarly, the system components represented by the rectangles and ovals range over hardware and software. Let us now turn to a more detailed account of what these structural descriptions are like.

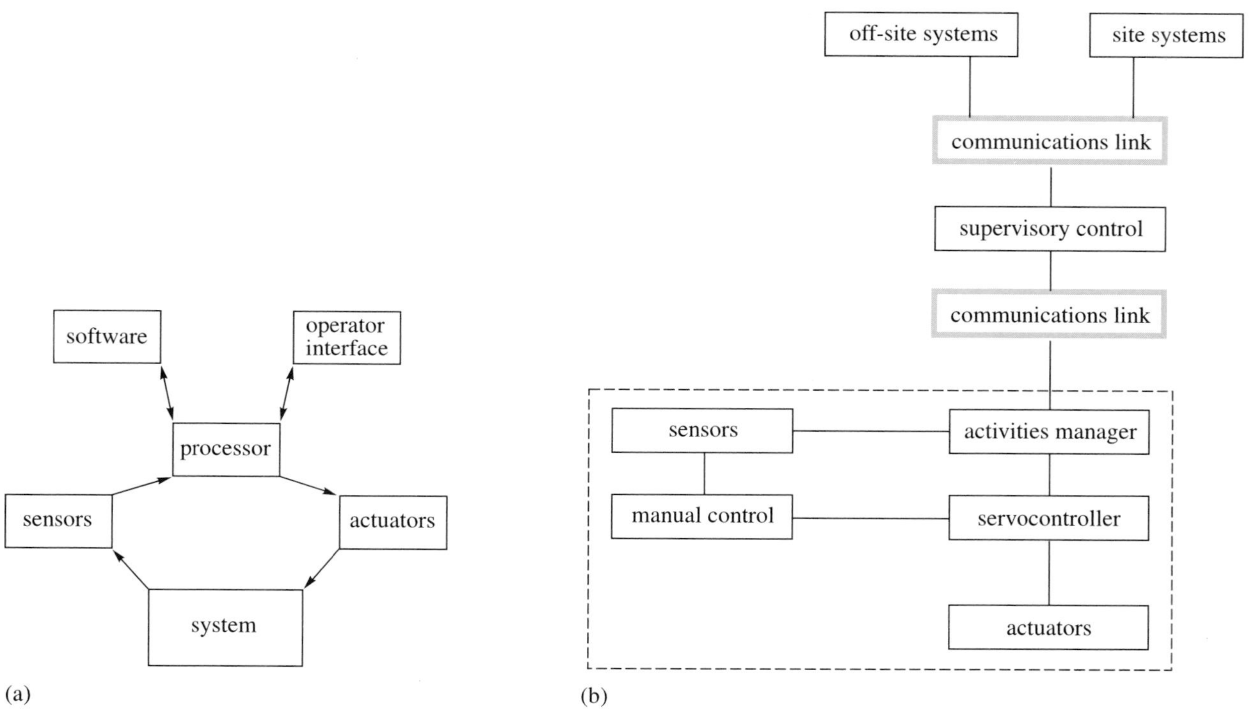

▲ ***Figure 9.3***
The Lancaster University Computerized Intelligent Excavator (photo), and architectural descriptions of its mechatronic design: (a) the general configuration of a mechatronic system; (b) control system hierarchy; (c) (overleaf) electronic system; (d) hydraulic system configuration; (e) software modules.

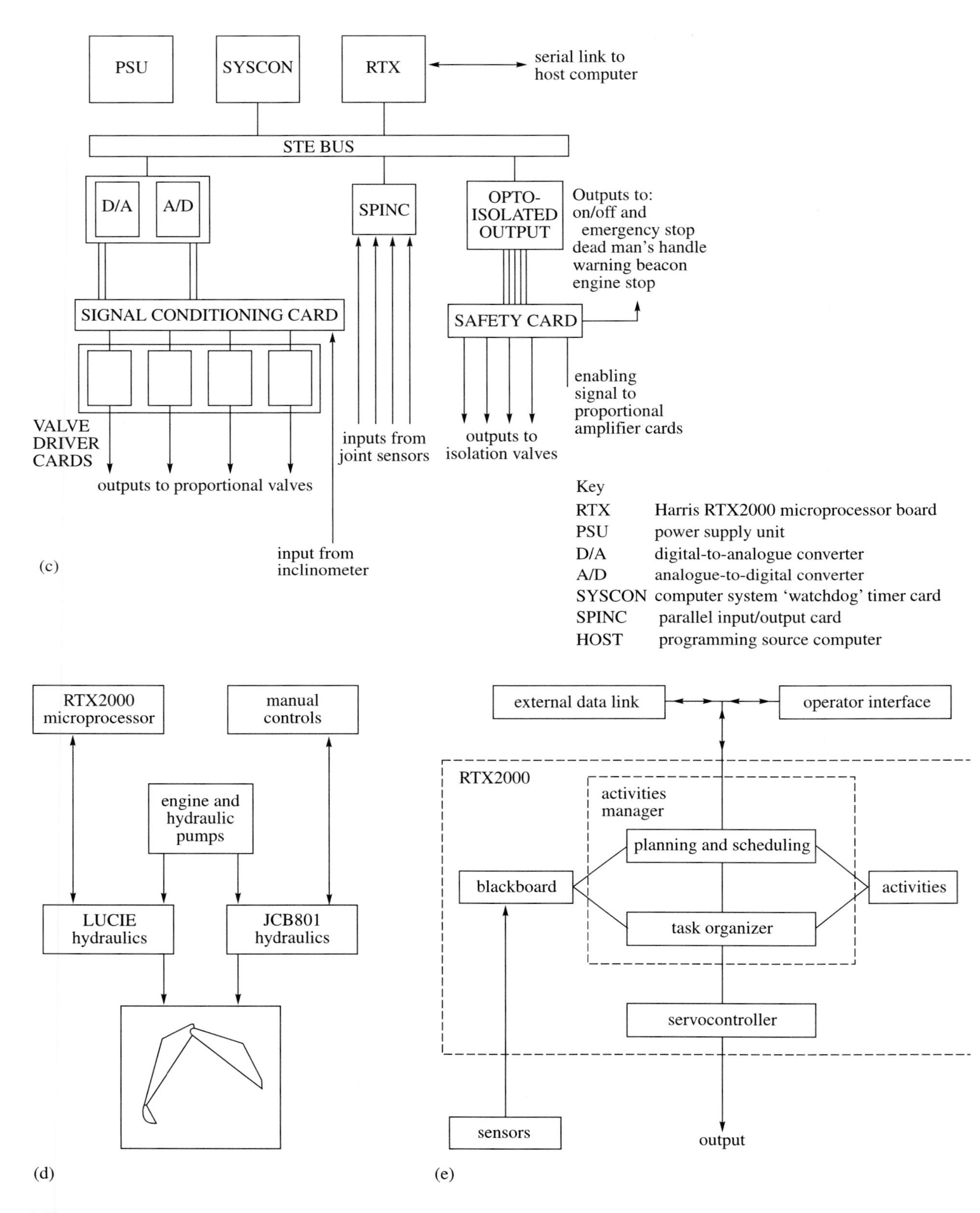
PSU
SYSCON
RTX
serial link to host computer
STE BUS
D/A
A/D
SPINC
OPTO-ISOLATED OUTPUT
Outputs to:
on/off and emergency stop
dead man's handle
warning beacon
engine stop
SIGNAL CONDITIONING CARD
SAFETY CARD
enabling signal to proportional amplifier cards
VALVE DRIVER CARDS
outputs to proportional valves
inputs from joint sensors
outputs to isolation valves
input from inclinometer
Key
RTX Harris RTX2000 microprocessor board
PSU power supply unit
D/A digital-to-analogue converter
A/D analogue-to-digital converter
SYSCON computer system 'watchdog' timer card
SPINC parallel input/output card
HOST programming source computer
RTX2000 microprocessor
manual controls
engine and hydraulic pumps
LUCIE hydraulics
JCB801 hydraulics
external data link
operator interface
RTX2000
activities manager
planning and scheduling
blackboard
activities
task organizer
servocontroller
sensors
output

(c)

(d)

(e)

9.3 Approaches to structuring

9.3.1 What is a structure?

Usually a ***structure*** is taken to be a set of parts in relationships with each other. Both the parts and the relationships may be of many different kinds. Examples of the parts in architectures might be sensors, processors, memories, motors, mechanical transmission units, energy sources, etc. More abstractly, the parts may be modules of a system. Examples of kinds of relationships I shall be talking about are *spatial*, *temporal*, *control* and *communication* relations. At the basic level, structures are described in terms of physical dimensions of space and time.

The structure is often described by enumerating the parts and the relations between them. This approach works well when the structure is simple. However, when the structure is complex, enumerative descriptions overwhelm us with detail: we cannot see the wood for the trees. We need some aids to cope with complexity. Two such aids are often used: *abstraction* and *modularity.*

Abstraction

The main idea of ***abstraction*** is that we leave out details and concentrate on the essentials. We emphasize what is general rather than specific. Because of this, the size of our description will be reduced and will be more tractable.

We shall discuss the notion of abstraction using an example: the concept of a ***state*** of a machine. This concept is used in all engineering areas: the mechanical engineer uses position and velocity to describe the state of a component; the electrical engineer uses voltage and current to describe the state of an electronic circuit element. Also, the software engineer uses data values of program variables to describe the state of a computation. If we did not want to be too specific about the physical nature of the variables and their values, we could just use a mathematical abstraction of states described in terms of symbolic variable names and symbolic values.

As discussed in the last section, in the early stages of design there is a lot to be said for stating specifications in an abstract form, without making a commitment about the physical implementation to be adopted eventually. The reason for making the least commitment in the early stages is to leave some leeway for consideration of other factors. For example, it may be possible to optimize (improve) the design in the light of various constraints that may apply by considering *trade-offs* through exchange of technologies. These constraints may relate to reliability, energy consumption, ecological issues, aesthetics, cost, manufacturability, to name just a few examples.

Modularity

Another approach to dealing with complexity is ***modularization***. We merge sets of components into larger units and describe structure in terms of these units. This enables us to concentrate attention on structures of manageable size. The internal structure of a module is hidden from view, simplifying the system description. When the time comes, we apply the same approach to the innards of the modules, again keeping the size tractable. This kind of approach leads to a hierarchy of embedded descriptions.

Let us look at an example. A car engineer might describe the structure of a car at the level of nuts, bolts, levers and cogs. There are thousands of these in a modern car and the description would be tedious and incomprehensible. Instead, the engineer merges hundreds of these into a module and calls it the engine, another group will be the transmission module, yet another the steering system, and so on. But the engine in turn can be described in terms of modules: the water pump, the valves and their control, the pistons and their linkages, etc. (Figure 9.4). The water pump can be further described in terms of components. We see a sequence of descriptions within descriptions: embedding. We see increasingly smaller detail.

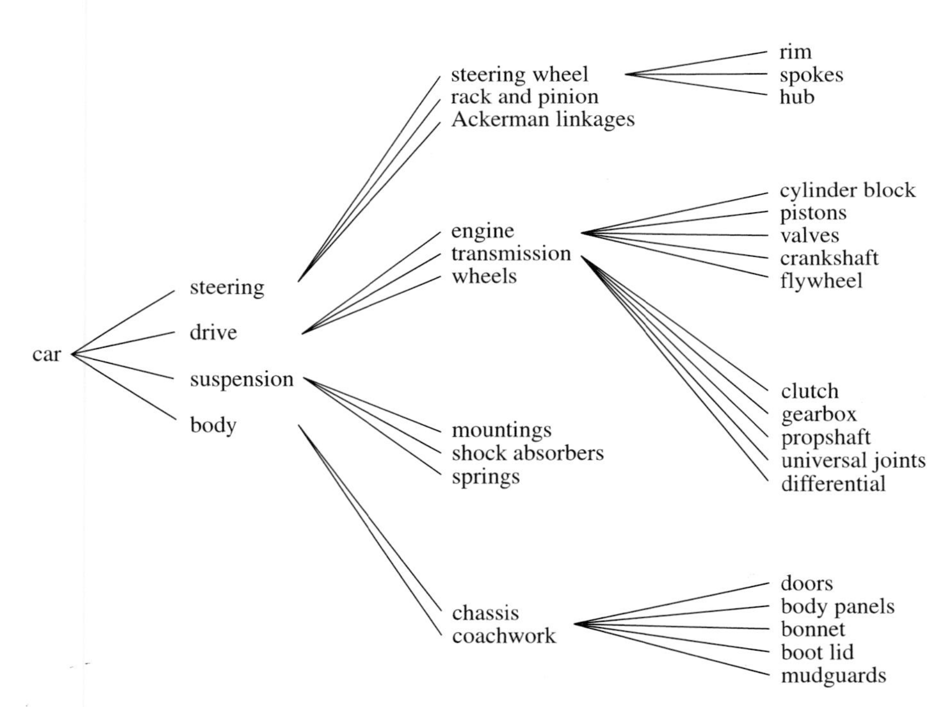

Figure 9.4
A car: hierarchy of embedded descriptions and functional decomposition.

The process of constructing descriptions in this form can also be applied to its function rather than to its physical components. It is then often called *functional decomposition*. The overall function of the object is successively divided (decomposed) into a set of constituent functions. In the case of the car, the overall function is self-locomotion under constant human control (though the car of the future may not be under constant human control – it may be autonomous some of the time). This is divided into the functions of converting the energy supply into kinetic form, transmitting the kinetic energy to the road wheels, providing steering functionality, and so on.

9.3.2 Modularization

The case for and against modularization

Modularization is desirable in large systems in order to control complexity in design, description, documentation intelligibility, fault finding, maintenance and repair. It can have a beneficial effect over the whole life-cycle of the product.

In small systems modularization may be undesirable. It can impose overheads in resource usage and thus increase costs. Modularization for reparability is of no use in a small, cheap product which is best treated as a 'throw-away' object when it becomes faulty. Similarly, modules imply interfaces which can be costly: for example, connectors.

It is thus important to apply modular treatment only where it is appropriate. A mixed approach may be better in some cases. It could be beneficial to design conceptually in modular form but to fabricate the product without modules in an integrated form.

The issue of modularity is particularly important in mechatronics in relation to the integration of different technologies. To illustrate this issue, consider the question of whether the computational elements of a mechatronic design should be segregated into a single 'computer' as a module, as opposed to distributing and applying computer technology in many places in the system or product. Recent developments in integrated-circuit semiconductor technology have led to new possibilities of deploying computing resources in the form of multiple single-chip microcomputer systems physically much nearer where they are needed, instead of a centralized, single, computer subsystem.

Consider the design of a small model AV system that could be used for teaching mechatronics. We can contrast two different architectures: distributed and centralized (Figure 9.5). In the centralized architecture a single microcomputer, perhaps an IBM-compatible personal computer, is used as the sole source of computing capability and is dealing with such functions as sensing, environment modelling, motor control and user interface management. In the distributed architecture, separate microprocessors are dedicated to each of these functions.

The use of several such subsystems in a product leads to a distributed computer architecture. The advantages and disadvantages of such designs, of course, need to be carefully evaluated in the design process. It is a common finding in computer technology that the hardware cost of a distributed architecture is lower than that of a centralized one for the same performance specification. On the other hand, the software development costs for embedded processors (that is, for processors that are integral parts of modules in the overall system hardware) are

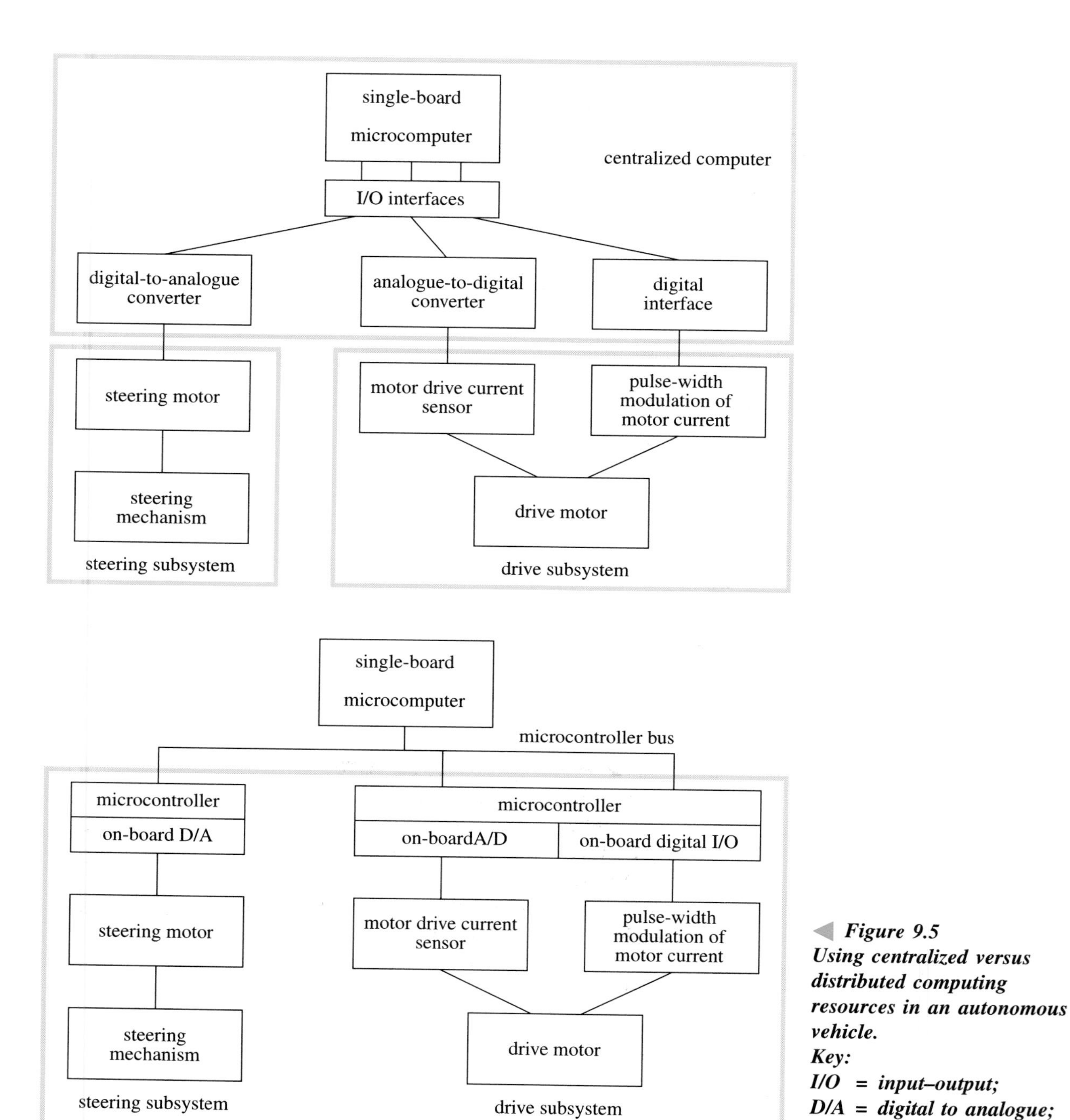

Figure 9.5
Using centralized versus distributed computing resources in an autonomous vehicle.
Key:
I/O = input–output;
D/A = digital to analogue;
A/D = analogue to digital.

usually higher than software development for desktop microcomputers. The reason for this is that testing software for embedded processors is more difficult, time consuming and requires special equipment. Other important trade-off factors to think about include data communication overheads in a distributed system, sharing of costly resources, single-processor performance limits compared with multiprocessor systems, and reliability.

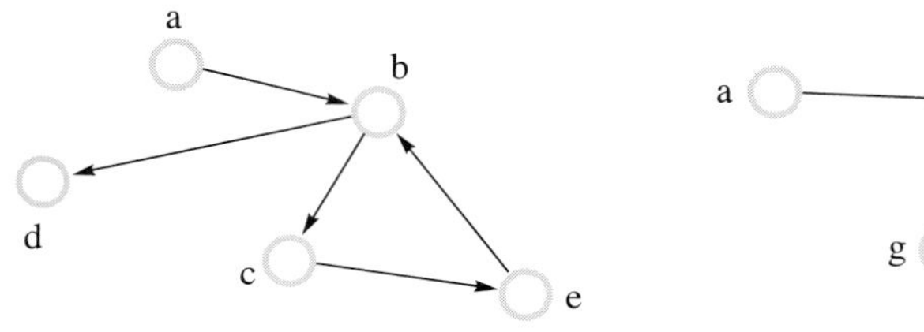

Directed graph
pathlength (a,e) ={3,6,...}
cycle (b,c,e,b) = true
reachable (e,a) = false

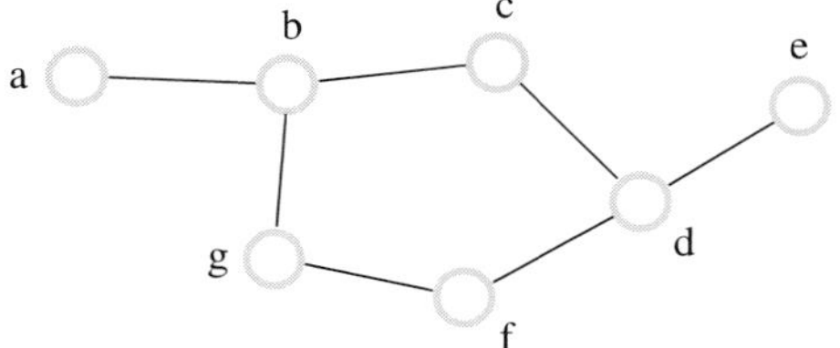

Undirected graph (network)
pathlength (a,e) = {4,5,9...}
cycle (a,b) = false

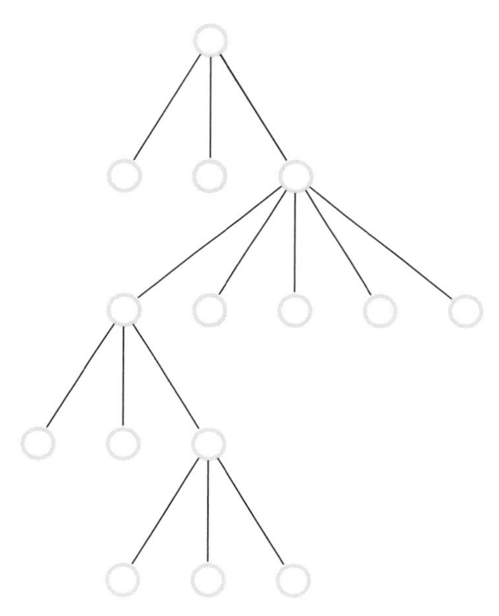

Tree graph
no cycles
single root node

Figure 9.6
Some simple concepts in graph theory: (a) directed graph; (b) undirected graph (network); (c) tree graph.

Connectivity between modules

The relationships between modules can often be conveniently described in graph-theoretic structures: networks and trees. In such a description the nodes represent modules and the connecting arcs represent relations of some kind (Figure 9.6). It is common to use arcs to denote, among other things, causal physical relationships (force, voltage), information transfer (signals, data, inputs, outputs), control (commands, execution sequence), the flow of material objects, and the membership in a set.

Hierarchy, heterarchy and even more complex structures are often seen in architecture descriptions. Usually these terms are not formally defined, but loosely speaking they relate to the graph-theoretic concepts of trees (hierarchy) and networks (heterarchy).

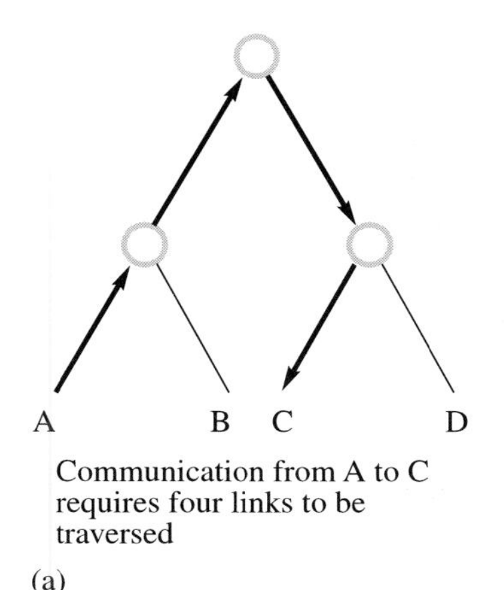

Communication from A to C requires four links to be traversed

(a)

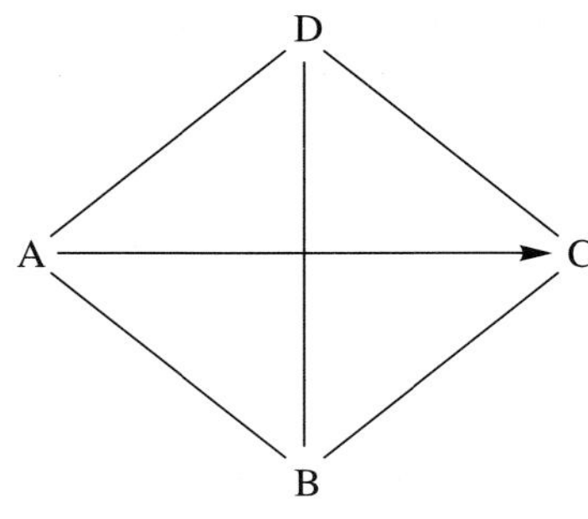

Communication between any pair of nodes requires only one link to be traversed

(b)

Figure 9.7 Communication in (a) a tree hierarchy and (b) a fully connected network.

Simon (1969) has argued that hierarchical structures are a good way to control complexity in systems. But hierarchies can be too rigid an organization in some cases. They enforce communication in a vertical direction between subordinate and superordinate modules, and inhibit or make costly the communication between modules at the same level horizontally, which must pass up and down the levels of the hierarchy.

The rigidity of hierarchical architectures led to the introduction of more flexible forms, called *heterarchical*. Modules are not as rigidly constrained in their interactions as they are in hierarchies. Communication networks are good examples of heterarchical organization: participants communicate with each other as the need arises, addressing each other directly. Figure 9.7 shows the differences in communication between a hierarchy and a network.

Flexible communication structure is not the only consideration, however. If there is a need for searching by a module, more efficient search strategies are available for trees than for networks. This again illustrates the idea of trade-off. In this case it is between communication cost and search cost.

Functional modules: vertical and horizontal

There has been much discussion recently in robotics, AI and cognitive science about the need for distinguishing horizontal and vertical modules in architectures for intelligent systems (Fodor, 1983; Brooks, 1986). The terminology is metaphorical: no spatial dimensions are involved and the distinction is quite abstract.

A *horizontal module* provides some functionality of an intelligent system from input to output. Such a module is capable of generating behaviour that is appropriate to some state of the environment, entirely on its own. It is thus a self-contained element for producing appropriate output to a specific input. An example of such a module would be an alarm module that stops a vehicle (the behaviour) whenever a tactile sensor at the front of the vehicle detects an obstacle (the input).

A *vertical module* cannot generate behaviour on its own. It can only contribute some functionality which must be combined with others in order to produce behaviour. An example of such a module is a memory module, which offers its services to a variety of other modules for storage of information. This form of modularity can be used to reduce duplication of identical facilities in an architecture. The required facility is separated out and is provided only once, as a 'server' of that facility. We shall discuss this issue in more detail in the next section.

Granularity and coupling in distributed systems

From the architectural point of view modularity can be organized at various degrees of granularity and coupling.

Granularity refers to the size and complexity of the components. In digital electronics this can vary from logic circuit elements to complete mainframe computer systems. In mechanical engineering this can vary between single mechanical components like linkages and gears to complete mechanisms or machines.

The degree of *coupling* characterizes the way the component elements communicate with each other. In digital electronics this can vary from direct wired connections in the case of logic circuits to wide-area networks operating through satellite links between computer systems. Intermediate degrees of coupling are exemplified by computer bus structures like the Multibus or VME (Versa Module Europe) bus. Such bus structures are commonly used to connect together the printed circuit boards from which electronic systems are assembled. In mechanical engineering the coupling may be described in terms of energy transmission components like drives which may be implemented in the form of levers, shafts, gearboxes, hydraulic couplings, force fields, etc.

When the resources of a system are partitioned and are at physically separate locations, we talk about ***distributed systems***. At the present time distributed systems are emerging as a major architectural design option in many applications. Many current mechatronic products utilize distributed architectures using embedded microprocessor subsystems.

9.4 Temporal issues and architecture

So far I have discussed architecture mostly in terms of spatial structure or at least spatial metaphors (recall that distinctions like horizontal versus vertical modules are not spatial). Temporal structuring is just as important as spatial structuring. A lot could be said about temporal structures in general, but here my concern is with temporal structures in relation to architectures.

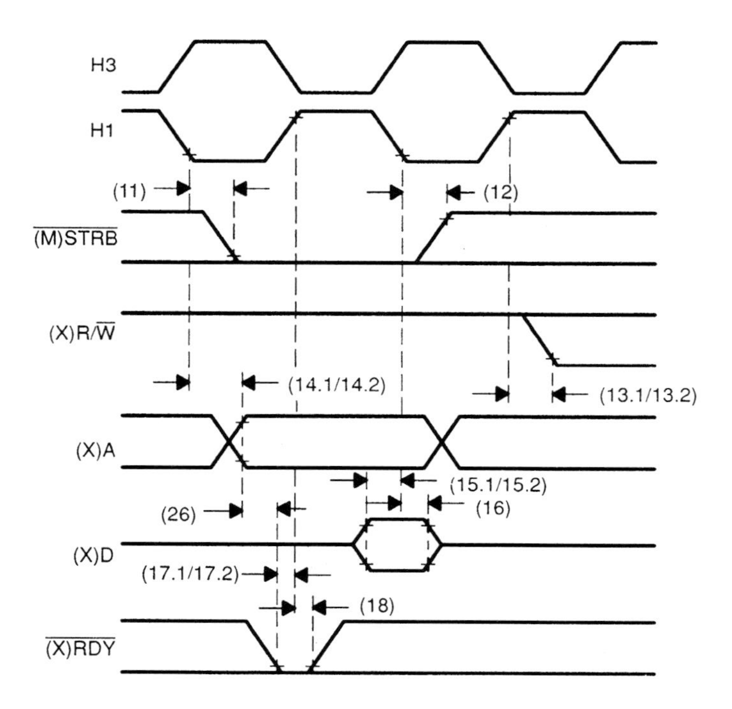

Figure 9.8
A timing diagram.

Earlier I have defined structure as a set of parts in relations with each other. In a temporal structure the parts are states or changes of states (also called events) and the relations are temporal, e.g. before, after, during, simultaneously.

As a graphical representation, timing diagrams are often used to depict temporal structure. In the simplest form a two-dimensional Cartesian coordinate system is used, with the horizontal axis representing time and the vertical axis representing states. As an illustration, Figure 9.8 shows the timing diagram of the memory-read operation in the TMS320C3x device. Note carefully that the states of a number of different signal lines are shown on the same diagram, so that the temporal relationships between them can be easily shown and annotated with measure information.

Timing diagrams are just one of many kinds of graphical tools for showing temporal structure. Others are occurrence networks, Petri nets and state charts. These are more graph-like in nature. Each has its advantages for emphasizing some aspect of temporal structure. Petri nets have been designed for the study of concurrency and parallelism in distributed systems. I shall not go into the details of these notational systems, however, but a closely related notation, state transition graphs, will be discussed later.

9.4.1 Control flow and data flow

In the discussion of architectures, temporal structure often makes its appearance in the form of ***control*** relationships. Informally, control defines what should happen and when. More specifically, by control I mean how the state of a system component is determined as a function of the states of other system components and as a function of time. (One might also say that this is the *causal structure* of a system, but philosophers might object. Causality is a complicated concept.)

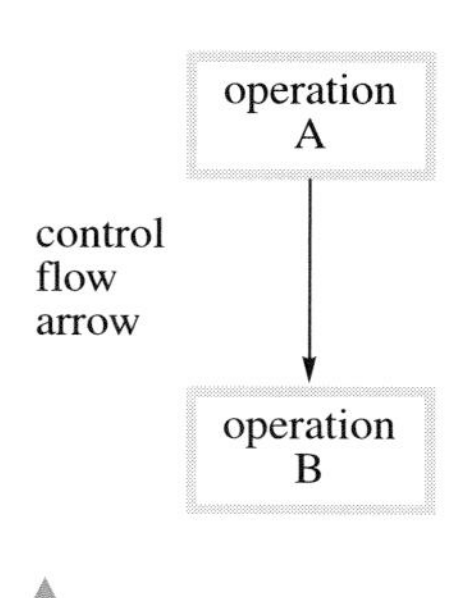

In software engineering, the expression ***control flow*** is used to describe how the locus of activity passes from one module to another in a temporal sequence. Here the locus of activity means the section of computer code which is currently being executed by a processor. That section of code is then *in control* of what the processor does from moment to moment. This kind of information is usually portrayed in the form of a *control flow diagram* (Figure 9.9). These are widely used in architecture descriptions.

In control system engineering, control is associated with a module (called the *controller*) if that module determines the state of some other module (often called the *plant*) as a function of time.

Thus we can see that the concepts of control and the temporal structure of events are intimately related in architecture descriptions.

Control flow is only one kind of temporal description. We are also interested in the availability of information at various places in the system as a function of time. What is known where and when? This kind of view of the system is described in a ***data flow*** diagram (Figure 9.10). This shows how the modules in an architecture communicate information between each other. For example, a communication network is a data flow diagram.

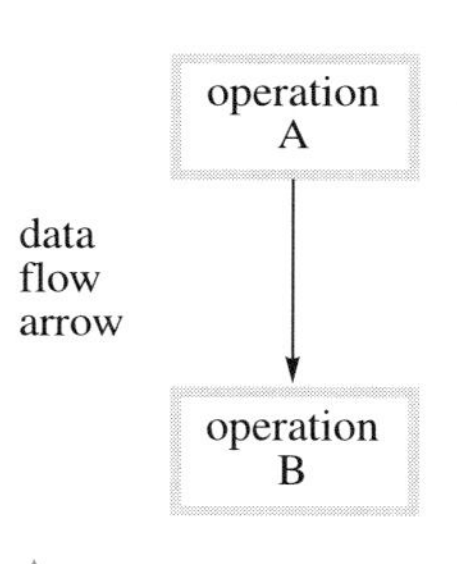

In mechatronics, the notion of data flow needs to be generalized to the description of flows of many other kinds of entities, like energy, objects, liquids and gases. This is quite natural, since a common abstraction in terms of space and time underlies the notion of flow in all these cases.

Descriptions of control flow and data flow give complementary kinds of information about a system. Both can be combined in a single diagram, but the results tend to be a bit confusing. Computer-aided design (CAD) systems can usually generate both kinds of information as and when needed. The design of such facilities, especially in a visual form, is an important research area in the field of human–computer interaction. It can be expected that the availability of specific mechatronic CAD tools will have a major impact on the mechatronic design process.

9.4.2 Automata and state transition graphs

Yet another way of using temporal information in designing a system is seen in ***state transition diagrams***. This technique has been widely used in many fields of engineering but is particularly appropriate in the informational aspects of design. Its most detailed elaboration has occurred in ***automata theory***, which deals with the abstract analysis of information processing machines. The abstract analysis is

only concerned that automata receive inputs from their environment, can be internally in one of a number of different states, and produce outputs as a function of their current state and current input. The inputs cause the *automaton* to change state.

The visual representation is again based on graphs. The nodes represent the states of the system and the arcs represent transitions between the states. The data communication aspects are also represented by associating the inputs and outputs with the arcs (Figure 9.11). The same information can also be represented in tabular form, as shown in Table 9.1.

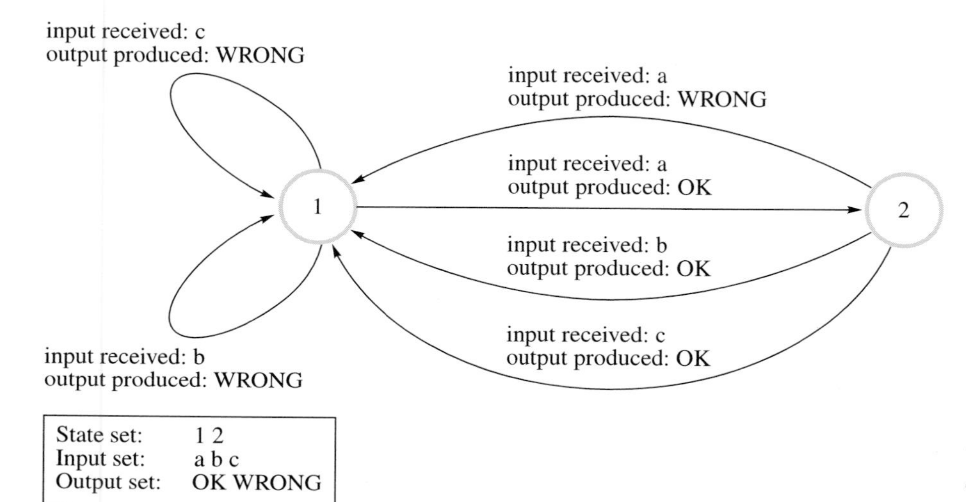

Figure 9.11
State transition graph representation of an automaton

TABLE 9.1 TABULAR REPRESENTATION OF AN AUTOMATON

State set	State transitions			Outputs		
	a	**b**	**c**	**a**	**b**	**c**
1	2	1	1	OK	WRONG	WRONG
2	1	1	1	WRONG	OK	OK

9.4.3 Serial and parallel processing

A major characteristic of systems is whether they utilize serial or parallel processing. This distinction also revolves around the temporal structuring of events.

- *Serial processing* implies that events do not overlap: they follow each other in a sequence, sequentially.
- *Parallel processing* implies that events do overlap and may occur concurrently with each other.

Clearly, parallel processing can lead to a reduction of processing time, since the same interval of time can be used for the execution of several operations or events. For this reason, parallel processing is often used in systems that have to operate in ***real time***. A real-time system is one which is guaranteed to perform its task within a time interval that is less than or equal to some stated upper bound.

The gain in time that comes from parallel processing has to be paid for in processing elements: higher parallelism implies more hardware. Parallel processing can therefore be seen as a trade-off between time and equipment resources. Ever increasing demands for greater processing power to be deployable in real time has produced a technological trend towards increased concurrency. However, concurrent systems are harder to design, construct and debug, in addition to their greater cost. The careful assessment of the benefits and costs of concurrent systems is a major design consideration.

9.4.4 Multiple processes and tasks: role of operating systems

In the software component of complex mechatronic systems the management of multiple processes is often segregated into a module. This module in turn is often a part of a computer or microcomputer operating system. Operating systems help in the utilization and control of the software and hardware resources of a computer. They enable the application software to have disciplined and coordinated access to these resources. An important example of such resource control is the scheduling of the processor or processors in the system. It is customary to divide the overall computation into concurrently executable *tasks* and to organize the execution of these tasks as a set of concurrent *processes*.

Multi-tasking operating systems are now widely used even in quite small microprocessor systems. They deal with the allocation of processor time, they control access to shared peripheral devices, manage interrupts, start, stop, create and destroy processes, and handle the synchronization and communication between processes.

9.5 Case studies of architecture designs

In this section I illustrate many of the concepts that have been introduced so far in this chapter by describing some architectures for intelligent machines which have been proposed in the literature.

This section will show how designers are addressing the issue of generating intelligent behaviour from a machine. The architectures described are designs that attempt to reconcile a number of apparently conflicting requirements by constructing *layered architectures* (introduced in Section 1.8 of this volume).

The first study is from a group of workers at the US National Bureau of Standards (NBS) who have been working originally in the field of robotics, at first in flexible manufacturing systems (Albus *et al.*, 1979), and more recently in space research (Lumia *et al.*, 1990). The earlier paper has had a considerable influence on the robotics and AI fields. Interestingly, Albus' design ideas were strongly influenced by a study of the human nervous system.

The second is from Rodney Brooks at MIT who has developed Albus' ideas further, into the notion of the *subsumption architecture* (Brooks, 1986).

You should keep in mind, therefore, that these contributions represent snapshots from the history of the subject. There is still considerable controversy surrounding the issue of architectures for intelligent machines. This controversy is sometimes referred to as the *cognitive* versus *reactive* architecture controversy. Reactive systems are so called because they react to events as if by reflex action, without doing a lot of abstract reasoning. The main advantage of reactive systems is that they can respond quickly. However, they tend to lack robustness in the face of varying environmental conditions. Cognitive systems are more robust, but at the expense of speed. This is yet another trade-off! The case studies described here are both on the cognitive side of this division. This controversy is reflected in Brooks' work. Having proposed the layered subsumption architecture in 1986, more recently he has changed direction and he now advocates 'flat' *reactive architectures*, i.e. the elimination of abstract cognitive information processing from the architecture.

9.5.1 The hierarchical architecture of Albus

This architecture has been designed in the context of robotics in flexible manufacturing systems. For robots to operate effectively in the partially unconstrained environments of manufacturing, they must be equipped with control systems that have measurement and sensory capabilities. The architecture described consists of parallel control and measurement hierarchies. The control hierarchy decomposes tasks into subtasks, and the measurement hierarchy analyses data from sensors. At each level the control hierarchy sends expectations to the measurement hierarchy, which returns computed values of the deviation between the observed and expected data. The control hierarchy uses this information to modify its task decomposition strategies so as to generate sensory-interactive goal-directed behaviour. The system has been partially implemented on a research robot using a network of microcomputers and a real-time vision system mounted on the robot's wrist.

Clearly, sensory measurement capabilities are necessary for robots to operate effectively in unstructured environments; but more than just measurement data is required. The data must be processed and analysed and the results introduced into the robot control system in real time so that the response is goal directed, reliable, and efficient. This is a problem in which complexity grows exponentially with the number of sensors and with the number of branch points in the control program.

The problem of controlling a sensory-interactive robot is similar in many respects to that of controlling any complex system such as an army, a government, a business, or a biological organism. The command and control structure for such systems is invariably a hierarchy wherein goals or tasks selected at the highest level are decomposed into sequences of subtasks which are passed to the next lower level in the hierarchy. This same procedure is repeated at each level until, at the bottom of the hierarchy, a sequence of primitive actions is generated, each of which can be executed with a single operation. Sensory measurements are fed back into this hierarchy at many different levels; the aim is to alter the task decomposition so as to accomplish the highest level goal in spite of uncertainties or unexpected conditions in the environment.

The advantage of hierarchical control is that complexity at any level in the hierarchy can be held within manageable limits irrespective of the complexity of the entire structure.

Part of the robotics research program at the US National Bureau of Standards, where this architecture was produced, is an investigation of methods for designing and implementing real-time sensory and control functions in a *hierarchical structure*. The project is developing engineering procedures for partitioning tasks into subtasks, assigning subtasks to logical modules, designing hardware to implement this logic, and writing software.

Hierarchical task decomposition

In general, the behaviour of a robot results from a temporal series of primitive actions which generate drive signals to actuators producing forces and movements. Each primitive action is typically specified by an instruction such as (`GOTO` point) which resides in a memory device. A sequence of instructions (i.e. a program) which executes a complete task can be represented as a sequence of states in a state diagram.

In the absence of sensory input, the sequence of primitive actions is fixed, both in its order and its timing. However, most present day robots allow external interlock signals to synchronize the timing of the robot's program with external machinery.

More sophisticated robots have the capacity to use sensory interlocks to activate conditional branches within the program. Conditional branching enables a robot to select one of several different programs depending upon sensed conditions in the environment. In some cases, such a robot can be programmed to cope with simple error conditions.

More sophisticated yet are robots that can incorporate values of sensed variables into their data structure so as to track moving objects and operate upon or manipulate parts which are not precisely aligned to a known reference. When this is represented graphically, the result is that the states corresponding to program instructions become regions instead of points in state space. The extent of these regions corresponds to the range of the possible responses to the sensed variables.

As the length and complexity of branching in robot programs increases, it becomes tedious to write them as linear strings of instructions. Furthermore, it soon becomes evident to anyone actually writing such programs that the same substrings occur repeatedly. The obvious solution to this is to define macros, or subroutines, consisting of often-used substrings. Of course, it is always possible to partition any particular task program into a set of macros. If one can define a set of macros which can partition all programs within a certain class of tasks, it becomes possible to write programs for that class of tasks completely in terms of macros. The set of macros then becomes an instruction set for a higher level programming language. Strings of macros represent robot programs written in the higher level language.

Of course, strings of macros can themselves be partitioned into consistently recurring groups to form second level macros, and recurring groups of second level macros can be defined as third level macros. In principle this process can be repeated any number of times to create a hierarchy wherein each level breaks a higher level input command into a sequence of subcommands to the next lower level. In this way a high-level task is decomposed through a succession of hierarchical levels, until at the lowest level a string of action primitives produce the forces and motions to accomplish the high-level task.

A four-level hierarchical decomposition for a task 'Assemble AB' is illustrated in Figure 9.12. On the left is a hierarchy of computing modules, each representing a library of macros at the corresponding level. Each H module receives two sets of inputs: (a) a list, or vector, of command variables C; and (b) a list, or vector, of feedback variables F. Each macro produces a state trajectory, and the H modules themselves are ***state machines***. Each H module samples its inputs C and F and its own internal state at some time $t=k$ and computes an output P. The output P becomes input to other H modules at time $t=k+1$. Part of the F vector to any H module consists of variables indicating the state of lower level modules. The rest of the F vector consists of sensory information from the environment filtered through a sensory processing hierarchy to that level.

Whether or not such an approach is practical depends upon how many macro names are required at each level to cover a class of tasks. At present this is still an open research issue. Clearly the list of names is task dependent. There are, however, a limited number of task types and a limited number of task decompositions required at each level by any single task. For example, the number of different types of elemental moves required for robots (or even humans) to perform routine mechanical assembly is not large. Systematic time and motion studies done for human workers reveal a surprisingly small number of elemental movements corresponding to first level macros required for mechanical assembly of small components. There are only a few different types of movements such as reach, grasp, lift, transport, position, insert, twist, push, pull, and release. A list of parameters with each macro can specify where to reach, when to grasp, how far to twist, how hard to push, and so on.

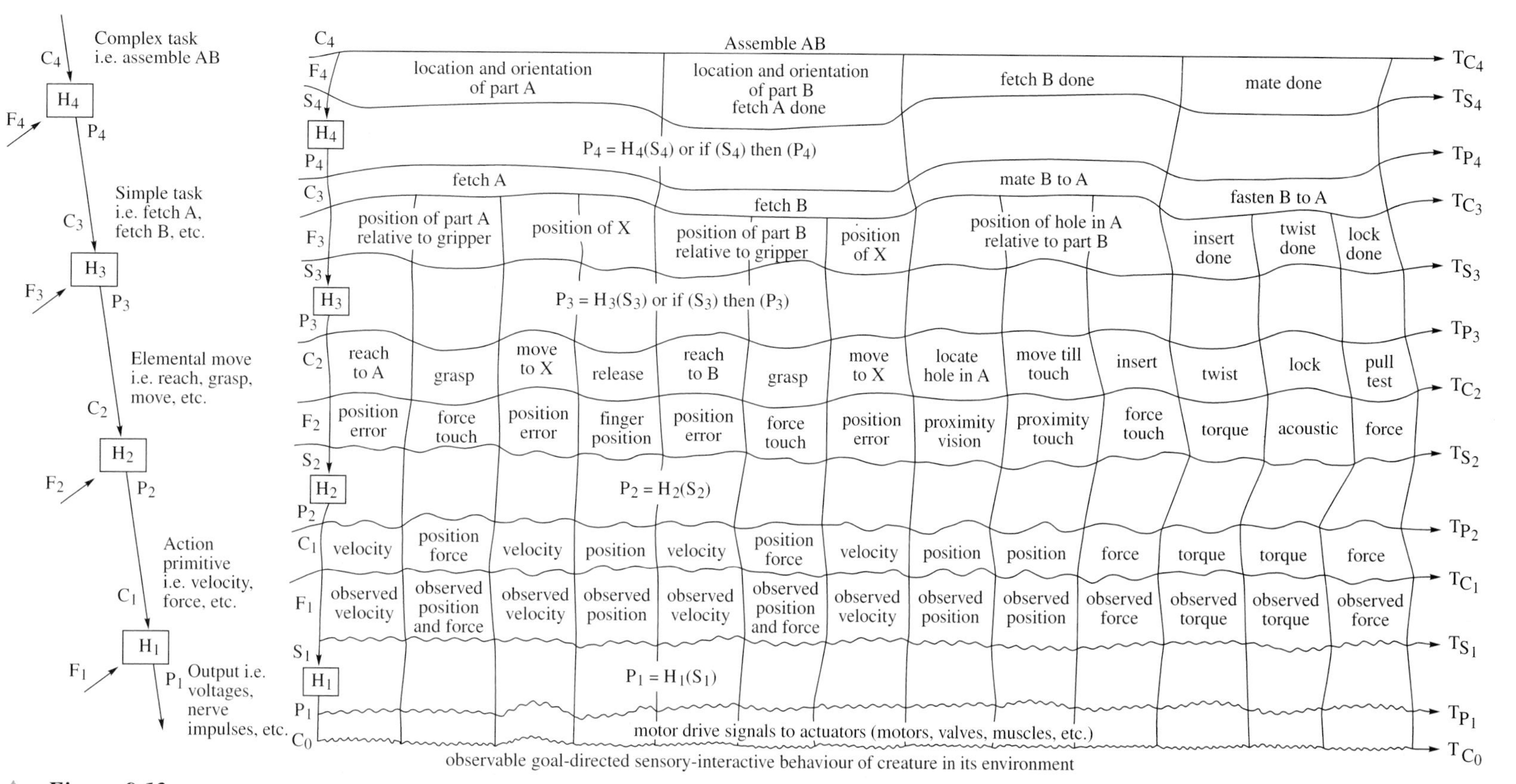

Figure 9.12

Hierarchical decomposition of a task.
Key:* C= *command variable;* H= *computing module (state machine);
F=*feedback variable from sensors and lower levels;* P= *output.*

Incorporating real-time sensory feedback

The type of hierarchical task decomposition described above is nothing more than good top-down structured program design applied to robot tasks. Each macro represents a relatively short sequence of instructions and a limited and well-structured set of branches. Programs at each level tend to be readable, understandable, and easy to debug and test for correctness.

The advantages of such a hierarchical decomposition extend far beyond programming convenience, however. The sophisticated real-time use of sensory measurement information for coping with uncertainty and recovering from errors requires that sensory data be able to interact with the control system at many different levels with many different constraints on speed and timing. For example, joint position, velocity, and sometimes force measurements are required at the lowest level in the hierarchy for servo feedback. This type of data requires very little processing, but must be supplied without time delays of more than a few milliseconds.

Visual depth (proximity) and information related to edges and surfaces are needed at the primitive action level of the hierarchy to compute offsets for gripping points. This type of data requires a modest amount of processing and must be supplied within a few tenths of a second.

Recognition of part position and orientation requires more processing and is needed at the elemental move level where time constraints are of the order of seconds. Recognition of parts and/or relationships between parts which may take several seconds is required for conditional branching at the simple task level.

Attempting to deal with this full range of sensory feedback in all its possible combinations at a single level leads to extremely complex and inefficient programs. Sophisticated analysis of measurement data, particularly vision data, is inherently a hierarchical process. Only if the control system is also partitioned into a hierarchy can the various levels of feedback information be introduced to the appropriate control levels in a simple and straightforward way.

Measurement as an active process

In robotics the primary purpose of sensory feedback is to control action, but there are many different kinds of action. Different sensory information is required for different tasks and even different portions of the same task. As time varies and conditions change, different sensors, different resolutions, and different processing algorithms may be needed. The speed requirements of real-time control do not permit all resolutions and all processing algorithms to be applied all the time. Thus, real-time allocation of measurement and processing resources must be done.

Furthermore, measurement data often must be compared with expectations and hypotheses in order to detect missing objects or events, or to detect deviations from desired trajectories. Thus, the control system must have the capacity to tell the measurement system what to expect and when to expect it. This means that there should be links from the control hierarchy to the measurement hierarchy as well as the other way around. Figure 9.13 illustrates such a structure.

These links may simply convey to the sensory processing modules what action is being taken so that the appropriate processing algorithms can be applied to the incoming data. However, in more sophisticated robot systems these links may also include data retrieved from memory or generated by mathematical models of the external environment. The entire set of links (including memory and mathematical models) represent a world model. Thus, the world model is itself a multilevel hierarchical entity.

As the task execution proceeds, the links from the control hierarchy comprising the world model produce sequences of expected data to the sensory processing modules at each level. Under ideal conditions these expected values will accurately predict the observed data. In other situations the processing modules at each level will detect deviations between the predictions and observations and produce error signals. These are the F vectors which modify the task decomposition at each level so as to maintain the robot performance within permissible limits.

In general, any deviation from the ideal task performance will be detected at the lowest level first. If the lowest level H function is capable of coping with that

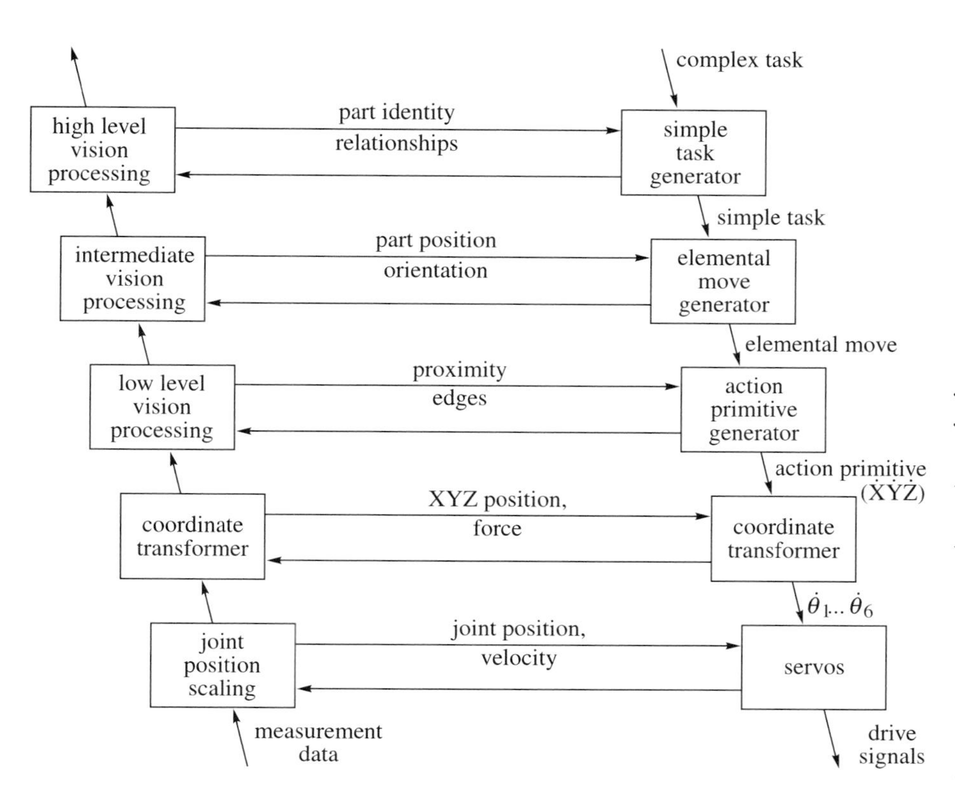

Figure 9.13
Links between control and measurement. The links from left to right provide feedback for control. The links from right to left provide context and expectations for data processing. Here the lowest level of Figure 9.12 has been split into two levels: one for coordinate transformations between work space (or end effector space) and joint space, the other for servo computations on the joint positions and velocities.

error, no higher level action need be taken. However, if the error persists or grows in spite of the action of the lowest level H function, then it will be detected at the next higher level where a different macro may be selected. By this means it becomes possible to implement very sophisticated multilevel error correction procedures in a relatively straightforward manner.

System implementation

The laboratory at NBS has implemented the type of cross-coupled measurement/control hierarchy described above. It was chosen to implement the hierarchy in a network of microcomputers because it was believed to be the best way to achieve low cost and upward compatibility. However, such a logical architecture can also be implemented on a single minicomputer. In fact, the first version of the NBS hierarchical control system was implemented in this way.

A later system was a network of microcomputers with the architecture shown in Figure 9.14. Time is sliced into 20 ms increments. At the beginning of each increment each logical module reads its set of input values from the appropriate locations in common memory. It then computes its set of output values which it writes back into the common memory before the 20 ms interval ends. Any of the logical modules which do not complete their computations before the end of the 20 ms interval write extrapolated estimates of their output accompanied by a flag indicating that the data are extrapolated. The process then repeats.

Each logical module is thus a state machine whose output depends only on its present inputs and its present internal state. None of the logical modules admits any interrupts except the clock interrupt which signals the beginning and end of the 20 ms computation intervals. This simple modular structure enormously simplifies the writing and debugging of software.

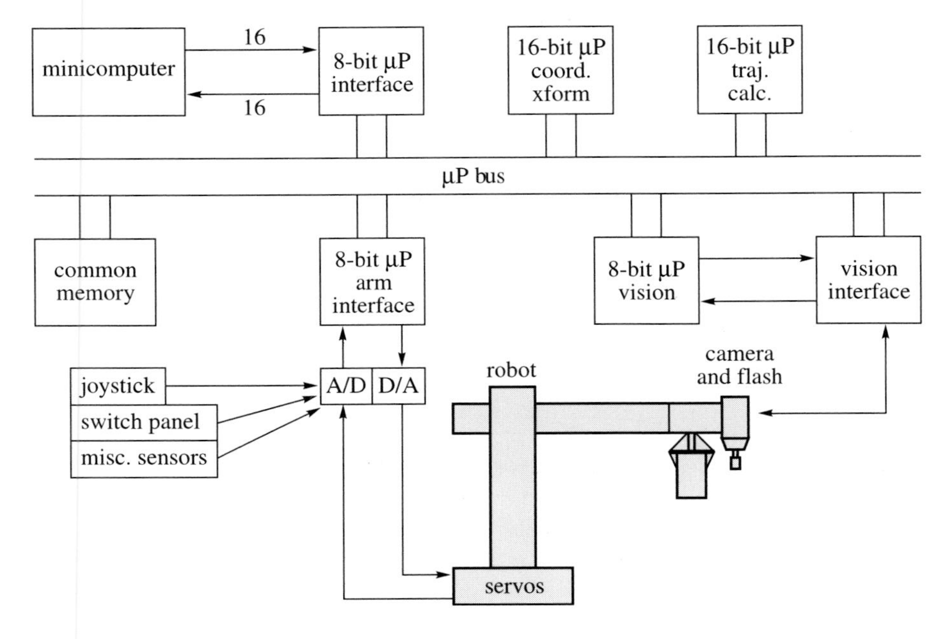

Figure 9.14
The architecture of the system implementation. μP = microcontroller.

In the last few years this architecture has been adopted as the NASA/NBS Standard Reference Model for Telerobot Control System Architecture (NASREM) in space research (Lumia *et al.*, 1990). The objective of this work is the construction of a flight telerobotics servicer, which will support assembly and maintenance operations on a space station.

9.5.2 The layered architecture of Brooks

Brooks' architecture is for controlling mobile robots. The system is intended to control a robot that wanders around in the office areas of a laboratory, building maps of its surroundings. The system is built up from layers to let the robot operate at increasing levels of competence. The layers are made up of asynchronous modules which communicate over low bandwidth channels. Each module is an instance of a fairly simple computational machine. Higher level layers can *subsume* the roles of lower levels by suppressing their outputs. However, lower levels continue to function as higher levels are added. The result is a robust and flexible robot control system.

A control system for an intelligent, completely autonomous, mobile robot must perform many complex information processing tasks in real time. It operates in an environment where the environmental conditions are changing rapidly. In fact, the determination of those conditions is done over very noisy channels since there is no straightforward mapping between sensors (e.g. television cameras) and the actions required.

Requirements

There are several requirements of a control system for an intelligent autonomous mobile robot. They each put constraints on possible control systems we might build and employ. Brooks offers the following list of such design requirements:

- *Multiple goals* Often the robot will have multiple goals, some conflicting, which it is trying to achieve. It may be trying to reach a certain point ahead of it while avoiding local obstacles. It may be trying to reach a certain place in minimal time while conserving power reserves. Often the relative importance of goals will be context dependent. For example, getting off the railroad tracks when a train is heard becomes much more important than inspecting the last ten track ties of the current track section. The control system must be responsive to high-priority goals, while still servicing necessary 'low-level' goals – e.g. in getting off the railroad tracks the robot still has to maintain its balance so it doesn't fall down.
- *Multiple sensors* The robot will most likely have multiple sensors (e.g. television cameras, encoders on steering and drive mechanisms, and perhaps infra-red beacon detectors, an inertial navigation system, acoustic rangefinders, infra-red rangefinders, access to a global positioning satellite system). All sensors have an error component in their readings. Furthermore,

often there is no direct analytic mapping from sensor values to desired physical quantities. Some of the sensors will overlap in the physical quantities they measure. They will often give inconsistent readings – sometimes due to normal sensor error and sometimes due to the measurement conditions being such that the sensor (and subsequent processing) is being used outside its domain of applicability. The robot must make decisions under these conditions.

- *Robustness* The robot ought to be robust. When some sensors fail it should be able to adapt and cope by relying on those still functional. When the environment changes drastically it should still be able to achieve some modicum of sensible behaviour, rather then sit in shock, or wander aimlessly or irrationally around. Ideally it should also continue to function well when there are faults in parts of its processor(s).
- *Additivity* As more sensors and capabilities are added to a robot it needs more processing power, otherwise the original capabilities of the robot will be impaired relative to the flow of time.

Starting assumptions

Brooks' design decisions for a mobile robot are based on nine factors:

- Complex (and useful) behaviour need not necessarily be a product of an extremely complex control system. Rather, complex behaviour may simply be the reflection of a complex environment. It may be an observer who ascribes complexity to an organism – not necessarily its designer.
- Things should be simple. This has two applications:

  (a) When building a system of many parts one must pay attention to the interfaces. If you notice that a particular interface is starting to rival in complexity the components it connects, then either the interface needs to be rethought or the decomposition of the system needs redoing.

  (b) If a particular component or collection of components solves an unstable or ill-conditioned problem, or, more radically, if its design involved the solution of an unstable or ill-conditioned problem, then it is probably not a good solution from the standpoint of robustness of the system.

- We want to build cheap robots which can wander around space inhabited by humans with no human intervention, advice or control and at the same time do useful work. Map making is therefore of crucial importance even when idealized blueprints of an environment are available.
- The human world is three dimensional; it is not just a two-dimensional surface map. The robot must model the world as three-dimensional if it is to be allowed to continue cohabitation with humans.
- Absolute coordinate systems for a robot are the source of large cumulative errors. Relational maps are more useful to a mobile robot. This alters the design space for perception systems.

- The worlds where mobile robots will do useful work are not constructed of exact simple polyhedra. While polyhedra may be useful models of a realistic world it is a mistake to build a special world such that the models can be exact. For this reason we will build no artificial environment for our robot.
- Sonar data, while easy to collect, do not alone lead to rich descriptions of the world useful for truly intelligent interactions. Visual data are much better for that purpose. Sonar data may be useful for low-level interactions such as real-time obstacle avoidance.
- For the sake of robustness the robot must be able to perform when one or more of its sensors fails or starts giving erroneous readings. Recovery should be quick. This implies that built-in self-calibration must be occurring at all times. If it is good enough to achieve our goals then it will necessarily be good enough to eliminate the need for external calibration steps. To force the issue we do not incorporate any explicit calibration steps for our robot. Rather we try to make all processing steps self calibrating.
- The interest is in building *artificial beings* – robots which can survive for days, weeks and months, without human assistance, in a dynamic complex environment. Such robots must be self-sustaining.

Levels of competence

There are many possible approaches to building an autonomous intelligent mobile robot. As with most engineering problems they all start by decomposing the problem into pieces, solving the subproblems for each piece, and then composing the solutions. Brooks claims to have done the first of these three steps differently to other groups of workers. The second and third steps also differ as a consequence.

Typically, mobile robot builders have sliced the problem into some subset of:

sensing,
mapping sensor data into a world representation, and planning,
task execution,
motor control.

This decomposition can be regarded as a *horizontal* decomposition of the problem. The pieces form a chain through which information flows from the robot's environment, via sensing, through the robot and back to the environment, via action, closing the feedback loop (of course, most of the above projects also included internal feedback loops). An instance of each piece must be built in order to run the robot at all. Later changes to a particular piece (to improve it or extend its functionality) must either be done in such a way that the interfaces to adjacent pieces do not change, or the effects of the change must be propagated to neighbouring pieces, changing their functionality too.

Brooks has chosen instead to decompose the problem *vertically* as the primary way of slicing up the problem. Rather than slice the problem on the basis of the internal workings of the solution he slices the problem on the basis of desired external manifestations (behaviour) of the robot control system.

To this end he has defined a number of *levels of competence* for an autonomous mobile robot. A level of competence is an informal specification of a desired class of behaviours for a robot over all environments it will encounter. A higher level of competence implies a more specific desired class of behaviours.

Brooks uses the following levels of competence:

1 Wander aimlessly around without hitting things.
2 Explore the world by seeing places in the distance which look reachable and head for them.
3 Build a map of the environment and plan routes from one place to another.
4 Notice changes in the static environment.
5 Reason about the world in terms of identifiable objects and perform tasks related to certain objects.
6 Formulate and execute plans which involve changing the state of the world in some desirable way.
7 Reason about the behaviour of objects in the world and modify plans accordingly.

Notice that each level of competence includes as a subset each earlier level of competence. This is the reason why this architecture is called a *subsumption architecture*. Since a level of competence defines a class of valid behaviours it can be seen that higher levels of competence provide additional constraints on that class.

Layers of control

The key idea of levels of competence is that we can build layers of a control system corresponding to each level of competence and simply add a new layer to an existing set to move to the next higher level of overall competence.

We start by building a complete robot control system which achieves level 1 competence. It is debugged thoroughly. That system is never altered. Next, a second control layer is built. It is able to examine data from the level 1 system and is also permitted to inject data into the internal interfaces of level 1, suppressing the normal data flow. The second layer, with the aid of the first, achieves level 2 competence. The first layer continues to run unaware of the layer above it which sometimes interferes with its data paths. The same process is repeated to achieve higher levels of competence (Figure 9.15).

In such a scheme we have a working control system for the robot very early in the design process – as soon as the first layer is built. Additional layers can be added later, and the initial working system need never be changed.

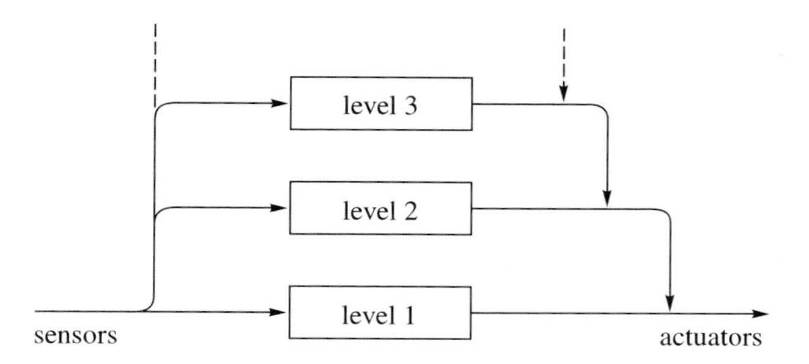

◀ *Figure 9.15*
Layers in the architecture.

Brooks claims that this architecture naturally lends itself to solving the problems of mobile robots delineated earlier. The fulfilment of the requirements can be discussed under the same headings as before.

- *Multiple goals* Individual layers can be working on individual goals concurrently. The suppression mechanism then mediates the actions that are taken. The advantage here is that there is no need to make an early decision on which goal should be pursued. The results of pursuing all of them to some level of conclusion can be used for the ultimate decision.
- *Multiple sensors* In part we can ignore the sensor fusion problem as stated earlier using a subsumption architecture. Not all sensors need to feed into a central representation. Indeed certain readings of all sensors need not feed into central representations – only those which perception processing identifies as extremely reliable might be eligible to enter such a central representation. At the same time, however, the sensor values may still be being used by the robot. Other layers may be processing them in some fashion and using the results to achieve their own goals, independent of how other layers may be scrutinizing them.
- *Robustness* Multiple sensors clearly add to the robustness of a system when their results can be used intelligently. There is another source of robustness in a subsumption architecture. Lower levels which have been well debugged continue to run when higher levels are added. A higher level can only suppress the outputs of lower levels by actively interfering with replacement data, and in cases where it cannot produce results in a timely fashion the lower levels will still produce results which are sensible, albeit at a lower level of competence.
- *Additivity* An obvious way to handle additivity is to make each new layer run on its own processor. We will see below that this is practical as there are in general fairly low bandwidth requirements on communication channels between layers. In addition, we will see that the individual layers can easily be spread over many loosely coupled processors.

The structure and operation of layers

In this architecture each layer is built from a set of small processors which send messages to each other.

Each processor is a *finite state machine* with the ability to hold some data structures. Processors send messages over connecting communication channels (Brooks calls these 'wires'). There is no handshaking or acknowledgement of messages. The processors run completely asynchronously, monitoring their input wires and sending messages on their output wires. It is possible for messages to get lost if a new one arrives on an input line before the last was inspected. This actually happens quite often. Input lines have single element buffers. There is no other form of communication between processors; in particular, there is no shared global memory.

All processors (called modules) are created equal in the sense that within a layer there is no central control. Each module merely does its thing as best it can.

Inputs to modules can be suppressed and outputs can be inhibited by wires coming from other modules. This is the mechanism by which higher level layers subsume the role of lower levels.

Each state is named. When the system first starts up all modules start in the distinguished state named 'NIL'. A state can be specified as one of four types:

1 *Output* An output message, computed as a function of the module's input buffers and instance variables, is sent to an output line. A new specified state is then entered.
2 *Side effect* One of the module's internal variables is set to a new value, computed as a function of its input buffers and variables. A new specified state is then entered.
3 *Conditional dispatch* A predicate (Boolean function) on the module's internal variables and inputs is computed and, depending on the outcome, one of two subsequent states is entered.
4 *Event dispatch* A new state is entered, determined by the events which have taken place in the recent past history of the inputs to the module.

These four types of states are used by Brooks to define a programming language. Programs written in this language determine the behaviour of each of the modules in the system. The language is a derivative of LISP, a programming language widely used in artificial intelligence research and development work.

Communication

Figure 9.16 shows the best way to think about these finite state modules for the purposes of communications. They have some input lines and some output lines. An output line from one module is connected to input lines of one or more other modules. One can think of these lines as wires, each with a source and destinations.

Input signals can be suppressed and replaced by the suppressing signal. Output signals can be inhibited. The mechanism for this can be described as follows.

An extra wire can terminate (i.e. have its destination) at an output site of a module. If any signal travels along this wire it *inhibits* any output message from the

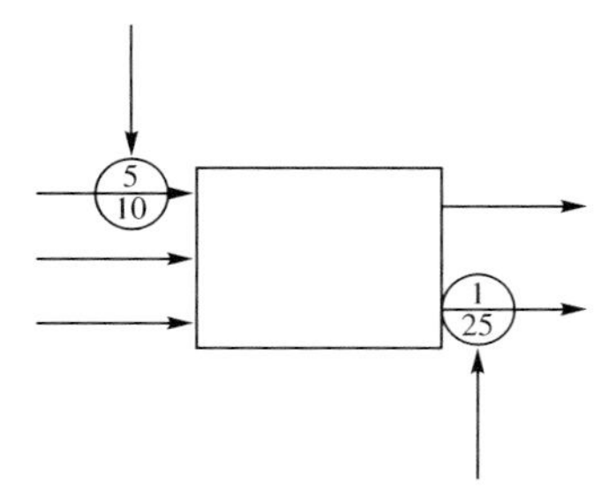

Figure 9.16 **Wires carry the communication between modules. The numbers are time durations.**

module along that line for some predetermined time. Any messages sent by the module to that output during that time period are lost.

Similarly, an extra wire can terminate at an input site of a module. Its action is very similar to that of inhibition, but additionally, the signal on this wire, besides inhibiting signals along the usual path, actually gets fed through as the input to the module. Thus it *suppresses* the usual input and provides a replacement.

For both suppression and inhibition the time durations are written inside a circle, as shown in Figure 9.16.

9.6 Conclusion

In this chapter I have discussed what architectures are and how they can be used in the mechatronic design process. I have indicated that architectures are concerned with the high-level structuring of designs, both in spatial and temporal terms. The use of architectures in design is to deal systematically with a number of requirements that arise. The various major features of architectures are responses to such requirements. Some of these requirements and the architectural responses to them can be summarized as follows:

- The need for hierarchy to control complexity.
- The need for heterarchy to control flexibility.
- The need for distributed systems to control cost and operating speed.
- The need for modularization to control intelligibility, manufacturability and maintainability.

I have also described some architecture proposals that are particularly relevant to mechatronics, especially when the emphasis is, as in this book, on *intelligent* machines.

It is important to recognize that the layered architectures of these proposals are applicable equally to the computational, mechanical, electronic, or other subsystems of a mechatronic system. Any particular layer of such an architecture may be implemented in one or more of the relevant technologies. This will indeed be the case particularly in the layers that interface directly to the environment, i.e. perceptual input and motor output. But even the cognitive subsystem may on occasion incorporate or be based entirely on mechanical, hydraulic or optical computational elements. It is important therefore to realize that an abstract approach to architecture design is possible and has some benefits. The choice of the implementation technology can be treated as a separate, though interacting, concern.

References

Albus, J. S., Barbera, A. J., Fitzgerald, M. L., Nagel, R. N., VanderBrug, G. J. and Wheatley, T. E. (1979) 'A measurement and control model for adaptive robots', in *Proceedings of the 10th International Symposium on Industrial Robots*, SME, Dearborn, Michigan.

Bradley, D. A. and Seward, D. W. (1992) 'The Lancaster University computerized intelligent excavator programme', in *Mechatronics – The Integration of Engineering Design*, Proceedings of the University of Dundee Conference, Mechanical Engineering Publications Ltd, London.

Brooks, R. A. (1986) 'A layered intelligent control system for a mobile robot', in *Proceedings of the 3rd International Symposium on Robotics Research*, MIT Press, Cambridge, Mass.

Fodor, G. (1983) *The modularity of mind*, MIT Press, Cambridge, Mass.

Lumia, R., Fiala, J., Wavering, A. and Albus, J. (1990) 'Kinematics and dynamics in a hierarchically organized robot control system', in Taylor, G.E. (ed.) *Kinematic and Dynamic Issues in Sensor Based Control*, Springer Verlag.

Simon, H. (1969) *The Sciences of the Artificial*, MIT Press, Cambridge, Mass.

CHAPTER 10
DESIGN

Paul Wiese

10.1 Introduction

Designing mechatronic products is more difficult than designing current technological products, because it involves the combining the elements of design of a number of different engineering disciplines, such as mechanical, electrical and electronic optical engineering. Except in the simplest mechatronic products, it is unlikely that one person can carry out every aspect of the design. It is important in the team situation, therefore, that every member of the team understands enough of the related technologies to make meaningful judgements on the necessary compromises that have to be made during the design process. The chief designer has to be conversant with each branch of technology – its approaches, methods and jargon – to lead and guide the team to a properly integrated solution. The role of the chief designer is therefore effectively that of team leader in a multidisciplinary team.

Before we look at each aspect of design in some detail, let us first see where the design of mechatronic products fits into the outside world.

Design is the fundamental starting point for any new product. Because design is about the achievement of objectives in an uncertain world, it always has to be based on a number of assumptions; hence the requirements the designer is working to are always of an ill-defined nature. For a mechatronic product this can be particularly difficult because of the variety of design approaches available, compounded by the interactions between the different technology disciplines.

Recent years have seen the world economy become increasingly competitive. Markets have reacted to increased competition in different ways – manufacturers might have to emphasize lower price, better quality, frequent changes in the product, or combinations of all three.

Manufactured goods are made on or by processing machines and the design of these goods and processing machines has had to react to the marketplace requirements. Additionally, social changes in the world have necessitated changes in design of process machines as the cost of labour has increased and the social consequences of long hours of tedious work have become apparent. The result has been the requirement for new types of process machines employing new concepts, higher speeds, modularity, productivity, rapid size change and social acceptability.

In many design areas these new requirements cannot be achieved easily by the evolutionary design trends of past years. Many developments of old designs reach a limit in their performance, and a new approach is the only way that further improvement can be made. It is here that the mechatronics approach is at its most powerful – in showing the way for further improvement when an existing approach has reached its limit.

As an example of how a limit in a process machine can be reached, consider a simple product-packaging machine. The processing of the product (e.g. the wrapping of a cosmetic product) is carried out with the product stationary in space and using several separate operations. Each operation is carried out at a discrete position (workstation) in the machine dedicated to that single operation. The machine therefore has two different types of function to perform: (1) to carry out the appropriate wrapping actions over several stages, and (2) to move the product (initially unwrapped, and then in each semi-wrapped state) from one workstation to the next. The maximum speed of the machine is then a direct function of the time taken to undertake each wrapped action, and the time the machine takes to move (to *index*) from one workstation to the next. The latter necessitates the acceleration and deceleration of the mass of the moving parts of the machine (in most cases the mass of the product itself is comparatively negligible). The wrapping process determines the minimum time to achieve the wrap, and inertia of the machine and the sustainable acceleration the product can accept determines the maximum indexing speed. For a given configuration, the maximum machine speed is fixed and can be mathematically determined. To achieve higher speeds a totally new processing approach is required.

Some new market requirements appear more onerous than before but in fact can assist design. Better product quality is a case in point, because it means improved consistency. To achieve better consistency requires greater precision in the machine, but it results in better control of that machine, since the manufacturing variables in the product are reduced.

The most successful way that today's market requirements are being achieved is by the design of modular mechatronic products. The addition of artificial intelligence enables the process machine operation to be changed and the machine to be designed around the process requirements, rather than around mechanical constraints in the machine itself (such as the mechanical drive system) or human control requirements. The result is a significant improvement in performance, whether measured as productivity, efficiency, process speed or repeatability.

One further important point has to be stressed. Just as many machines of today have evolved by steady development over many years, many aspects of design have hidden virtues and problems which have become accepted by designers and are incorporated without further thought. For instance, in a mechanical geared drive, the variation in gear pitching due to machining tolerances and the resulting backlash problems are well understood and covered by design rules available to everyone. Substitution of well-proven techniques by new mechatronic solutions may have obvious measurable advantages, but in the substitution the designer may become painfully aware of some of the hidden advantages in the existing technologies of today. The substitution of the mechanical gear drive by an independently controlled electric motor may provide much greater flexibility in gear ratios but the lower rotating inertia of such a system may prove troublesome on acceleration and deceleration, and the flexibility of the mechanical coupling between motor and drive shaft may provide an unwanted variable missing in the older geared type of system. Colloquially, it is called swings and roundabouts. Mechatronics is not all swings!

10.2 The mechanics of design

Let us now look at the way a design team should operate, and how to apply this to the design of mechatronic products. The design team should note the need for:

- a design plan or design strategy for the future based on the type of products being designed;
- a phased approach to the final design when appropriate;
- formal product specifications as sheet anchors for the design process;
- good awareness of manufacturing costs and their relationship to sale price;
- an active and continuous awareness of the human–machine interface problem in its many facets.

Too many design teams operate without a plan or some idea of a strategy towards the design of particular products over, say, the next ten or twenty years. Few products remain forever static at their initial design level, and it is essential for the sales team to have a product that can be updated and developed to generate new sales potential at regular intervals. Examples of this are seen in everyday life. The motor car is an example – it is difficult to think of any car that has remained static at the original design level. The Ford Cortina went through Marks 1 to 5 over a 22-year period, for instance.

Any worthwhile design team should be able to generate a technical strategy towards their future 'Project X'. This Project X can incorporate technical needs and possibilities which at present cannot be met, but which can be envisaged. This is simply illustrated in mechatronics, because we can envisage a machine with complex neural intelligence networks, sophisticated sensors and transducers, etc., even when we know it will take several more years to develop the technologies to the required level.

Chapter 1 of Volume 1 referred to the European car manufacturers' research programme, Prometheus, which has specific objectives for the year 2010.

▲ ***A design concept from General Motors Corporation***

Although Prometheus was set up for research, each of the car manufacturer's design teams in the joint programme will have a *design plan* so that they can incorporate the research into their future products. I will call this approach ***strategic design***, and, if carried out successfully, it enables intermediate phases of the design programme to be split off at times appropriate to the marketplace requirements, whether these are sales led or technology led. This ***phased approach*** may indeed take us in a different direction than towards our ultimate Project X, but then the whole strategic plan needs to be flexible, and the direction can be altered as market forces dictate.

Before we turn to the more formal aspect of design (the necessity for specifications), let us look at how the design process itself has changed and is changing. If we look at design in mechanical engineering, we can see how the importance of specialists has increased as technology has progressed. In the earliest days, the design office would be a small self-contained office, which issued drawings to the manufacturing side. The legal design authority was held by the chief designer, and little outside expertise was needed. Figure 10.1 represents this simplistic form.

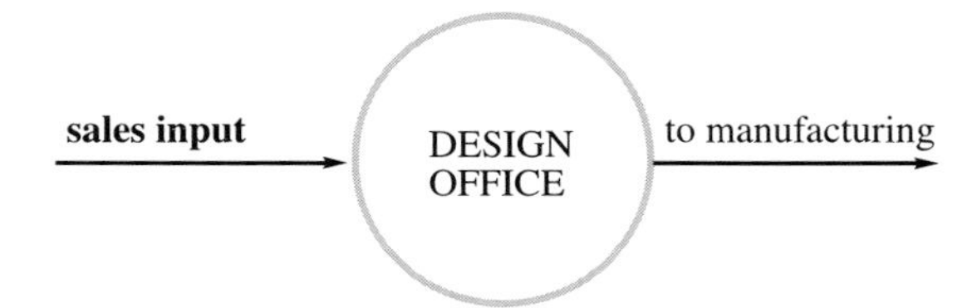

◀ ***Figure 10.1***

As technologies developed, some expert information had to be obtained from outside specialists, as represented in Figure 10.2.

As the technologies developed further, this outside expertise was frequently grouped into separate specialist departments within the organization, such as a stress office, a thermodynamics group, and a computing group – again shown diagrammatically in Figure 10.3.

The design office was receiving inputs from the specialist departments, and its work was becoming mainly one of blending and synthesizing, while still being responsible for original design work.

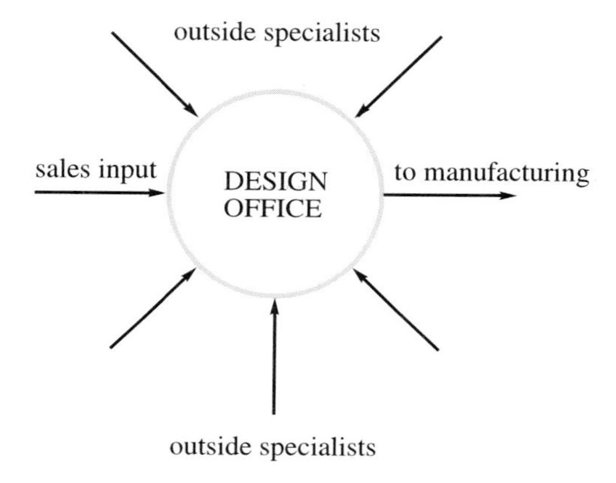

Figure 10.2

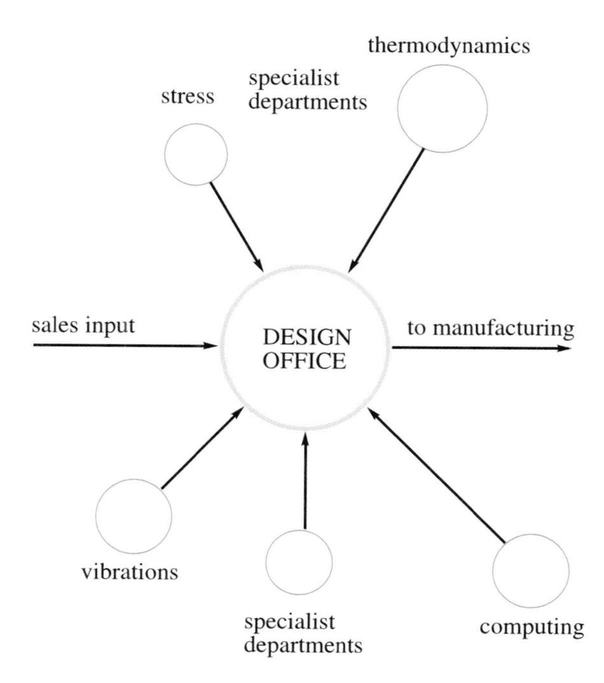

Figure 10.3

With the rapid increase in computing power in the last fifteen years, the expertise in the specialist groups has increased dramatically, whereas the role of the design office has diminished. Much of the original synthesis is now done in the specialist areas, which communicate with each other before providing the design office with the idealized solution to their specialized problems. The role of the 'design office' is now more one of assembling the specialists' solutions into the whole. Of course, the design office or design role still encompasses all the activities described above, but the important point is that the specialized technological work is carried out separately rather than being a part of the work of individual designers.

A further important step has been the emergence of ***concurrent engineering***, where the manufacturing engineering is carried out in parallel with the design activity, instead of in series, after the design work is complete. The drive towards concurrent engineering has come from the need to dramatically reduce product development timescales and the absolute need to design within strict cost targets. The latter can only be achieved when the manufacturing methods and processes are planned in conjunction with the design process. Indeed, in the best examples of this practice, material is being machined before the design is complete, and before drawings are available in even rudimentary form. This is possible by use of advanced CADCAM (computer-aided design/manufacture) techniques. Concurrent engineering is a most important facet of the modern design process, in mechatronic product design as elsewhere. Consequently, when we look at the modern method of design, we see not only the specialist technology areas within design, but manufacturing engineering as well. This is shown in diagram form in Figure 10.4.

An easy way to differentiate between manufacturing engineering and the older discipline of production engineering is to remember that manufacturing engineering covers the design of the manufacturing process, so only has to be done once for a particular design. Production engineering is concerned with the actual making of the designed product in whatever quantities the sales or market demands. This approach quickly demonstrates that manufacturing engineering is part of the design process, whereas production engineering is not.

This scenario illustrates the need for team work and communications. Note that this example started with mechanical engineering, which in most mechatronic products will still be the lead technology. But to Figure 10.4 we now have to add the extra specialist and main groups in the related technologies – electronics, electrical engineering, software, optical engineering, etc. It is clear that the chief designer must now act as a team leader. One thing that has not changed is that the legal authority for the design remains with the chief designer, no matter how big or how important the specialist departments become.

Let us now look at some of the mechanisms of design.

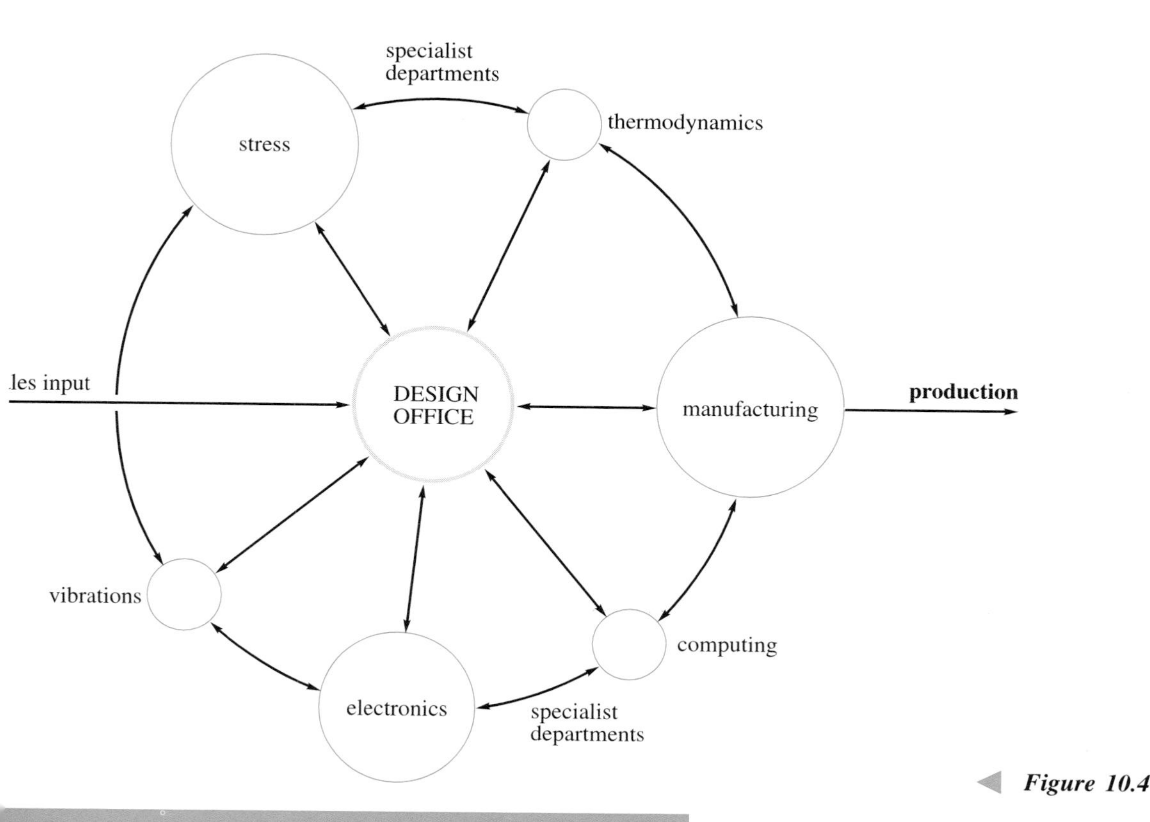

Figure 10.4

10.3 Product specifications

10.3.1 The requirements specification

The starting point for any engineering design is the ***requirements specification***. Designing without one is like running a meeting without an agenda – it can be done but usually results in chaos. In design, chaos only leads to disaster later, so the necessity for the requirements specification cannot be overemphasized.

The requirements specification is laid down by the 'customer', as to just what is required from the new product. The customer in an industrial context is usually the sales department in the same company, and the requirements specification covers the important parameters the new product must achieve to be attractive in the market place, i.e. the type of product that can be sold. The requirements specification must therefore contain the likely selling price or the manufacturing cost, as the cost to produce is a vital part of the design process. This is a most important point because typically 75% of the manufacturing cost of a product is dictated or influenced by the initial design.

For the designer, the requirements should be stated in as much detail as is possible, because there is always more than one design solution to any given problem. In a mechatronic product this is even more important because of the greater complexity of the design, and the involvement of several different engineering disciplines in the design solution. The conventional requirements will need to specify much wider goals and functions and to incorporate time-related sequence at overall and detail level. In particular the requirements specification needs to quantify acceptable performance and quantity levels, and the target efficiency of the whole mechatronic system, as well as its constituent parts.

Although the designer does not (and should not) draw up the requirements specification, it is the designer's job to ensure that it is as complete as possible, and that detail which will be needed in the design is there and correctly stated. The designer will be familiar with the extra parameters in a mechatronic product whereas the sales department will not be, so this point is of vital importance. It is in these grey areas that trouble subsequently arises if the design fails to meet its objectives – although all designers have experienced the mysterious changes that can appear in requirements *after* the design is complete!

10.3.2 The design specification

Once the requirements have been agreed, conceptual design can start. The designer should begin to compile a ***design specification*** for the product. In effect, this is the engineering response to the requirements specification, and details not only the expected performance from the product, but also the target manufacturing cost, the standards used in the design (both national and international), bought-out items, software details, etc.

The manufacturing cost is an essential part of the design specification and should be broken down into target costs for each module and submodule. To achieve minimum manufacturing cost, the designer will need to work closely with the manufacturing or production engineers and the buying office. This process is now called *concurrent engineering* (see Section 10.2). In many existing companies it is an unfortunate fact that designers do not always understand that the price at which a product can and will sell is determined by the market place and not by the manufacturing cost. There is a long list of innovative engineering designs which were unsuccessful because this simple economic fact was not understood. If you cannot design the product's manufacturing cost to be below the market-determined selling price, the product will be unsuccessful no matter how brilliant the design is thought to be. For the designer of a mechatronic product this is a difficult problem to overcome because of the complexity of any intelligent machine with costs involved in software, simulation, etc., as well as pure mechanical/electrical design.

A new product should be specified via a listing of schematic assembly drawings – either on computer-aided design (CAD) or hard copy – which show the complete product via its modules or submodules. The use of a list of detail drawings (the *master part list*) as part of the design specification is to be avoided, as component detail drawings do not and should not show assembly tolerances, tightening torques, software listings, etc. A design specification should be sufficiently complete to enable the product to be drawn in detail by an outsider, with the full performance targets stated. The listing of the schematic drawings is a powerful technique in establishing the basic initial design standard and the subsequent introduction of modifications via additional schematics. Hence the design specification can be used at any point in time to establish not only the original design standard but the subsequent modification history. As the use of CAD becomes more universal, the design can be defined by an electronic model, but an *audit trail* through subsequent modifications is essential for quality purposes.

An essential step before any actual design work begins is for the requirements specification and the design (engineering) specification to be reconciled. This is best done at a formal meeting, where any differences (usually shortfalls in required parameters) can be discussed and an agreed position reached. In a worst case it may mean that a design cannot be conceived which can meet the required objectives. More likely there will be shortfalls in some areas and if these are not accepted by the selling team at the outset, endless problems will arise later. Note that although the design specification is about performance parameters, project timescales also need to be outlined at this stage.

When agreement is reached, *both* specifications should be reissued to represent the agreed position. It is the designer's job to check the specifications for completeness, and to state where deviations may occur. Both specifications should be treated as formal documents which represent the formal statement of the design. It follows that updating and reissue of the specifications should also be carried out by a formal system.

10.4 Conceptual design

The initial conceptual design phase for a mechatronic product is probably more important than for existing, traditional-type products, since the design solution is seen through multidisciplinary eyes, and will incorporate new and probably untried techniques. At this stage the problems of communication between the various engineering disciplines involved in mechatronics can be at their most wearing. Each discipline has its own jargon and its own method of expressing design. The mechanical designer uses technical drawings with 2-D side and end elevations on hard copy drawings, or 3-D modelling on CAD. The electrical and electronic designers are thinking in terms of circuit diagrams, with design of printed circuit boards done automatically by CAD. The software designer is using flow diagrams and programmed short cuts peculiar to the language used for the program.

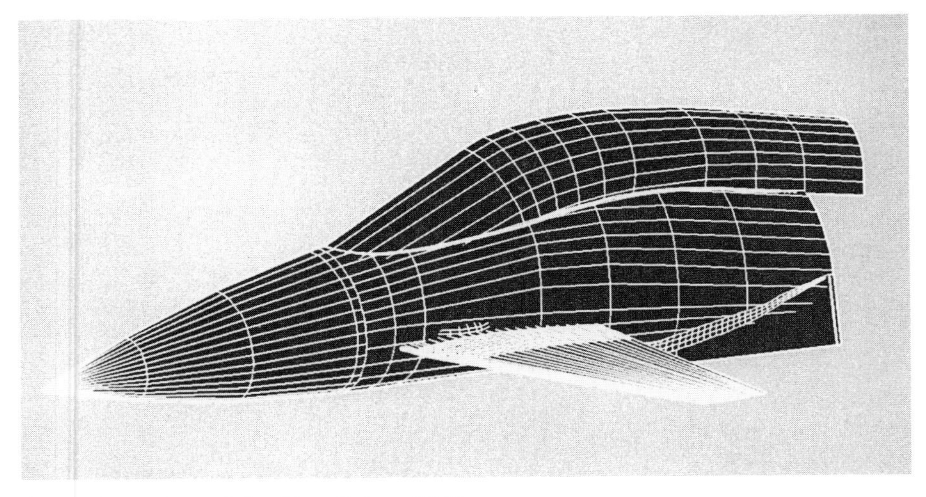

▲ ***A CAD view of the Eurofighter 2000***

While the design can be largely liberated from the traditional evolutionary approach most apparent in mechanical engineering, the cross-discipline communication problems in mechatronic design are deeper than just different jargon usage. At the same time the traditional evolutionary approach to design gives a solid formulation to launch a new design. Starting with a completely clean piece of paper can be compared to parking in an empty car park – any position may be suitable, but which will be best?

The balance between some aspects of existing technology and a mechatronic approach was touched on briefly in Section 10.1. In this conceptual design phase these parameters have to be considered fully. Any new mechatronic product will contain a large amount of existing technology, and several stages of product standard may be needed to be built and run before the final 'mechatronic product' can be designed. This is because the first requirement of any design is the establishment of accurate data on which to base the design. In a sophisticated aerospace-type environment, where it is impossible to design and make several interim/experimental aircraft just to accumulate design data, extensive use has to be made of rigs, wind tunnel models, etc.

In many industries it is not possible to adopt a full-scale test type of approach, since full operating conditions cannot be simulated in a factory environment. For instance, the designer of a large steam turbine generator for an electric power station cannot test the design in the workshop since neither the steam supply nor the means to dissipate many megawatts of electrical power is available. Similarly in the process industries, the availability and flow of materials cannot be sustained in a workshop for a long period. In both cases the first time the product runs under design conditions is in the customer's plant, in front of the customer, who naturally expects 100% performance immediately.

This lack of comprehensive operating data is a salient point in design since it explains the need for the 'right first time' approach to any design, and also helps explain the inherent conservatism of many designers. In choosing a mechatronic approach by putting intelligence into the machine, it is even more vital to have accurate operating data available on which to base the design. Designers often like to think they know everything about their own design, but it is almost impossible for them to obtain accurate operating data. In many industries the only data available is from operator experience, which is subjective at the best of times.

As an example, I was asked to improve the performance of a high-speed processing machine which made 8000 items per minute. Four machines had been built and were being tested in customer plants in different countries on a three-shift, 24-hour basis. No two machines operated the same way or to the same quality. It transpired that the original design allowed the operators to adjust or alter 107 variables on the machine. The machine processed organic materials which varied in characteristics, and each operator on each shift at each plant had his own pet settings to try to obtain optimum performance. What chance was there to obtain good operating data in this scenario with this number of variables? A protracted nine-month programme of strict parametric testing on a full statistical basis was necessary before the design characteristic could be fully evaluated and corrections made. The machine now produces the best quality product in its field in the world, at speeds of up to 10 000 items per minute – but the operator-adjusted variables have been reduced to single figure numbers!

In the process industries, the use of *statistical process control* (SPC) is giving some useful information on machine operating performance, but it is the recent introduction of microprocessor-controlled supervisory and data-collection equipment which is at last providing the needed information. Such a system is the Molins OASIS system which will monitor a machine continuously and inform both operator and plant management when faults occur, and most importantly, as adverse trends arise. An example of the power of this approach is shown at a plant in Sydney, Australia, where the early production units were installed and were monitored via a satellite/modem link in the UK. On one occasion the UK office were able to telephone the plant in Sydney and tell them that a certain function on their number two machine had stopped working. Neither the operator nor the plant management were aware the machine was not operating correctly!

The reason I have laboured this point is because for any machine with AI it is essential for the designer to have basic operating data at the concept stage. If this is unobtainable, then one or two interim stages in the product design may be unavoidable. This is evolutionary design in the accepted sense, but at the beginning of the 80/20 rule, not at the end. (The 80/20 rule says that 80% of the information required is available with 20% of the effort; the remaining 20% takes 80% of the effort.)

A further point to consider in the design of machines for the process industries in particular is that many of these machines process organic materials. Organic materials have properties which can vary with temperature, humidity and pressure in a more random manner than inorganic materials. Further, there is little documented data on organic materials, and even test methods are not standardized. These are just some of the variables that designers must cope with.

Let us now look at a few examples of mechatronic products and their conceptual design, and then look at one case in detail.

10.4.1 Aircraft

The complexity of a modern aircraft is a natural haven for a mechatronic approach, where the 'fly-by-wire' system of control is now in daily service. The mechanical control systems using hydraulics, pneumatics and 'wire-in-pulley' systems have been replaced by a computer-controlled electric actuator system, with no mechanical back-up.

The reliability of the mechanical type of control system has become largely understood by experience over the last 70 years of operation. The new system doesn't have the benefit of this experience, and in the meantime reliability is achieved by duplication and triplication of the systems.

Perhaps the best example of a mechatronic approach in aircraft is the European Fighter Aircraft, or EFA. Any aircraft controlled by a human being has to be designed to be aerodynamically stable. On any new aircraft type, extensive testing is carried out until the aircraft can be said to fly 'without vices'. This inherent stability, however, limits the extent of its manœuvrability – and all combat aircraft have to have extreme manœuvrability. With an aerodynamically unstable design the aircraft can change direction very quickly. However, the human brain cannot react fast enough to fly an aircraft designed as an unstable aerodynamic platform, and the control has to be carried out by computers, as in the EFA. In effect, the computer is asked to direct the aircraft from one unstable path to another in a continuous motion. It is said a human can control the EFA for no longer than 4 seconds. This is a true mechatronic machine, where the AI is faster than the human. This does not, however, imply it is more intelligent!

10.4.2 High-speed machines

Process plant machinery predominantly has to hold, shape and move materials, and market forces are constantly requiring increased output rates, whether these are achieved in tens, hundreds or thousands per minute. Typically, ***process machinery*** is that used to perform operations such as stamping, printing, weaving, packing, wrapping, bottling, spinning and many more besides. It does not normally cover machining operations but does cover complex lines incorporating many operations, as in an assembly line. ***High-speed machinery*** is a term currently in vogue to describe process machinery, whether the materials processed are organic or inorganic.

▲ ***The Eurofighter 2000***

The current need to find significant increases in output through speed, flexibility and accuracy has led to two major UK research initiatives in this area during the 1980s and 90s under the title 'The Design of High-speed Machinery'. As a result of these initiatives, important work in the mechatronic field has been completed and is currently being applied to specific machines. Let us look at this work and consider how it can be used on a typical machine. For our example, we will consider a high-speed box-making machine.

A box-making machine, known technically as a flexo/folder/gluer (which describes the three stages in box making, as we shall see later), is the last machine in the line in a corrugated board plant, and comprises the 'converting' end of the line where the board is converted into boxes. The output of the machine is corrugated board boxes in collapsed form for easy transport. At the next stage, in a customer's plant, a box-erecting machine will take these collapsed boxes and form them into the familiar crown or printed box for filling with the customer's product. The output speed of a box-making machine of this type depends on the *blank* size and hence the machine size but is typically 25 500 sheets per hour for a machine handling blanks up to 1700 mm by 660 mm. This type of machine is shown diagrammatically in Figures 10.5 and 10.6.

How could we improve the performance of the machine by a mechatronic approach? Because of the complexity of the machine and the technologies associated with each process industry (in this case the corrugated board industry) it is impossible to give a comprehensive answer in this chapter, but I hope that an outline will illustrate the approach needed.

First, what is the machine's output and what areas can be improved? The output is collapsed corrugated board boxes, and the areas that could be improved are:

output speed
overall efficiency
quality
set-up time
flexibility for different box shapes
reduced labour input.

The above operations are carried out automatically by the machine against fixed settings in each module which are specific to the box being produced. The machine is controlled by a microprocessor which stores up to 250 box details, and the machine can be called to set to any one of these 250 variants on demand by the operator. The machine will then automatically reset to the new setting on each of 28 axes. However, if the machine is incorrectly set up initially or if the settings are changed or wear takes place, then incorrect size boxes result. To avoid the problems, we have to ask the following questions at each stage.

Stage 1 Is the blank square and in register (i.e. correctly positioned along the axis of the machine)?

Stage 2 Are the nip (or driving) rolls correctly set for the thickness (or calliper) of the blank?

Stage 3 Is the printing correctly positioned, of good quality, and with colour in alignment (in register)?

Stage 4 Are the creases correctly positioned and to the correct depth? Are any slots (die cuts) in the blank clean and in position and has the die cut scrap been removed?

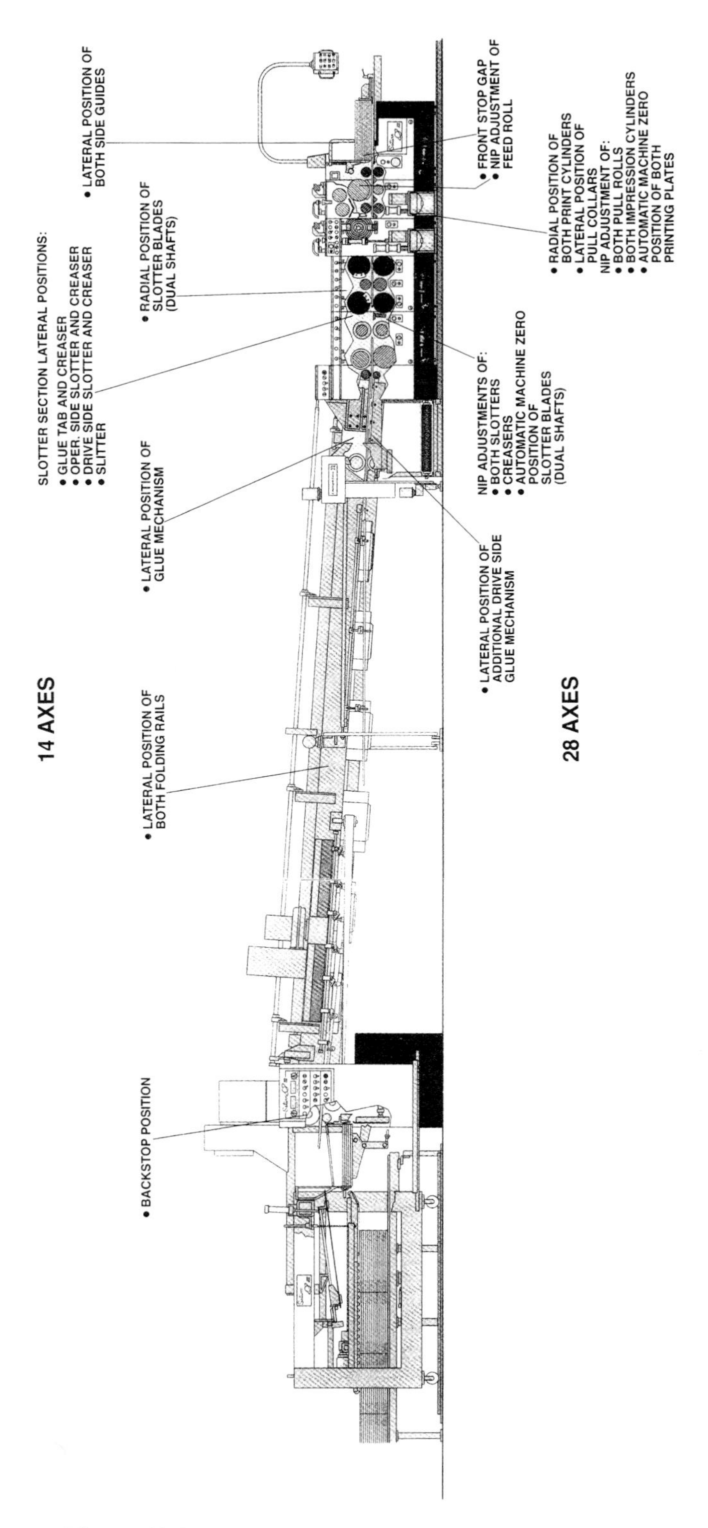

Figure 10.5
An automatic box-making machine.

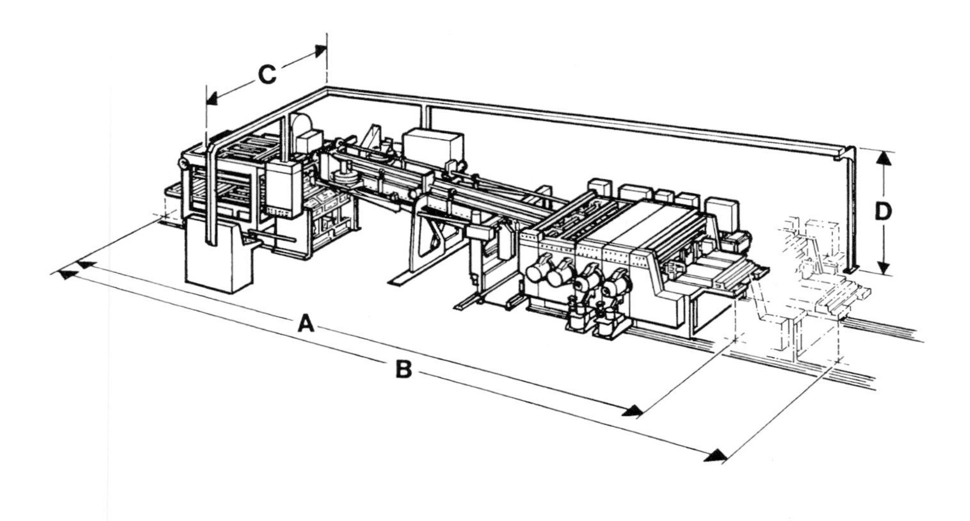

TYPE NUMBER	**2668**	**3797**	**5113**
MAXIMUM SPEED (SHEETS/HR)	27,000	18,000	15,000
DIMENSIONS:			
LENGTH CLOSED (A)	44′ 9½″ (13.652 M)	52′ 8 1/16″ (16.054 M)	56′ 1 1/16″ (17.096 M)
LENGTH OPEN (B)	53′ 5⅜″ (16.291 M)	62′ 9⅜″ (19.136 M)	65′ 11⅞″ (20.114 M)
WIDTH w/o D.C. (C)	19′ 2″ (5.842 M)	20′ 9″ (6.325 M)	20′ 9″ (6.325 M)
HEIGHT (D)	10′ 9⅜″ (3.286 M)	10′ 10¼″ (3.308 M)	10′ 10¼″ (3.308 M)
WIDTH w/D.C. TO O.S.	N/A	35′ 5″ (10.795 M)	39′ 1⅞″ (11.935 M)
WIDTH w/D.C. TO D.S.	27′ 0 1/16″ (8.230 M)	35′ 0⅛″ (10.668 M)	36′ 9⅛″ (11.205 M)
ELECTRICAL POWER:			
INPUT KVA	93	93	93
VOLTAGE	460 V, 3Ph, 60 Hz	460 V, 3Ph, 60 Hz	460 V, 3Ph, 60 Hz
MAIN DRIVE MOTOR	SPLIT (2) 25 HP	50 HP	50 HP
CONTROL	110 V, 1Ph, 60 Hz	110 V, 1Ph, 60 Hz	110 V, 1Ph, 60 Hz
PNEUMATIC:			
PRESSURE	80psi (5.6kg/cm^2)	80psi (5.6kg/cm^2)	80psi (5.6kg/cm^2)
VOLUME – NORMAL	50 SCFM	50 SCFM	50 SCFM
VOLUME – SET-UP	100 SCFM	100 SCFM	100 SCFM
WATER:			
PRESSURE	40psi (2.5kg/cm^2)	40psi (2.5kg/cm^2)	40psi (2.5kg/cm^2)
TEMPERATURE	110°F (43°C)	110°F (43°C)	110°F (43°C)
CYLINDER CIRCUMFERENCE OVER DIES	27 ¼″ (692mm)	37 ½″ (952mm)	50″ (1270mm)
MAXIMUM PRINT AREA	63″ × 24″ (1600mm × 609mm)	93″ × 35 ½″ (2362mm × 902mm)	106 ½″ × 48″ (2705mm × 1219mm)

Figure 10.6 Specifications for the flexo/folder/gluer.

Stage 5 Has sufficient glue been laid down, and in the correct position? Is the glue viscosity correct?

Stage 6 Are the folds made correctly and square to one another, i.e. no *fishtailing* (malalignment).

Stage 7 Are the glued joints correctly bonded?

Stage 8 Is the stacking of the blanks correct, i.e. in register? Are the correct number of blanks in each stack?

Of course, the above is only a first look at possible problems in the machine; there are many more problem areas, particularly when dealing with different paper types through the machine (e.g. virgin paper or recycled, line/medium weight). However, we can see that each stage can be treated separately and the machine made self-adjusting. For instance, at stage 1, photoelectric cells or a vision system can determine whether each blank is square and in register, or skewed. A skewed blank at stage 1 will ensure incorrect treatment at all subsequent stages unless we put in a corrective procedure. A simple camera system will check for correct glue lay-down, and another for fishtailing after the folding section. Print ink and glue viscosity can be measured and so on.

Before the design process can start on making the machine fully self-correcting, basic data are needed on how the existing machine behaves in a day-to-day production environment. Accordingly, devices such as sensors, cameras and transducers have been set up on the machine to obtain on-line instantaneous information and trends, and this information can be analysed in different ways to suit the needs of the operator, supervisor, maintenance department or the designer.

The information also needs conversion into data that can be used to monitor and correct the machine operation. For instance, if the registration of the blank at stages 1 and 2 is tending progressively to go out of alignment, servomotors can be fitted to correct the position against instructions from the photocell sensors or camera system, based on data contained in the pre-set limits in the data bank. The addition of further intelligence would enable each blank to be individually positioned to ensure the most consistent and optimum requirements (such as creasing position). Again, as the product varies, the machine can be designed to recognize the change and compensate for it. An example would be a change in calliper (thickness) of the blank which could be accommodated automatically. With existing machines bad boxes would result.

The microprocessor type of supervisory equipment enables all these variables to be logged and analysed for operation machines. The designer then knows the actual material variables and the rate of wear and debris accumulation that the machine will have to contend with.

10.5 Embodiment of design concepts

The mechatronic product field is so wide that it is impossible to give definite design rules. In view of the different technologies involved, the designer needs to be able to keep a clear mind to the several different approaches possible by considering each technology solution in turn. This in itself requires the lead designer to be able to understand enough of all the technologies involved to make a decision and then to use specialists in the technologies, even if this involves going outside the group for assistance.

10.5.1 The human–machine interface

Different products may have a unique solution or several competing ones. In assessing competing solutions the designer needs to look right through to the final ***human–machine interface***. It is most important to consider this point at the outset.

All machines require human involvement, either on the operating side or the maintenance side, or both. The social consequences of long hours of tedious work are now recognized. A new complex design of process machine may completely deskill the operator, such that the operator's life becomes one of boredom instead. An approach is then to make one operator responsible for several machines, which makes good economic sense. For the operator, the change is to one of added responsibility and the aim here is to match the operators' skills to the machine needs. This must be considered at the design stage, however difficult this may be, when the product is to be sold over a wide market sector with vastly differing human skill levels.

Some highly skilled occupations require a skill level to cope with emergencies rather than day-to-day routine operations. This is most evidently true in potentially hazardous situations. Let's take a modern aircraft as an example. Much of the routine flying of an aircraft can – and is – done automatically. Automatic landings are now commonplace (although the pilot may decide to carry out a manual landing 'to keep his/her hand in'). The necessity for a pilot to be present and highly skilled (besides being a comfort for the passengers) is to be able to take charge in an emergency.

Another example is the shift charge engineer in an electricity power station. Again, the person's experience and skill level (and salary expectations) will be well in excess of what is required for day-to-day running of the station. The function is to be able to take sole command of the situation when the unexpected problems or emergencies occur in such a complex plant. We can add artificial intelligence to a machine but we cannot yet anticipate every emergency.

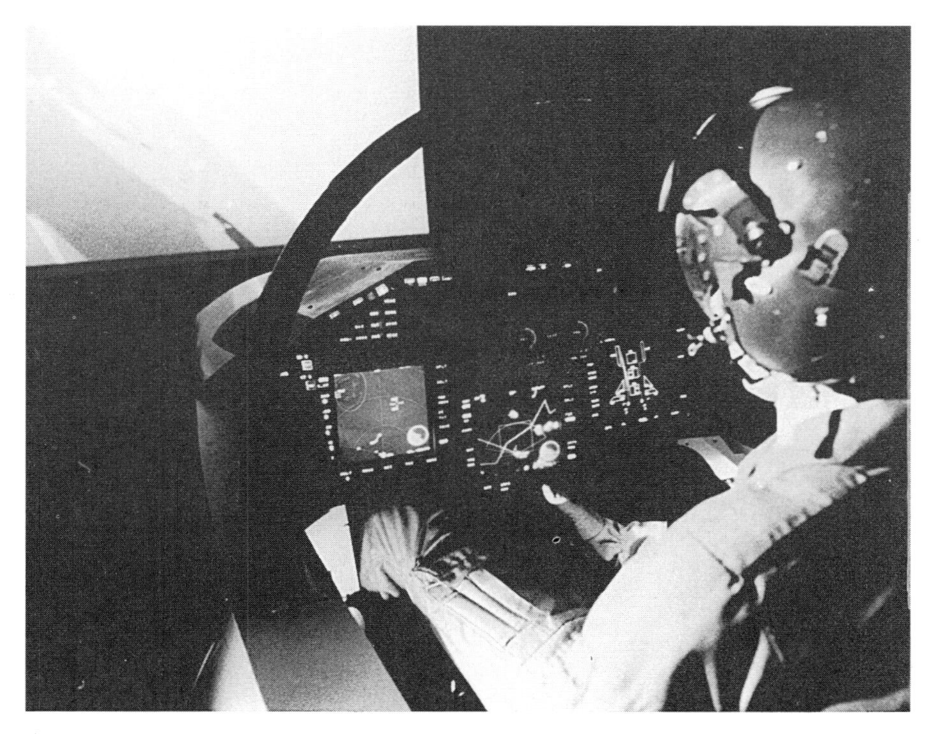

The Eurofighter 2000 cockpit

The designer must also consider the situation where AI could prove problematic to the human operator. Suppose that an airliner is about to fly into a mountain. The pilot may attempt a manœuvre to avert the catastrophe but in doing so takes the aircraft outside its normal flying envelope, right to the limits (or even beyond) of airframe stressing. Would AI allow this to happen or prevent it? How does AI anticipate the mountain catastrophe situation?

Emergencies for a mechatronic product may well be of a nature vastly different from those that occur with conventional plant. Much of the improvement gained by a mechatronic approach to a design problem can be illustrated by comparing the synchronization of two drive trains (or shafts), first using a conventional mechanical geared system and secondly a mechatronic, independently controlled, electric motor approach. The gears will have a fixed ratio which can only be changed by fitting new gears. The electric motors can be programmed to work at virtually any gear ratio, and the change in ratio can be made with the system running. However, if a power failure occurs, the geared system will run down but will still be synchronized, whereas the independent motors will almost invariably not remain in synchronism. In many mechanisms this will lead to very dangerous and damaging collision situations.

Whereas power failure cases lend themselves to fairly straightforward fault analysis, the random effect of spikes on the electrical supply voltage, rapid transients in voltage and other electrical noise effects, such as electromagnetic radiation, are less easy to diagnose or anticipate. This problem has led to new restrictions being placed on designers by the Health and Safety at Work Act, where mechatronic machines operate in the vicinity of humans, and where mechatronic solutions are being used in the control of dangerous processes, such as in chemical and nuclear plants. The whole field of fault analysis, safety and reliability is not completely defined for all areas of mechatronics at the time of writing. Research into these areas is in progress, but at present the problem has to be considered deeply at the design stage, especially as to how it relates to the human–machine interface.

Mechatronics has enabled designs to be achieved in areas that would be impossible using previous technologies. For instance, a purely mechanical camcorder is difficult to conceive. As Charles Babbage found with his attempts to build the first computer, sometimes the technology has to be developed to enable ideas to be brought to a practical state. In consumer 'electronics', many of the products are mechatronic rather than electronic, and it is instructive to look at how the designer has achieved the aims. As we are looking here at conceptual design rather than actual detail design examples, we can note how new markets have been devised (rather than developed) by the use of mechatronics.

When the final customer is the general public, the designer has to be very aware of the human-machine interface, and it would be fair to say that much of today's consumer electronics are too complex for the public to understand. Designers do not always understand that instruction books are rarely read from cover to cover, and quite frequently not at all. (Ask yourself how many of your friends know and can use all the facilities of a modern microwave cooker or video recorder.) What the designer frequently has achieved is the creation of attractive products in terms of capability for the required function, cost and appearance but has not solved the human–machine interface problem.

While appearance is frequently thought of as the domain of the industrial designer, in mechatronics the whole casing of a product is much more integrated into a whole, rather than a device which can have an external pretty box to cover it. The design and function of product *housings* is an area of its own. In its simplest form, the housing – and indeed the machine framework – may be the minimum to achieve several functions separated in space. Consider the design of a crane as used on many high-rise building construction sites. The 'housing' is little more than a steel skeleton to connect the pulleys for the lifting area, the driver's cab, and the counterweight. However, using a skeleton rather than a solid-sided box construction also solves the problem of side wind loads and wind aerodynamic forces on the crane, so the housing is fairly functional. Similarly, the streamlined casing of a camcorder looks attractive but the camcorder also feels good when it is held in the hand, because the design has achieved a correct weight balance. Some housings seem purely functional, but may confer hidden advantages. One major European aeroengine manufacturer gained worthwhile fuel consumption benefits compared with its commercial rivals by studying the aerodynamic design of engine nacelles in modern jet airliners. Today's car design has changed through a slow development of the original 'horseless carriage' concept, through the 'streamline' age, to a real study of aerodynamics to reduce drag and hence reduce fuel consumption. Many other examples are available in everyday life. The point is that design has moved away from the purely functional approach to housings and now sees it as an integrated part of the whole and one of the most important areas of human–machine interface.

10.5.2 Design trade-offs

Throughout the conceptual design phase, different solutions to each problem must be considered and the optimum solution chosen. There may often be a question of a ***trade-off*** (the optimization of a machine through exchanging one technological system for another). To achieve the right balance, it is important to ensure that all the considerations and all variables are included in the analysis. Let us first look at a simple example in everyday life.

Consider the question of operating a city commuter bus service. Initially, each bus had a two-person crew – a driver and a conductor, with totally separate duties. The bus company then analysed the situation, and decided economies could be made and costs reduced by using a 'one-person' bus, and asking the driver to collect the passengers' fares when they boarded the bus. This undoubtedly reduced costs *to the bus company*, but has produced the situation where long lines of traffic are held up while the bus is stationary in the road, the 'driver' not driving the bus but collecting fares. Passengers can be forced to stand outside the bus at the mercy of the weather while queuing to pay the fare, fuel is being consumed by all the queuing vehicles behind the bus, and possibly the police are trying to keep traffic moving whilst these mobile chicanes (the buses) block the road. Did the bus company really analyse this situation correctly, or just parochially? In fact, their solution was correct for their set of circumstances, but incorrect for the community at large. The point to see here is that the criteria to use as judgement need to be accurately defined. The bus company were correct, using their own criteria, but what if the taxation authorities had trebled their road tax on these buses to take account of the extra congestion caused? Then the bus company's set of criteria would be changed and they would have to add an extra external variable into their analysis.

Let us now look at some examples in mechatronics. In information processing for instance, we need to consider the trade-off between, for example:

centralized versus distributed processing,

hard-wired versus programmed logic,

digital versus analogue electronics,

propriety versus specially designed integrated circuits.

In motion execution systems, consider:

timing by mechanical cams and linkages versus electronic clocks,

feedback by mechanical governors versus electrical tachogenerators,

remote control by mechanical linkages and drives versus electronic controls.

In the major area of energy conversion subsystems, consider:

> a single prime mover generating mechanical power (usually rotative), and transmitting this power through shafts, gears, belts, chains and linkages, versus separate motors generating mechanical movement for each separate function;
>
> a local prime mover plus a generator to convert energy into electrical form, transmitting to separate motors at each output location (such as a diesel electric railway locomotive) versus a centralized power source and generator, transmitting electrically over a distributed network to local output sources (such as electric railway locomotives and workshop and shipyard cranes).

The analysis will depend on the variables and criteria for each case considered, hence it is impossible to give design rules to cover every eventuality. The important point is to ensure that the analysis is as thorough as possible and that any hidden criteria are recognized. Some cases are straightforward (choice of the power source for a modern airliner will not take long) but the design of a fully modular high-speed processing machine may lead to some very difficult decisions. Also, always consider the effect of the path chosen on outside variables. Will these outside variables then change in an adverse manner, negating the original scenario? (See the bus company example.)

10.5.3 Design trends

In considering design, especially design in a phased strategy, it is essential to take into account the trends in development of the specialized items used in mechatronic products. Trends such as the following are important and need consideration as to their use and phasing in the overall design.

> the change from analogue to digital electronics,
>
> the change from wired logic to programmed logic,
>
> the change from centralized to distributed information processing,
>
> the change from separate machines to interconnection of intelligent machines into systems,
>
> replacement of mechanical transmission systems by electronically controlled electrical motors,
>
> replacement of conventional sensors by smart sensors,
>
> increasing ability for machines to learn,
>
> increasing ability to make decisions under uncertainty,
>
> increases in the types and scope of sensors,
>
> increasing ability to measure at very high speeds.

While most, and probably all, of these current trends will be known to the reader, the total pros and cons for their adaptation must be studied at each stage. As we have said before, there are swings and roundabouts at every stage! A good example occurred in America in the early 1940s. Steam locomotives were being replaced by diesel-electrics on the American railroads, and these 'new technology' engines were equipped with supervisory gear on the diesel engines. It was found that the reliability of the diesel electric locomotives was very poor, and a fraction of that achieved by the steam locomotives. Investigation showed that the problem was not in the prime mover and transmission of the new engines, but in intermittent faults in the supervisory gear causing erroneous signals. Removal of the supervisory equipment restored the reliability figures to acceptable levels. In this case the reliability of the instrumentation had not been developed to the level needed for the whole 'product', despite the then perceived need for this type of equipment, and thus gave false availability and reliability figures.

10.5.4 Design for quality, reliability and performance

The parameters which affect quality, reliability and performance are many and closely interconnected. In situations where safety is important, duplication and triplication of electronic systems are not unusual, in fact the Health and Safety at Work Act demands it. The same approach is rare in mechanical engineering but should and must be considered, particularly in the design of mechatronic products. The classical mechanical engineering approach of over-engineered solutions, with safety factors high enough to ensure reliability, leads to heavy and expensive designs (particularly when rotating or reciprocating masses are involved).

The *quality* of a mechatronic system includes many factors, of which how well the system fulfils its objective is just one. Long life, robustness, user friendliness, environmental friendliness and performance are others. These criteria are built in at the design stage and are difficult to add in later. The present jargon describes this as 'fitness of purpose', which describes it exactly.

Performance is a measure of total system behaviour and system total behaviour. The actual performance must obviously meet the performance specification, because the performance specification sets the target performance required. If the achieved performance is outside the specification, the system has failed, however brilliant the concept. It is rather like taking an examination: the examiner always seems to ask awkward questions whereas we could answer perfectly if *we* set the questions.

Similarly, *reliability* needs to be seen. It is important not only that the product will perform for long time periods without breakdown, but also that the system continues to behave to its specification if all its inputs are within the specification. These aspects are built in at the design stage and illustrate why the product specification and design specification are so important, and why they need to be agreed formally. They should be seen as formal, contractual documents, even if they are not given to or made available to the end user.

Reliability and quality of the system also imply robustness, and a measure of how well the system will perform if some or even all its inputs move outside specification, momentarily or permanently.

10.5.5 Design approaches

As mentioned previously, the lead designer needs to approach the design of a mechatronic product by considering solutions from the viewpoint of each technology involved. This approach will vary in effectiveness depending on the product, but is worthwhile in every case. However, it is rare that a design is in a completely greenfield situation, with no preceding design to influence the thinking. For most commercial design, economics will dictate the use of some existing technologies and designs. In the motor car industry it is now common to use the same floor pan (a major pressing) for many models, even across several competing, but collaborating, manufacturers. In Section 10.4.2 the phased approach to the design of a mechatronic high-speed processing machine was touched on, and it is this phased approach that is most likely to be needed when mechatronics are introduced to new areas – which is often where the most exciting possibilities arise. In these cases the designer, and indeed the whole team or company, needs to look far enough ahead to realize the final design and the final benefits and commercial advantages available.

Let us take as an example the design for a high-speed processing machine and look at how a mechatronic design approach may need to be carried through in phases. The first requirement is to set down the objective of the redesign *in its final form*, so that the strategy can be devised to achieve this objective, and the number of phased designs established. More importantly, by clearly setting down the final objectives of the design, the designer can assess the commercial implications of the work and time schedules, and set manufacturing cost targets. In our example the requirement of the processing machine is to make the maximum number of good, saleable products in a given time. Hence the machine needs to produce consistent quality, and in particular to identify or reject any out-of-quality items.

It is most likely that the existing machines run at constant speed and deliver the items at an output stage without regard to their quality. For example, a bottle will still arrive at the output end whether it is full, partly filled or empty, and a box-folding machine will still fold boxes whether the folds are perfect, misplaced or badly formed.

In this simplified example let us say that our final objective (to deliver the maximum number of good items) will be achieved by designing a variable speed machine which can monitor the quality of product it produces at each stage, reject any poor product, and adjust itself to correct the formation of poor products. Further, the machine will monitor its own trends, and self-compensate for wear, movement of location stops, etc. It is most unlikely that any designer could achieve all these objectives immediately in one stage, if only because of the lack of operating data on which to base a design. So what is the approach? I suggest:

Phase 1: Obtain maximum operating data by equipping existing machines with extra sensors, transducers, and supervisory computers to establish basic operating data, trends, and machine performance.

Phase 2: Introduce modularity into the design with, say, independent electric drives to modules, computer control, and additional sensors and transducers with some close looping of secondary functions.

Phase 3: Improve the control side of the design by close looping main drive motors and introduce variable speed operation, with main functions reacting to adverse trends. Add a system to identify and mark poor and rejected products.

Phase 4: Final version, fully compensating variable speed machine, close looped with control-system strategy based on good quality product only at output end. Reject station for any poor quality items, complete management information system (MIS) printout with each delivered batch.

Note that the above is quoted as an example of how the phased approach may be introduced, and is not intended as a definitive approach to a particular machine design. However, the market place can frequently put a bar on the introduction of too sophisticated and too large a jump in technology, due to the natural apprehension and conservatism of customers. Being first in the field does not ensure commercial success – the British were the first to introduce jet airliners with the Comet, but the Americans made the commercial breakthrough with the Boeing 707 and Douglas DC8.

The concept of variable speed is important in the above example. The constant speed machine has been with us for many years because of the constraints imposed by using mechanical (gear) drive systems. We have been designing process machines to achieve good consistent quality at constant speed using variable input materials. What is needed is a machine that produces constant quality from variable materials. Variable speed is an important parameter in achieving this, which mechatronics lets us use.

10.6 Conclusion

I have shown that the design of mechatronic products is more difficult than using existing conventional technologies, because it involves different engineering disciplines which have to be integrated. The possibilities that mechatronics offers us are so exciting that this approach will become the normal approach very shortly for all products, and our horizons are only as far as our imagination. However, to achieve these new goals our design has to be formalized and restrained, so that the products will be acceptable and not too advanced for the market place.

The points I have tried to emphasize in this chapter can be taken as outline design rules. The trade-offs covered in Section 10.5.2 will enable specific design rules to be formulated for each specific design.

- Mechatronic design will usually be a team effort and it is important to realize not only the part played by each member of the team but also the differing approaches taken by the different engineering disciplines involved. However expert and enthusiastic you are, it is unlikely you can design the whole product single handed. In particular, the chief designer's function is to act as team leader, understand the different disciplines and their approaches to design, and make the decisions on the inevitable compromises.
- Formal specifications should be used to act as sheet anchors for the design. It is often the case that quality, performance and reliability are achievable when the design targets are quantified and agreed.
- An overall design strategy is needed, incorporating a phased approach to the design's final objective, leading to an actual achievement of the specification requirements of the day.
- There should be an active and continuous awareness of the human–machine interface problem in its many facets. Remember that however intelligent we may make a machine, human intelligence is greater, and also that human intelligence can be used against the machine design by those so disposed.
- Consider trade-offs and trends carefully, particularly if the product is designed to achieve a long operating life.

ACKNOWLEDGEMENTS

Grateful acknowledgement is made to the following sources for permission to reproduce material in this text.

Chapter 1

Figure 1.1: V.A.G. (United Kingdom Ltd); *Figure 1.2:* ECT; *Figure 1.3:* Soil Machine Dynamics Ltd.

Chapter 2

Figure 2.13: Adapted and reproduced by courtesy of Nikon UK Ltd.

Chapter 4

Figure 4.4: The Guardian: photograph: Tom Jenkins.

Chapter 7

Figure 7.11: Courtesy of Rexroth Pneumatics Ltd.

Chapter 8

Figure 8.14: Meade, M.L. and Dillon, C.R. (1991) *Signals and Systems: Models and behaviour*, Chapman and Hall, © 1986, 1991 by M.L. Meade and C.R. Dillon.

Chapter 9

Figures 9.1, 9.2, 9.8: Texas Instruments (1991) *TMS320C3x User's Guide*, courtesy of Texas Instruments Incorporated; *Figure 9.3:* Bradley, D.A. and Seward, D.W. (1992) 'The Lancaster University computerized intelligent excavator programme', in *Mechatronics – The Integration of Engineering Design*, Proceedings of the University of Dundee Conference, Mechanical Engineering Publications Ltd, London, © The Institution of Mechanical Engineers 1992; *Figures 9.15, 9.16:* Brooks, R.A. (1986) 'A layered intelligent control system for a mobile robot', in *Proceedings of the 3rd International Symposium on Robotics Research*, MIT Press.

Chapter 10

Figures 10.5, 10.6: The Langston Corporation; *Photographs: Section 10.2:* General Motors Corporation; *Sections 10.4, 10.5:* British Aerospace Defence Limited.

Cover

Courtesy of Dr. Paul Margerison.

INDEX TO VOLUME 1

Page numbers in bold indicate principal references to key terms, which are flagged in the text by the use of bold italic type.